SPACE AND THE
AMERICAN
IMAGINATION

SMITHSONIAN HISTORY OF AVIATION SERIES
Von Hardesty, Series Editor

On December 17, 1903, human flight became a reality when Orville Wright piloted the *Wright Flyer* across a 120-foot course above the sands at Kitty Hawk, North Carolina. That awe-inspiring 12 seconds of powered flight inaugurated a new technology and a new era. The airplane quickly evolved as a means of transportation and a weapon of war. Flying faster, farther, and higher, airplanes soon encircled the globe, dramatically altering human perceptions of time and space. The dream of flight appeared to be without bounds. Having conquered the skies, the heirs to the Wrights eventually orbited the Earth and landed on the Moon.

Aerospace history is punctuated with many triumphs, acts of heroism, and technological achievements. But that same history also showcases technological failures and the devastating impact of aviation technology in modern warfare. As adapted to modern life, the airplane—as with many other important technological breakthroughs—mirrors the darker impulses as well as the genius of its creators. For millions, however, commercial aviation provides safe, reliable, and inexpensive travel for business and leisure.

This book series chronicles the development of aerospace technology in all its manifestations and subtlety. International in scope, this scholarly series includes original monographs, biographies, reprints of out-of-print classics, translations, and reference materials. Both civil and military themes are included, along with systematic studies of the cultural impact of the airplane. Together, these diverse titles contribute to our overall understanding of aeronautical technology and its evolution.

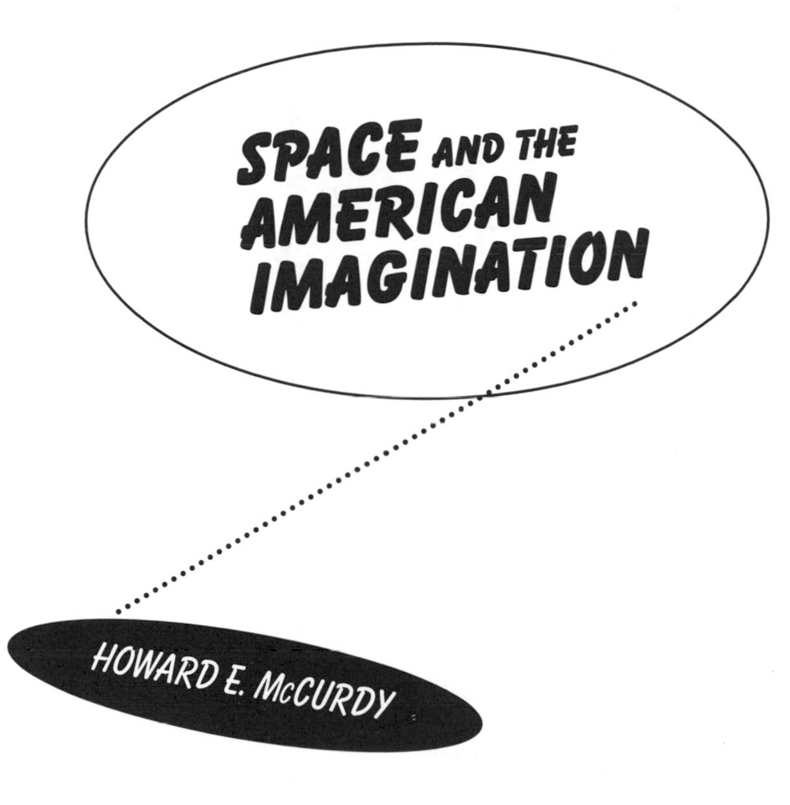

SMITHSONIAN INSTITUTION PRESS
WASHINGTON AND LONDON

To my classmates at Queen Anne High School
Seattle, Washington
Class of 1959

Copy Editor: Karin Kaufman
Production Editors: Jenelle Walthour and Ruth Thomson
Designer: Kathleen Sims

Library of Congress Cataloging-in-Publication Data
McCurdy, Howard E.
Space and the American imagination / Howard E. McCurdy.
p. cm.
Includes bibliographical references and index.
ISBN 1-56098-764-2 (alk. paper)
1. Astronautics—United States—Public opinion. 2. Mass media—United States—influence. 3. Astronautics—Government policy—United States. I. Title.
TL789.8.U5M338 1997
387.8′0973—dc21 97-16

British Library Cataloguing-in-Publication Data is available

Manufactured in the United States of America
04 03 02 01 00 99 98 97 5 4 3 2 1

♾ The paper used in this publication meets the minimum requirements of the American National Standard for Information Sciences—Permanence of Paper for Printed Library Materials Z39.48-1984.

CONTENTS

LIST OF ILLUSTRATIONS

ACKNOWLEDGMENTS

When I was not yet in my teens, a relative of mine, Grace Warren, wrote me a series of letters on the planets and extraterrestrial life. Her husband, Dana Warren, was a theoretical physicist at the Lawrence Livermore National Laboratory in California and an expert on cosmic rays. Together, they excited my imagination with extraterrestrial facts and fantasies. I visited them in the summer of 1956. The following year, the Soviet Union launched *Sputnik 1* and *Sputnik 2*. The excitement surrounding that event encouraged my classmates at Queen Anne High School in Seattle to form a rocket club (unauthorized by school authorities). I joined, along with one of my best friends, Joel Farley. Together we conducted crude but delightful experiments with homemade rockets. Appropriately motivated, I decided to pursue a career in science. I chose chemistry and went to Oregon State University to study the subject. William E. Caldwell, who had been my father's professor twenty years earlier, welcomed me into the science program. It took one year to discover that I was more interested in political science than science per se. My professors at Oregon State, the University of Washington, and Cornell University made public policy understandable. Cornell professor Paul P. Van Riper gave me my first research assignment involving the civil space program, an insignificant but nonetheless stimulating responsibility.

Starting out as a young professor, I was encouraged by the dean of public administration scholars, Dwight Waldo, to explore the relationship between imagination and public policy. He had written occasionally about fiction and public administration, and I did the same. The more I wrote, the more I learned about the ways in which imagination affected the course of government. I was impressed by the work of scholars such as Roderick Nash and Joseph Corn, who described how imagination shaped public attitudes and the government policies to flow from them.

This book has allowed me to combine my interests in science, imagination, and public policy, permitting me to revisit those wonderful images that entertained me in my youth and investigate in a scholarly way the government policies to emerge from them. I could not have completed this work without the inspiration of the people who encouraged these interests in the past.

Special thanks go to Roger Launius and Sylvia Kraemer at the National Aeronautics and Space Administration in Washington, D.C. The NASA History Office, which Dr. Launius directs, supported the research that resulted in this book with a special contract. Much of the material used in the study was gathered from the archives of the NASA History Office, masterfully assembled by the irreplaceable Lee Saegesser. Colleagues at the American University, my regular place of employment, and the University of Washington, where I served as a visiting professor during the 1995–96 academic year, listened and criticized as the book evolved. Professor Richard Berendzen, who provided frequent lessons in astronomy and astrophysics, deserves special thanks. Graduate students at both institutions checked sources and located materials; I particularly want to recognize the assistance of Vince Talucci, Mary Huston, and Richard Faust. To those who read parts or all of the manuscript, especially James R. Hansen and Robert Wohl, I extend my appreciation.

Preparation of this study required a wide-ranging knowledge of popular culture, space science, rocket technology, and public policy. Any errors that may have crept into the work in spite of repeated cross-checks are my own or the result of an overambitious imagination.

INTRODUCTION
THE VISION

Where there is no vision, the people perish.

—Proverbs 29:18

Political reality is different from real reality.

—Joel Achenbach, 1995

Since its beginnings, the U.S. space program has been motivated by a highly romantic dream. According to this vision of cosmic exploration, humans would leave the Earth's surface and explore the universe, just as their ancestors had crossed oceans to investigate foreign lands. Space stations would ring the earth; humans would colonize the Moon and Mars. Rocket scientists would develop spaceships that could move through the void at incredible speeds. Space-age technologies would transform life back on Earth, bringing wealth and power to the nations that controlled the next frontier, and space-age explorers would solve the mysteries of the universe and reveal the mind of God. Much of this could be achieved, according to the vision, before the twenty-first century dawned.[1]

Not by private enterprise alone would this dream be realized. A government agency was established to organize the march into space. Through tax-funded programs, public officials have labored to transform fantasy into fact. The government agency created to carry out this vision—the National Aeronautics and Space Administration (NASA)—conducts many activities, but none is so central to its cause as the realization of the spacefaring dream.

Some forty years into the venture, the reality of space exploration differs considerably from anticipated events. No one-hundred-person space stations ring the earth, no spaceships skip through the solar system at warp speeds, no advanced life forms have been found on Mars. The first phase of space exploration has been a disappointment, given the expectations and timetables advanced by first pioneers. The dream of glorious exploration contains more fiction than fact.

This has not deterred advocates of exploration. Not in space. As the millennium approaches, exploration advocates still press for human and robotic expeditions to Mars. Magazines and books still publish articles on the conversion of nearby planets into habitable spheres. Space flight engineers still strive to construct workable space stations and rocket ships that will cut the cost of space flight by the much-coveted "factor of ten."[2]

The vision of space exploration persists not because it is inherently real, but because it appeals so powerfully to human aspirations. Those aspirations do not float suspended in the public consciousness but are rooted in cultural traditions that extend back thousands of years. The spacefaring vision draws upon cultural traditions as forceful as the terrestrial exploration saga and the myth of the frontier. Its allure draws strength from the hope that these traditions will be replayed in a new domain, the realm of outer space. Space exploration is as much a re-creation of the past as a vision of the future. Cultural traditions help make the spacefaring vision strong, exciting, and, above all, entertaining. Space exploration has proved to be one of the most entertaining images of the twentieth century, by definition, holding the interest of its audience and giving pleasure.[3] Its capacity to entertain accounts in substantial measure for its appeal.

The attractiveness of the spacefaring vision also creates problems for the persons circulating the dream. That which makes the spacefaring image so powerful also makes it vulnerable. Government activities are rarely as entertaining as the images that motivate them, especially when those images are based on romantic interpretations of the past. To paraphrase J. B. S. Haldane, space is not only stranger than we imagine, it is stranger than we can imagine.[4] Although it is possible for a phenomena that exists in imagination for hundreds of years to come true, it is more likely that imagination will launch programs that require altered expectations.

As public policies are implemented, they invariably require some sort of reconciliation between vision and reality. The visions that give rise to government policy, if they are powerful enough to excite public interest, invariably encounter a contrary world. This is especially true in the realm of space, where cultural traditions are strong, the role of imagination pronounced, and the cosmos full of surprises. The creation of public policy under these conditions is not a tidy process. Real government policies need to be entertaining in order to attract public interest in an age of information saturation, but in order to work, those policies also need to be authentic. Modern government requires the reconciliation of

The grand vision of space exploration drew much of its strength from its capacity to excite and entertain. Advocates stirred public interest with images of space stations, exotic spacecraft, alien life-forms, lunar bases, and expeditions to Mars. This painting by Chesley Bonestell, which appeared in a 1954 issue of *Collier's*, depicts preparations for the return journey of an early expedition to Mars. (By Chesley Bonestell with permission of the Estate and Bonestell Space Art.)

imagination, popular culture, and real events. This book investigates the way in which this process has occurred in the realm of space exploration.

Imagination involves forming mental images of events or processes that are not actually present. Any mental image produced by the imagination may be called a vision, although in the public sphere the term typically refers to an image of considerable scope that broadly depicts events or processes well into the future. Images and visions are transmitted in various ways. Many reach their audience through works of fiction, which depict characters and events that are not real. Others are nonfictional and depict events that are presumed to be real but have not been experienced directly by the persons who view them. In the realm of space exploration, some of the most powerful images have been transmitted through nonfictional works gathered broadly under the heading of popular science. Books, movies, television, radio, newspapers, magazines, music, paintings, and the theater transmit images, both fanciful and real. The most pervasive images draw on the ideas, customs, and beliefs held by the public at large: what is generally know as popular culture. Persistent images become part of the popular culture themselves.

Imagination shapes public policy by creating expectations. Works of imagination anticipate events or outcomes yet to come, encouraging people to presume that the future will unfold in a particular way. Some works of imagination inflame fears, scaring people with portrayals of events that the audience would as soon avoid. Still others create mental pictures that help people understand how the world operates, causal images that explain why a problem exists or how it might be cured.

In most cases, images do not duplicate reality. They may anticipate reality, but they cannot capture it exactly. Some images are so fantastic that they can never come true, although that does not necessarily lessen their influence. Notwithstanding their veracity, what matters is the capacity of images to motivate behavior. Myths rooted in works of fantasy, such as fairy tales, can transmit values and shape popular beliefs in powerful ways.

In formulating public policy, politicians and policy advocates work with relatively simple models of society and the world. The ability of policy advocates to advance their agendas depends to a considerable extent upon their skill at constructing models that are familiar, easy to understand, and desirable. The models used to explain space exploration drew upon some of the most attractive ideas in American society. Outer space, according to its advocates, would become the "new frontier," repeating the historic events thought to have created the American dream. Drawing on the analogy of the airplane, advocates promised that space travel would become "as safe and inexpensive for our grandchildren as jet travel is for us."[5] Space explorers would discover alien life forms, just as terrestrial explorers in past centuries had cataloged new species on the earth. Science would unlock the secrets of the universe, transforming the human experience in the same fashion as previous discoveries, and space technology would create a cornucopia of consumer goods, just as invention created economic progress in the past. The attractiveness of the space exploration vision was based not so much on the anticipation of a radically new future as on the repetition of comfortable images from the past.

The familiarity of models or ideas such as these give them their power. The more deeply rooted such ideas are, the more powerful the resulting vision. A vision deeply rooted in preexisting notions can survive decades of disappointment as humans struggle to make it come true. Never mind that the underlying notions contain romantic ideals separated in many ways from the events that inspired them. Cultural beliefs such as the value of frontiers may contain more myth than substance, but that does not weaken their appeal. In the realm of ideas, the strength of an image is not determined by its inherent validity but by its ability to create a workable model in the minds of the people to whom it is designed to appeal.

The original vision of space exploration was also enhanced by its capacity to entertain. Works of imagination dealing with space exploration are among the most entertaining in American culture. They exalt the courageous explorer who overcomes unimaginable hardships to conquer new frontiers and promote the ultimate promise of new beginnings, where everything will be better than before. They contain some of the greatest mystery stories of modern times. From the first photographs of the back side of the Moon to the search for extraterrestrial life, expeditions into space both human and robotic have excited public interest with the mysteries of the universe. Altogether, the vision of space exploration is full of adventure and resurrection and fantastic dreams come true.

In an era of television and mass communication, the capacity of new ideas to excite and entertain is crucial to their survival in the governmental policy milieu. Ideas that do not excite and entertain are replaced by those that do. Imagination performs a critical role in sorting out policy proposals, helping to determine which proposals land on the governmental agenda, which get approved for financial support, and how well they persist as obstacles arise. In this respect, space exploration has proved to be the archetypical example of the modern public policy. The initial vision of space exploration was transmitted through works of imagination that encouraged audiences to dream of wondrous new worlds. For centuries, works of fiction enthralled readers with fantastic voyages to the planets and stars. Through a deliberately organized public relations campaign, using the medium of popular science, advocates of space exploration worked to convince Americans that these dreams could come true. When the actual space program began, television treated it as one of the greatest news stories of all time.[6]

Many commentators are critical of the tendency among people who shape public opinion to rely upon the tools of the entertainment industry to attract converts to their cause. The line between entertainment and information has become so blurred, they say, that the public can no longer distinguish between what is real and what is not. This charge is frequently directed at broadcast journalists, who

The reality of space exploration, during the early years of the venture, differed considerably from the romantic vision offered by advocates of cosmic flight. Human interplanetary travel did not occur on the timetable advanced by the first space-flight pioneers. Robotic probes to Mars, such as this 1976 *Viking* lander, did not reveal the lush and habitable planet of imaginative lore, but a dry and frigid sphere. (NASA)

in their effort to compete head to head with the entertainment industry create "infotainment" as a means of grabbing public attention. The criticism is also directed at producers of motion pictures and television programs who use factual events to create docudramas that bear little resemblance to reality. This process, the critics charge, undermines democratic processes because the perpetrators weaken the ability of people to make intelligent decisions about the future course of government policy.[7]

Public policy by its very nature is not entertaining. Meetings of Congress, military expeditions, and other government activities are not as entertaining as dramas or films. Space exploration involves hours of tedium broken by only occasional moments of drama or terror. Any effort to make public policy compete with the entertainment industry ultimately will fail. It will create gaps between expectations and reality and undermine trust in a government that cannot accomplish mythical feats.

Modern government, however, cannot operate without visions that excite and entertain. Repeatedly in the course of public affairs, works of imagination have won converts to a new cause: helping to promote conservation of natural resources, assisting in the movement to abolish slavery, fostering the regulation of food and drugs, and promoting space exploration. Advocates of new ideas frequently merge fact and fancy in works of imagination as a means of shaping public policy. Many visions and expectations compete for the attention of policy makers and only the most inviting images move to the top of the policy agenda. The public at large is bombarded by ideas and opinions. In an ideal society, expert judgment would separate fact from fantasy. In the real world, however, expert opinion is often divided. Experts do not know for certain whether an advocated policy will produce a desired effect or how future events will unfold. When this happens, imagination helps fill the void.

Imagination recasts issues in terms that nonexperts can understand, reducing complex problems to simple concerns, such as the struggle between good and evil. By framing issues in terms the average viewer can comprehend and by making those issues seem more interesting than they actually are, imagination enlarges the audience engaged in policy debates. Works of imagination invite the inattentive public to become involved in issues it would otherwise avoid, and in so doing, shifts the balance of power toward groups most adept at using imagination to build support for their ideas. Imagination allows people to visualize a particular activity taking place, a key ingredient in convincing policy makers to undertake it.

To become public policy, a vision must seem familiar, feasible, and desirable. The truth or validity of the vision is to a large degree irrelevant to the policy debate *because no one knows for certain how the policy will turn out.* The validity of a particular vision is determined by its compatibility with established cultural beliefs and its ability to attract a large audience through its capacity to excite and entertain. That which gives the vision its power, however, invariably ensures its demise. Inevitably, reality intrudes. Gaps between imagination and reality arise. Policies constructed on a fictitious foundation, however, do not spontaneously disappear; they persist even as facts intrude in discomforting ways. Reconsideration and abandonment of policies require an opposing reality and a countervailing vision around which the opposition can rally. Otherwise, the old policy is likely to linger. This is especially true for policies such as space exploration, where alternative visions are ill-formed, the opposition poorly organized, and the underlying technology benign.

Under such circumstances, politicians and policy advocates will not abandon the original dream. Politicians will be reluctant to kill visions for which underlying cultural norms remain strong. Advocates will not abandon their beliefs just because their prophesies fail. Rather than acknowledge defeat, they will work harder to make the original vision come true and reemploy their faith to build support for original aspirations. They will seek new explanations for old ambitions and new discoveries to spur old policies.

In the case of space travel, advocates have used a rich and multifaceted vision to promote their dreams. Their vision has drawn strength from works of imagination that excite and entertain, works that draw much of their power from preexisting cultural norms. Gaps inevitably have arisen between image and reality. The power of the original vision, however, has overcome this deficiency. The public in general (along with many government officials) have refused to abandon the spacefaring dream.

PRELUDE: THE EXPLORATION IDEAL

> The ability to imagine is the largest part of what you call intelligence. You think the ability to imagine is merely a useful step on the way to solving a problem or making something happen. But imagining it is what makes it happen.
>
> —Michael Crichton, *Sphere*, 1987

In the spring of 1959, less than eight months after the creation of the National Aeronautics and Space Administration (NASA), a internal NASA committee chaired by Harry J. Goett met to chart a long-range plan for the human exploration of space. Committee members agreed upon certain activities the civil space agency would faithfully pursue in the decades to follow: the development of spacecraft capable of orbiting Earth, a laboratory or station in space, and human expeditions to the Moon and nearby planets, preceded by robots. "The ultimate objective of space exploration," members concurred, "is manned travel to and from other planets."[1] Later that year, NASA officials ensconced the recommendations of the Goett committee within the agency's first comprehensive long-range plan. While the document embraced a number of robotic and remotely controlled missions, the long-range vision for the ensuing decades remained clear: "Manned exploration of the moon and the nearer planets must remain as the major goals."[2] Seventeen months later President John F. Kennedy approved the first significant step in this long-range plan with the 1961 decision to go to the Moon. Since then, NASA's mission, as well as the popular image of space flight, has been dominated by a grand vision of extraterrestrial exploration, with humans clearly in charge.

Human exploration of the solar system was only one of many possible directions in which the U.S. civil space program could have headed. An impressive program of scientific discovery could have been organized around satellite technology, robotic probes, instruments in space, instruments on the ground, or the emerging science of remote sensing. Granted, such technologies were not well understood at the beginning of the space age, but neither were rocketry and space travel. To conceive of space exploration in any form required an act of imagination. In this respect, human space travel possessed one great advantage over robotics and planetary probes: it was, given the history of imagining space travel and the cultural traditions upon which it drew, easier to visualize.

Proponents communicated the vision of space travel so effectively that any other approach to cosmic investigation seemed inconceivable. They created a set of expectations so wonderful and exciting that the expectations became cultural norms—assumptions so deeply embedded in American society that their validity no longer needed investigation.

For as long as humans have been able to conceive of heavenly bodies other than Earth, people have dreamed about traveling to alternative worlds and exploring them. Lucian of Samosata, one of the few Greek thinkers who believed in the existence of other worlds, describes an odyssey in which his characters travel to the Moon and encounter lunar inhabitants who move about on large, three-headed vultures. Seventeenth-century astronomer and mathematician Johannes Kepler wrote a fanciful work in which a journey to the Moon reveals that its craters are artificial constructions, built by the local inhabitants in order to produce shaded places from which they can escape the rays of the Sun. Public interest in stories about extraterrestrial travel flourished after Galileo Galilei revealed in 1610 that the traveling points of light in the night sky were planets like the earth, though different in size. Acceptance of the plurality of worlds doctrine naturally encouraged people to ponder what humans might find if they journeyed to and explored these Earth-like realms.[3]

People who wrote such stories did not agonize long over the means of transport to and from these extraterrestrial bodies, solving the problem with little more than an active imagination. Lucian transported the crew of his sailing ship via the power of a giant whirlwind. In his 1638 story *The Man in the Moon*, Domingo Gonsales (pseudonym for the English bishop Francis Godwin) relied upon a flock of wild swans capable of carrying large loads to transport him to the lunar surface. Cyrano de Bergerac wrote two seventeenth-century novels describing yet another voyage to the lunar surface. One of his methods of transportation depended on bottles filled with early morning dew, Bergerac observing that dew seemed to rise upward when exposed to the rays of the morning sun. In 1775 Louis-Guillaume de la Follie produced levitation through electric power, which the hero of his novel generated by turning cranks and gears attached to two large sulphur balls. Joseph Atterley (actually American professor George Tucker) relied upon an antigravity substance called *lunarium* to transport his crew to the Moon in

1827, as did H. G. Wells in his 1901 novel *First Men in the Moon*. The centrifugal force produced by a giant flywheel hurled the components of an Earth-orbiting station into space in Edward Everett Hale's 1869 novel *The Brick Moon*. In an influential 1897 novel, *Auf Zwei Planeten*, German author Kurt Lasswitz likewise employed a gravity-defying material to transport Martian explorers to Earth. Suspended in space at a point directly above the North Pole, the Martians' descent to the earth's surface was expedited by the erroneous observation that descent would prove effortless because the planet does not move at the point of its poles.[4]

The most fantastic schemes for extraterrestrial travel avoided physical devices altogether, transporting humans by telepathic means. In the fantastic *Somnium*, Kepler falls asleep and dreams that spirits can speed humans to the lunar surface through shadows of the earth and Moon during eclipses of each. The dream method of transportation was also favored by Edgar Rice Burroughs, who in the early twentieth century thrilled a generation of young readers with an adventurous sequence of novels set on the planet Barsoom (the local name for Mars). The hero of his novels, John Carter, falls into a trance outside the mouth of an Arizona cave, awakening a moment later on that distant sphere.[5]

Such methods of transport required readers to suspend disbelief. Having overcome the boundaries of probability, readers could enjoy the contemplation of conditions on unreachable lands. Transport problems resolved, the main purpose of the story could proceed. Fortunately for the authors of such stories, a broader collection of travelers' tales had already set the precedent for fanciful methods of conveyance. The earthbound fantastic voyage had been a central part of Western literature ever since Homer wrote the *Iliad* and the *Odyssey* in the eighth century B.C. The tradition of epic journeys on the earth provided a far more sturdy foundation for the contemplation of space travel than science fiction stories alone.

Never mind that many transportation schemes required a suspension of the laws of physics. Miraculous events are a staple fare in travelers' tales, in both the methods of transport and the nature of conditions in the places at which the travelers arrive. In Shakespeare's *The Tempest*, the exiled Duke of Milan uses magical powers to create a terrible storm that blows the king of Naples and his ship's crew onto the island where the play is set. The executor of this scheme, an airy spirit named Ariel, culls the king's ship out of its fleet and brings the vessel unharmed into a quiet cove. This magical tale provided the inspiration for the classic 1956 science fiction film *Forbidden Planet*.[6] Because Shakespeare's play challenges the audience's imagination, the audience is allowed to suspend belief and contemplate the possibility that control of the weather and other seemingly miraculous events are within the reach of humans. Occurrences that contradict contemporary standards of understanding occur with natural regularity in distant places.

Like the characters in *The Tempest*, Lemuel Gulliver in *Gulliver's Travels* and the famous Robinson Crusoe all reach their destinations through terrible storms. In his classic children's story *The Lion, the Witch and the Wardrobe*, C. S. Lewis, who was also a writer of science fiction, has his characters find the land of Narnia by climbing through the back of a enormous wardrobe in a professor's country home. Alice travels by means of a rabbit hole and looking glass, and Dorothy commutes from her Kansas home to Oz in a ferocious tornado.[7]

Because no one has ever been to the places depicted in these tales, readers are allowed to exercise their imaginations and speculate on the plausibility of the wonders described within. No one has ever visited the mythical islands described by Homer in the *Odyssey*, nor the land of the Lilliputians in *Gulliver's Travels*. So far as we know, no one-eyed giants or diminutive people inhabit the islands of the earth, but who can say with certainty that stranger creatures do not exist in places as yet unexplored?

Not constrained by the requirements of reality, the authors of such tales were free to embellish their stories with highly romantic ideals. In a typical exploration fantasy, an earthly traveler is called upon to resolve a struggle between good and evil, or at least to demonstrate intellectual superiority among misguided people. After saving the Lilliputians from a naval invasion by neighboring Blefuscu, Lemuel Gulliver proceeds to rescue the conquered nation from slavery at the hands of the Lilliputian emperor. Four human children help rescue the inhabitants of Narnia from the tyrannical rule of the evil White Witch in *The Lion, the Witch and the Wardrobe*, a story full of magic, including the resurrection of the good lion Aslan from the dead. Like explorers on real terrestrial expeditions, Odysseus is called upon to overcome unimaginable hardships on his ten-year journey home in *The Odyssey*. For centuries, explorers in fact and fiction have been treated like outstanding athletes, capable of superhuman acts of strength and bravery. The romantic qualities of these stories invite readers to hope that at least some of the elements in travelers' tales might be true.

During the late nineteenth and early twentieth centuries, a number of individuals were motivated by stories such as these to take what previously had been considered romantic fantasy and make it reality, sometimes preparing works of imagination themselves as a means of promoting their cause. Until the mid-nineteenth century, nearly all extraterrestrial tales employed methods of exploration that were scientifically impossible. Inevitably, science and invention exposed the more fanciful assertions as the falsehoods they were. Scientific observations, for example, confirmed that the Moon had no atmosphere, upsetting the plot lines in a number of travelers' tales. For the most part, this did not bother the audience for such stories, as most readers were happy simply to be entertained. A growing number, however, conscious of advances in science and technology, found these shortcomings troublesome, as standing in the way of actually completing the fantastic voyages described in these tales. Fascinated by the prospect of space flight, they identified inaccuracies and attempted to correct them.

The so-called fathers of modern rocketry—Robert Goddard, Hermann Oberth, and Konstantin Tsiolkovskii—were all inspired by errors of imagination. "My interest in space travel was first aroused by the famous writer of fantasies Jules Verne," noted Tsiolkovskii. "Curiosity was followed by serious thought."[8] Among the outpouring of fictional works on space travel during the mid-nineteenth century, the work of the French writer Jules Verne was distinct. In his early twenties, struggling to establish himself as a successful writer in mid-nineteenth-century Paris, Verne developed the technique of inserting scientific explanations into simply told travelers' tales, enabling readers to acquire scientific information while enjoying an otherwise romantic adventure. By the time he released his first spacefaring story, Verne had already produced novels on crossing Africa by balloon, journeying to the center of the earth, and exploring the North Pole. In 1865 he pub-

The pioneers of space flight were motivated to undertake their studies by works of imagination. In some cases they produced such works themselves. The Russian space pioneer Konstantin Tsiolkovskii wrote science fiction stories as he struggled to devise a practical method of space transportation. (NASA)

lished *De la Terre à la Lune* (*From the Earth to the Moon*), a sequel, *Autour de la Lune* (*Round the Moon*), appearing five years later. Verne took special care to make his stories appear as plausible as possible. In the two lunar adventures, he relied upon mathematical formulas to calculate the velocities and transit requirements necessary to carry his spacefarers toward the Moon. He provided his spacefarers with devices for the replenishment and purification of air and even stumbled across the use of rocket power as a means of breaking the velocity of the space capsule as it fell toward the Moon. The actual means of departing the Earth, however, stumped Verne. Lacking an appropriate technology to propel his travelers to an escape velocity of twenty-five thousand miles per hour, Verne fell back upon the entertaining but scientifically implausible scheme of shooting the space capsule out of an enormous cannon.[9]

By 1875 Verne's lunar adventures had been translated into Russian. Konstantin Tsiolkovskii, born in 1857, had by this time moved from his village home to Moscow, where he struggled to educate himself at the Chertovsky Library. An earlier bout with scarlet fever had left him nearly deaf, a condition that caused Tsiolkovskii to withdraw from formal schooling and isolate himself in a world of books and dreams. "The idea of travelling into outer space constantly pursued me," he later wrote. Like many people of his day, Tsiolkovskii could not devise a realistic method of space travel. He toyed with a number of propulsion schemes, including centrifugal force, perpetual motion, and antigravity devices, and wrote science fiction stories in an effort to popularize his ideas. After repeated mathematical calculations, Tsiolkovskii settled on the use of multistage rockets, a conclusion explained in his *Investigation of Universal Space by Means of Reactive Devices*, first issued in 1911. The treatise was a serious work of imagination in itself, introducing readers who could obtain copies of this obscure work to orbiting space stations and methods for colonizing the cosmos.[10]

Hermann Oberth was born more than a generation later, in 1894, in Transylvania, a site more famous for its vampire tales than space fantasies. While attending the Bischof-Teutsch Gymnasium in his hometown of Schässburg, the young Oberth also discovered the novels of Jules Verne. He read Verne's lunar adventures "at least five or six times and, finally, knew [them] by heart."[11] Abandoning his original intent to follow his father into medicine, Oberth instead pursued his boyhood interest, earning advanced degrees in mathematics and physics and writing as his doctoral dissertation a treatise on the problem of extraterrestrial flight. Like Tsiolkovskii, Oberth adopted rocket propulsion as a substitute for Verne's cannonball approach. He arranged for publication of his dissertation by Rudolf Oldenbourg, a Munich publishing house. Once past the lengthy presentations on launch angles and trajectories, propellants, and valves, readers of *Die Rakete zu den Planetenräumen* (The Rocket into Planetary Space)

French novelist Jules Verne helped inspire serious investigations of space flight by injecting scientific information into otherwise romantic adventure tales. In his 1870 story *Autour de la lune* (*Round the Moon*) he inadvertently stumbled upon the use of rocket power as a means for controlling the trajectory of his space capsule on its route toward the Moon. (NASA)

were treated to an imaginative discussion of a human expedition into space and the establishment of an Earth-orbiting space station. The book was an immediate success, moving through three editions between 1923 and 1929. To help promote his ideas, Oberth traveled to Berlin to provide scientific advice for a film on space flight being produced by Fritz Lang. The 1929 movie, *Frau im Mond* (English title, *By Rocket to the Moon*), generally considered the first realistic treatment of space travel in movie form, depicts a number of principles set forth in Oberth's book and used in the actual flight forty years later, including the rollout of a multi-stage rocket ship from a large vehicle-assembly building.[12] The movie allowed Oberth and his associates to spread the gospel of space flight among the German intelligentsia. As one of Oberth's disciples explained, "The first showing of a Fritz Lang film was something for which there was no equivalent anywhere. . . . It is not an exaggeration to say that a sudden collapse of the theater building during a Fritz Lang premiere would have deprived Germany of much of its intellectual leadership at one blow."[13]

Verne's method for departing Earth involved an implausible scheme, the use of an enormous cannon to propel a space capsule to escape velocity. German rocket pioneer Hermann Oberth read the adventures during his youth so frequently that he knew their details and deficiencies "by heart." Like Tsiolkovskii, Oberth settled on the use of rocket power as the most reliable means of propelling humans beyond Earth. (NASA)

Like his European counterparts, Robert Goddard's interest in rocketry was motivated by works of imagination. According to his wife, Goddard read *From the Earth to the Moon* several times and absorbed himself by penning comments and corrections in the margins.[14] As a young person, Goddard read a serialized version of *War of the Worlds* in the *Boston Post*, which, he said, "gripped my imagination tremendously."[15] In his autobiography, Goddard recalls climbing a backyard cherry tree and dreaming of a voyage to Mars, an event he quietly celebrated thereafter as his "anniversary day."[16]

Goddard believed that in order to escape the inevitable cooling of the Sun, humans would eventually leave Earth and migrate to planets in other solar systems. "The navigation of interplanetary space must be effected to ensure the continuance of the race," he wrote at the age of thirty. He further believed that the Moon could be used as a launching pad for trips into interplanetary space and contemplated the construction of space factories on the Moon and planets designed to produce rocket fuels.[17] Although imagination inspired him to think wondrous ideas, it rarely caused him to promote them. Due to personal shyness and a discomforting encounter with the national press, Goddard was notoriously reluctant to advance his vision of space flight in public.

In 1919 Goddard, then thirty-seven years old, published *A Method of Reaching Extreme Altitudes*, a treatise on the use of rockets to propel objects into space. Sickly and reclusive, he had spent most of his life in Worcester, Massachusetts, where he earned three college degrees, accepted a teaching position, and experimented with rocket propulsion. The pamphlet, published by the Smithsonian Institution, is a rather sober discussion of the mass required to propel objects to altitudes beyond those attained by high-altitude balloons. Toward the end of the pamphlet, Goddard advanced a proposal from an earlier paper he had prepared on "The Navigation of Interplanetary Space" while recuperating from a bout with tuberculosis in 1913. He suggested that a small rocket of sufficient power (without humans on board) could be launched from Earth's surface and travel to the Moon, where its collision with that body would be confirmed by the explosion of a pyrotechnical device large enough to be viewed through telescopes on Earth.[18]

Public reaction to *A Method of Reaching Extreme Altitudes* caused Goddard to recoil from further promotional ideas. His proposal for a "moon rocket" was lampooned in the national press, where it was front page news. Editors at the *New York Times* pointed out (incorrectly) that a rocket would not work in a vacuum because it would have nothing "against which to react." Goddard knew this to be false, having already tested rocket thrust in vacuum tubes. The sensational national publicity distressed Goddard, and, as historian Frank Winter has observed, he "became even more press-shy and secretive."[19] His future writings concentrated on turbines, pumps, stabilization, and fuels. In 1926, from a field near Auburn, Massachusetts, he completed the first successful launch of a liquid fuel rocket, an event of abundant importance. Goddard requested that the Smithsonian Institution, his main source of financial support, not publicize the results. Four years later, in an effort to avoid further publicity and seek better weather, he moved his launch operations to a remote area near Roswell, New Mexico.

Tsiolkovskii, Goddard, and Oberth all turned to rocket power as the most efficient means to achieve a constant source of propulsion that could achieve the velocities required to escape the gravity of Earth and steer transit craft through space. In developing the science of rocketry, each made profound contributions. In shaping the image of how rocketry would be used for extraterrestrial flight, their contributions were most unequal.

Tsiolkovskii's work had practically no effect on the U.S. space program. He was an obscure figure within Russia, and even less well known outside. Although Goddard's work was widely recognized within U.S. scientific circles, his refusal to participate in even the most elementary promotional efforts assured that his

America's leading rocket pioneer, Robert Goddard, was inspired to undertake his studies by the writings of Jules Verne and H. G. Wells. In his autobiography, Goddard remembers climbing a backyard cherry tree and dreaming of a voyage to Mars, an event he subsequently celebrated as his "anniversary day." In 1926, from a field near Auburn, Massachusetts, Goddard became the first person to successfully launch a liquid-fuel rocket. (NASA)

influence on the popular conception of space in the United States would be nearly as slight as Tsiolkovskii's. As strange as it might seem, Oberth had the greatest influence on the popular conception of space travel, although it arrived through the work of his followers.

The absence of any eminent U.S. scientist to serve as a spokesperson for the awakening space movement left the American wing in the hands of two groups devoutly committed to the romantic exploration vision: German expatriates and science fiction fans. Both found reenforcement for their astonishing views within newly organized rocket societies. Responding to the increasing volume of extraterrestrial narratives, both fictional and nonfictional, partisans of space flight throughout Europe and the United States founded societies devoted to the conquest of space. They were inspired by an explosion in the quantity of science fiction as well as the appearance of serious investigations into the dynamics of rocketry. Tsiolkovskii's followers founded the Society for Interplanetary Travel in 1924, which collapsed less than one year later due to internal disputes. The Verein für Raumschifahrt, popularly known as the German Rocket Society, began its work in

Because Robert Goddard was notoriously shy, promotion of space flight in America fell to science fiction fans. One of the leading promoters, Edward Pendray (*right*), was a reporter for the *New York Herald Tribune* who wrote science fiction stories on the side. A collection of science fiction writers, gathered in Pendray's Manhattan apartment in the spring of 1930, formed the American Interplanetary Society for the purpose of promoting the spacefaring dream. (National Air and Space Museum)

1927. The American Interplanetary Society (later renamed the American Rocket Society) was founded in 1930 in New York City, and the British Interplanetary Society (BIS) appeared in 1933. Similar groups arose in Austria (1926), Canada (1936), and France (1938), as did allied groups in the pioneering countries.[20]

Leaders of the rocket societies were steadfastly committed to the attainment of human space flight, a dedication that went considerably beyond the general curiosity in astronomical phenomena prevalent at that time. Using the technology of the telescope, nineteenth-century astronomers and their popularizers had generated substantial public interest in the nature of the cosmos. Books and magazines regularly reported the latest findings, and by the 1930s, associations devoted to the investigation of astronomical phenomena were already commonplace in Europe and America. The *Société Astronomique de France* was formed in 1887 and the American Astronomical Society in 1899.

The rocket societies distinguished themselves from the astronomical groups by promoting human space flight as the primary means of cosmic investigation. The founders of the American Interplanetary Society announced in the opening sentence of their first bulletin that their principal aim was "the promotion of interest in interplanetary exploration."[21] The official statement announcing the formation of the society set out the philosophy in more detail: "It is our intention to build this society into a national organization with financial and other resources of such importance we can offer real inducement and stimulation to American scientists . . . in the development of rockets, rocket cars and other proposed methods of traveling in space and communicating with the planets."[22] Writing about the creation of the German Rocket Society, of which he was a founding member, the prolific science writer Willy Ley explained that its creators were "dedicated to promoting the idea of space travel." That "driving ambition," Ley said, served as the inspiration for their work.[23] The word astronautics (the scientific term used to describe the investigations of the rocket societies) means "navigating the stars." It was invented by a French science fiction writer.[24]

The leaders of the rocket societies proclaimed a gospel of remarkable power. Humans, they said, would carry out expeditions of discovery in space as ambitious as those of earlier explorers on Earth, maintaining the spirit of adventure and discovery they had inspired. Scientists would build large rocket ships capable of transporting people into space, and once in space, humans would construct space stations and assemble spacecraft capable of voyages far beyond Earth. Expeditions to the Moon and nearby planets would ensue, followed by permanent bases and settlements. Explorers would probe the mysteries of the universe, locate strange creatures, and make miraculous discoveries. It was a remarkable vision, powerfully attractive to people so inclined to believe.

It was also at that time wholly fantastic. Practical experience with the science of rocketry during the 1930s was primitive at best, and the production of a workable vehicle capable of launching even the smallest object into space lay a quarter-century away. The lack of practical experience, however, did not deter visionaries; to the contrary, it encouraged them to make bolder claims.

In the late 1930s members of the British Interplanetary Society formulated plans for a rocket trip that would land humans on the Moon and return them safely to Earth. The members were utterly serious and developed a number of technical devices in pursuit of their goal, such as plans for a "coelostat" to assist in navigating the spacecraft along its path, a "carapace" to protect the crew from atmospheric heating, and fall-away rocket tubes that could be jettisoned once the ship had spent its solid fuel. Various details appeared in two issues of the *Journal of the British Interplanetary Society* in 1939 and attracted widespread attention in the British press. Arthur C. Clarke, a twenty-one-year-old amateur astronomer and civil servant, calculated the escape velocities. That same year, in one of the first publications in what would be an extraordinarily prolific writing career, Clarke published an article titled "We Can Rocket to the Moon—Now!"[25]

The German Rocket Society was the most influential of the advocacy groups. Its members conducted field tests, as did those in other rocket clubs, but for the Germans, this was more than an exciting hobby. By 1932 they had conducted 87 flights and more than 270 static tests with real rockets. Experimenters had made impressive progress with the mechanics of liquid-fuel propulsion and the techniques of regenerative cooling. Beginning that year, the first of several members went to work for the German army, where they eventually produced the world's first large liquid-fuel launch vehicle, the V-2. After World War II, 125 members of the German rocket team emigrated to the United States, forming the nucleus of the group that built the large rockets used to propel the first Americans to the Moon.[26] Of equal importance to their technical accomplishments was their ability to promote the space flight dream. The Germans produced not only better rockets but also experts better able to communicate a technically sophisticated vision of space flight. Two of the society's members, Willy Ley and Wernher von Braun, became the principal spokespersons for the spacefaring dream after moving to the United States.

Ley was born in Berlin, Germany, in 1906. His father was a wine merchant and his mother the daughter of an official in the German Lutheran Church. As a student in Berlin and East Prussia, Ley developed a broad-ranging interest in paleontology, astronomy, and physics. In 1926 he obtained a copy of Oberth's treatise on space flight, which excited Ley so much that he gave up his plans for a career in the earth sciences and devoted himself to the promotion of space flight. Ley had an unusual proclivity for communication, a talent he put to use as a lecturer and writer of popular science books. At the age of nineteen, he rewrote Oberth's book in a style compatible with popular consumption. The following year he helped organize the German Rocket Society and accepted the position of vice-president (Hermann Oberth agreed to serve as president), from which he helped build the membership to more than one thousand, with sufficient funds to operate a rocket proving ground with its own staff of mechanics and engineers on the outskirts of Berlin.[27]

In 1933 the National Socialist Workers' Party took control of Germany. The work of the German Rocket Society shriveled, due in part to a lack of funds and the increasing suspicions of the new government. "The value of the sixth decimal place in the calculation of a trajectory to Venus interested us . . . little," said the army captain placed in charge of developing long-range rockets for the German military.[28] Some of the society's members migrated to the government's military rocket development program at Peenemünde, a remote island in the Baltic Sea. Ley emigrated to the United States. In 1935, with help from friends in the American Rocket Society (it had changed its name the previous year), Ley fled to New York City, where he eked out a living writing articles and books on zoology. As V-2 rockets began to fall on London and other targets in the fall of 1944, Ley returned to his first interest, publishing *Rockets: The Future of Travel Beyond the*

After emigrating to the United States, Willy Ley (*right*) found the American space-flight movement firmly in the hands of science fiction writers. In his native Germany, Ley had helped found the influential German Rocket Society. Joined after World War II by German rocket engineer Wernher von Braun (*left*), the two visionaries became the most outspoken advocates in America for the spacefaring dream. (National Air and Space Museum)

Stratosphere. The book was an instant success. Ley's ability to link the development of rocketry to space travel established him as the principal popularizer of space flight in the United States at that time.

Like others before him, Wernher von Braun became involved in rocketry as a result of reading works of imagination. Recalling his reaction as a young man to one such story, von Braun said, "It filled me with a romantic urge. Interplanetary travel! Here was a task worth dedicating one's life to! Not just to stare through a telescope at the moon and the planets but to soar through the heavens and actually explore the mysterious universe! I knew how Columbus had felt."[29] Introduced to the members of the rocket society at the age of eighteen by Willy Ley, von Braun quickly ingratiated himself with the society's top staff. A talented engineer, von Braun was also remarkably charismatic and a relentless promoter of his own career. That summer he helped launch small, liquid-fuel rockets from the society's Raketenflugplatz testing grounds in northern Berlin. Two years later, having completed work on his bachelor's degree, von Braun went to work for the German army's new rocket program. The army captain who recruited the twenty-year-old von Braun commented on "the energy and shrewdness with which this tall, fair young student with the broad massive chin went to work."[30]

Von Braun played a key role in the development of the German rocket center at Peenemünde. He was technical director for research and development within an overall organization that by the war's end had produced more than six thousand V-2 rockets, twenty-nine hundred of which fell on targets in Great Britain and the Continent. As the Third Reich collapsed, von Braun led one hundred top rocket personnel to a small village on the Austrian border so that they could surrender to the U.S. Army. Interrogators were amazed at the assertion that this thirty three-year-old had developed the V-2 rocket and had even prepared plans for a missile that could reach New York.[31]

As a group, the Germans were no more technically adept and no more obsessed by nationalistic concerns than their counterparts in other countries, nor did they possess a deeper tradition of science fiction. What distinguished the leading members of the German Rocket Society was their unusual talent for communicating the gospel of space flight. The society's top engineers and scientists were drawn from intellectual classes whose members were accustomed to communicating with social and political elites. Oberth's father was a physician, Ley grew up in middle-class Berlin, and von Braun's father was a Prussian aristocrat, the Baron Magnus von Braun, a landowner in the province of Silesia and an important public official in pre-Nazi governments. Wernher's mother was a well-educated woman from the Swedish-German aristocracy with strong interests in biology and astronomy.[32]

Once in the United States, Ley and von Braun found much of the interest in space travel lodged in hands of science fiction fans. "No one had the slightest interest in the subject except science fiction magazines," Ley complained after his arrival in the United States in 1935.[33] A polyglot of science fiction writers had begun meeting some five years earlier at the Manhattan apartment of G. Edward Pendray, a reporter for the *New York Herald Tribune*. Pendray, a strange-looking man with a pronounced goatee, had come to New York from Wyoming in 1925 (he was born in Nebraska in 1901) and was something of a feature in the local literary scene, having penned a number of science fiction stories under the pseudonym of Gawain Edwards. His wife, Leatrice Gregory, was a widely syndicated newspaper columnist.

The writers who met at Pendray's apartment were active contributors to *Science Wonder Stories*, one of the cheap pulp paper magazines that specialized in chimerical adventure tales. "The most popular theme of science fiction," Pendray explained, "was interplanetary (or interstellar) travel." Being writers rather than engineers, their "imaginations could outpace dull practical considerations with a velocity comparable to that of light."[34] At one meeting in the spring of 1930, a group of twelve attendees formed an association designed to promote their fantastic schemes. The "principal moving spirit," according to Pendray, was David Lasser, managing editor of the magazine in which their stories appeared. A Baltimore native, Lasser had studied engineering at the Massachusetts Institute of Technology from 1920 to 1924. After a brief career as an engineer and technical writer for a variety of New York firms, he signed on as managing editor for Gernsback Publications. The following year, he helped found the American Interplanetary Society and was elected its president. Pendray became vice president, while Gregory served for a time as the society's librarian and sewed parachutes for use in the group's experimental rocket program.[35]

As a consequence of these developments, the principal explication of space flight in the United States during the twenty-five-year period leading up to the beginning of the space age fell to people with a highly romantic view. Pendray became the principal spokesperson for the group after Lasser left Gernsback Publications in 1933 for a career in the trade union movement. His position was gradually supplanted by that of Ley and von Braun, who added technical credibility to the fantastic schemes.

As if to merge the two trends, von Braun began work on a novel describing an imaginary mission to Mars, completing it while helping to develop the U.S. Army's rocket program after World War II. The novel did not enlist a commercial publisher, but the appendix to the novel, which outlined the technical requirements for the voyage, did. The ideas contained in *The Mars Project* were part of an increasingly public campaign that pushed von Braun to the forefront of the effort to realize the dreams of lunar and planetary pioneers.[36]

One additional factor helped reenforce the attractiveness of the spacefaring dream. The founding of the rocket societies and the promotion of their point of view occurred at a time when the last great era of terrestrial exploration seemed to be coming to a close. Advocates of interplanetary flight used popular interest in terrestrial exploration, along with the myths that had grown up around it, to promote space travel. The idea that the traditions of terrestrial exploration could continue in a new realm excited many people to follow the dream.

To the public, it seemed that few areas worthy of exploration remained on the surface of the earth. The "golden age" of polar exploration ended, by most accounts, in 1929, with the first airplane flight over the South Pole. The search for the great Northern Passage had ended a quarter-century earlier, marked by the completion of Roald Amundsen's tortuous three-year expedition through the ice-bound seas of northern latitudes, ending in 1905. Completion of the Yellowstone expedition of 1870 and Powell's navigation of the Grand Canyon in 1869 seemed to conclude more than three hundred years of major North American expeditions. The heroic age of African exploration had drawn to a close with confirmation of the source of the Nile in 1876 and completion of the full-scale expedition by Henry Norton Stanley down the Congo River in 1877. The golden age of Alpine mountaineering had ended with the conquest of the Matterhorn in 1865. The science of discovery did not cease with the conclusion of these events, but the ability of newspaper owners and other publishers to generate popular interest in new expeditions declined.

Expeditions such as these were treated as a form of mass entertainment by the publishing industry. They were used to sell newspapers, as in the case of the *New York Herald*, whose editors dispatched the journalist Henry Morton Stanley to lead an expedition to Central Africa in a search for the British missionary David Livingston. Public interest in expeditions and the reports from them were used to raise funds. Early members of the National Geographic Society, who numbered barely two hundred, decided in 1888 to issue a popular magazine in order to spur interest in their affairs. The rise of the so-called pulp magazines, generally marked by the appearance of Frank Munsey's *Argosy* in 1896, was due in large part to the general interest in adventure stories set in foreign lands. The public desire to personally, and safely, experience the great expeditions helped create interest in travel and promote the family vacation, a trend promoters of the national park movement and their allies in the American railroad industry made effective use of.[37] Whether the last great era of exploration had actually closed was of less importance than the dwindling supply of mysterious terrestrial lands in which to set entertaining tales.

As a form of mass entertainment, expeditions were mythologized in ways that conveyed a romantic view of the world and the place of humans within it.

Recounted in lectures and the popular press, exploration tales became a vehicle for expressing the beliefs of the times, validating, by comparison to practices in foreign lands, belief in the superiority of Western institutions and values. Expeditions also helped reconfirm the belief that humans were capable of making great sacrifices and enduring extreme hardships in order to achieve presumably great ends.

Proponents of space travel suggested that interplanetary travel would continue the adventure of exploration in an endless realm. Said Ley in introducing the story of space flight to readers of the first edition of *Rockets*:

> It is the story of a great idea, a great dream, if you wish, which probably began many centuries ago on the islands off the coast of Greece. It has been dreamt again and again ever since, on meadows under a starry sky, behind the eyepieces of large telescopes in quiet observatories on top of a mountain in the Arizona desert or in the wooded hills near the European capitals. . . . It is the story of the idea that we possibly could, and if so should, break away from our planet and go exploring to others, just as thousands of years ago men broke away from their islands and went exploring to other coasts.[38]

Philip E. Cleator, founder of the British Interplanetary Society, wrote in 1936 that "the spirit of adventure, the lure of the unknown, will attract man to Mars just as surely as they caused him to penetrate into the wilds of Africa and the solitude of the Polar regions."[39]

By introducing space travel as an extension of terrestrial exploration, advocates of space flight found themselves promoting the traditional image of "small ships and brave men" sailing off into the unknown. It was hard to imagine space exploration in any other way. When members of the British Interplanetary Society prepared their plans for an expedition to the Moon, they assumed that the rocket ship and its crew would not be able to communicate directly with people back on Earth, a presumption derived from the seagoing expeditions of centuries past. BIS members toyed with schemes such as a flashing light to signal the arrival of the crew on the Moon.[40] The image of spacefarers as lonely men (and, occasionally, a woman) cut off from terrestrial contact persisted through the 1950s.[41]

An essential ingredient of any motivating vision, whether terrestrial or otherwise, is the ability to imagine distant events and places without the benefit of knowing exactly what is going on there. This allows people to transfer diverse expectations to the vision, drawn from popular beliefs, precisely because the real circumstances are unknown. The illusion of a Great Southern Continent, excited by Marco Polo's 1477 report of a land, rich in wealth, too far south for the Great Khan to conquer, was, in the words of Daniel Boorstin "embellished precisely because it could not be disproved."[42] Once experience reveals reality, expectations invariably fall, but the underlying vision rarely dies. Rather, people transfer the vision to a new realm. The dream moves on.

Myths supporting previous earthbound expeditions had fallen under the weight of frequent visitation to the places imagined. Beliefs about the Great Southern Continent collapsed after explorers discovered Antarctica to be a cold and barren land. The South Pacific was not an earthly Eden, free from Christian notions of sin, a myth that persisted well into the twentieth century, and Africa was not the "dark continent," the source of animal instincts and original sin, any more than other terrestrial locations.[43]

Popular conceptions of exploration, however, fulfill important human needs. Old myths die hard. As each of the old frontiers yielded to investigation, humans transferred their expectations to new realms. Interest in Alaska, the polar regions, and the Himalayas kept the promise of exploration alive through part of the twentieth century. Before long, however, romantic images of even those places wore away. Outer space appeared just in time to maintain the dream. As the middle of the twentieth century approached, exploration had nowhere to go but up.

Having formulating their initial vision, advocates of space exploration faced a daunting task. They had to find a group of people willing to finance their schemes, and, in spite of the occasional fantasy to the contrary, that proved extremely difficult. Money that appeared in fiction vanished in the real world. (In *Frau im Mond* industrialists hand out money for a lunar expedition in the belief that the Moon is rich in gold. The intrepid explorers do not disappoint their backers, for once on the Moon, they discover a cave full of precious metal.) Sensible advocates recognized the need for government support: only governments, with their extensive treasuries, possessed the resources necessary to implement elaborate space plans. Placing space exploration on the public policy agenda, however, was as hard as actually traveling to the Moon. The public did not believe space flight was real, political leaders knew it was too expensive, and military leaders, who had the funds to lavish on rocketry, were more interested in practical applications than in exploring the universe. In the United States, where exploration advocates encountered especially high levels of disbelief, converting popular fantasies into real policies required a concerted, well-planned campaign.

MAKING SPACE FLIGHT SEEM REAL

Man will conquer space *soon*. What are we waiting for?

—*Collier's,* 1952

In late 1949 George Gallup conducted a poll in which he asked Americans to imagine what sort of scientific developments would take place by the year 2000. Eighty-eight percent of respondents believed that a cure for cancer would be found within the next fifty years. Sixty-three percent envisioned a future in which trains and airplanes would be run by atomic power. When asked whether "men in rockets will be able to reach the moon within the next 50 years," however, doubters prevailed. In spite of the outpouring of books and articles on space exploration during the previous two decades, only 15 percent of those polled believed that this reality would actually occur.[1]

Imagination matters when societies contemplate new ventures. People must have the ability to visualize a solution to the phenomenon with which the society grapples and possess confidence in the attainability of the goal. In his excellent work on the origins of the U.S. space program, Walter McDougall pays homage to the importance of imagination in determining the shape of America's response to the challenge of space. He identifies three forces that were required to get the space program underway: an economy rich enough to afford this expensive endeavor; the appropriate technology, particularly the development of rocketry; and imagination, or what McDougall calls "culture, the realm of symbolism." Once these forces started "pushing in the same direction," McDougall argues, a large, government-supported space program "was automatic."[2]

By 1949, when Gallup tested the belief in space flight, the United States hardly lacked a popular culture devoted to space exploration. Opinion leaders, however, had done little to eliminate public disbelief. Far more than in Europe, where engineers and rocket scientists played a leading role in the promotion of space flight, space flight in America remained the province of people interested in fantastic tales. Science fiction played such a large role in the American image of space travel that public skepticism remained high even after rocketry became a practical science. To the American public, space travel was intriguing but infeasible. Before government support for actual space missions began, advocates had to convince Americans that space flight was possible. This required feats of imagination as impressive as those contained in fictional tales.

Fantastic stories had dominated American interest in extraterrestrial phenomena for some time. In 1835 the *New York Sun* had titillated readers with reports that the famous astronomer Sir John Herschel had observed a large number of creatures on the surface of the Moon through a specially constructed telescope. Herschel was a real person, who had actually arrived in Capetown to make astronomical observations of the southern sky. In a series of stories, which were wholly fictional, readers of the *Sun* learned that Herschel had located moon bison, bat-persons, and a unicorn.[3] Beginning in 1869, Edward Everett Hale had serialized a fictional account of an earth-orbiting space station in the *Atlantic Monthly*. He called it "The Brick Moon," a reference to the material out of which the orbital facility was constructed.[4] By the 1870s Jules Verne's book *From the Earth to the Moon* had become available in the English press. Two decades later the English writer H. G. Wells published *War of the Worlds*, serializing the story in the American magazine *Cosmopolitan*. At the Pan-American Exposition in Buffalo in 1901, visitors could take a simulated trip to the Moon in a winged spacecraft. The ride reappeared at Coney Island the following year, where its popularity inspired promoters to build Luna Park, a rival amusement facility.[5] By one estimate, sixty million visitors passed through the turnstiles at Luna Park during its first five years.[6]

Fans of space fantasy could expand their imaginations by reading dime-store magazines, an important medium for popular culture in early twentieth-century America. The American publisher Frank Munsey turned *Argosy* into a cheaply priced, pulp-paper magazine in 1896. *Argosy* carried adventure stories about the West, Africa, sea travel, and an increasing number of science fiction stories.[7] In 1926 Hugo Gernsback launched the first pulp magazine entirely devoted to science fiction, *Amazing Stories*. It spawned a series of competitors, including *Astounding Stories,* and ushered in what many have called the "golden age" of science fiction in the United States.[8]

As the market for science fiction in the pulps expanded, so did the interest of Hollywood movie producers. Swashbuckling Flash Gordon jumped from the comic strips to movie matinees in 1936, followed shortly by Buck Rogers, who made a detour through the radio waves on his way to the silver screen.[9] Television

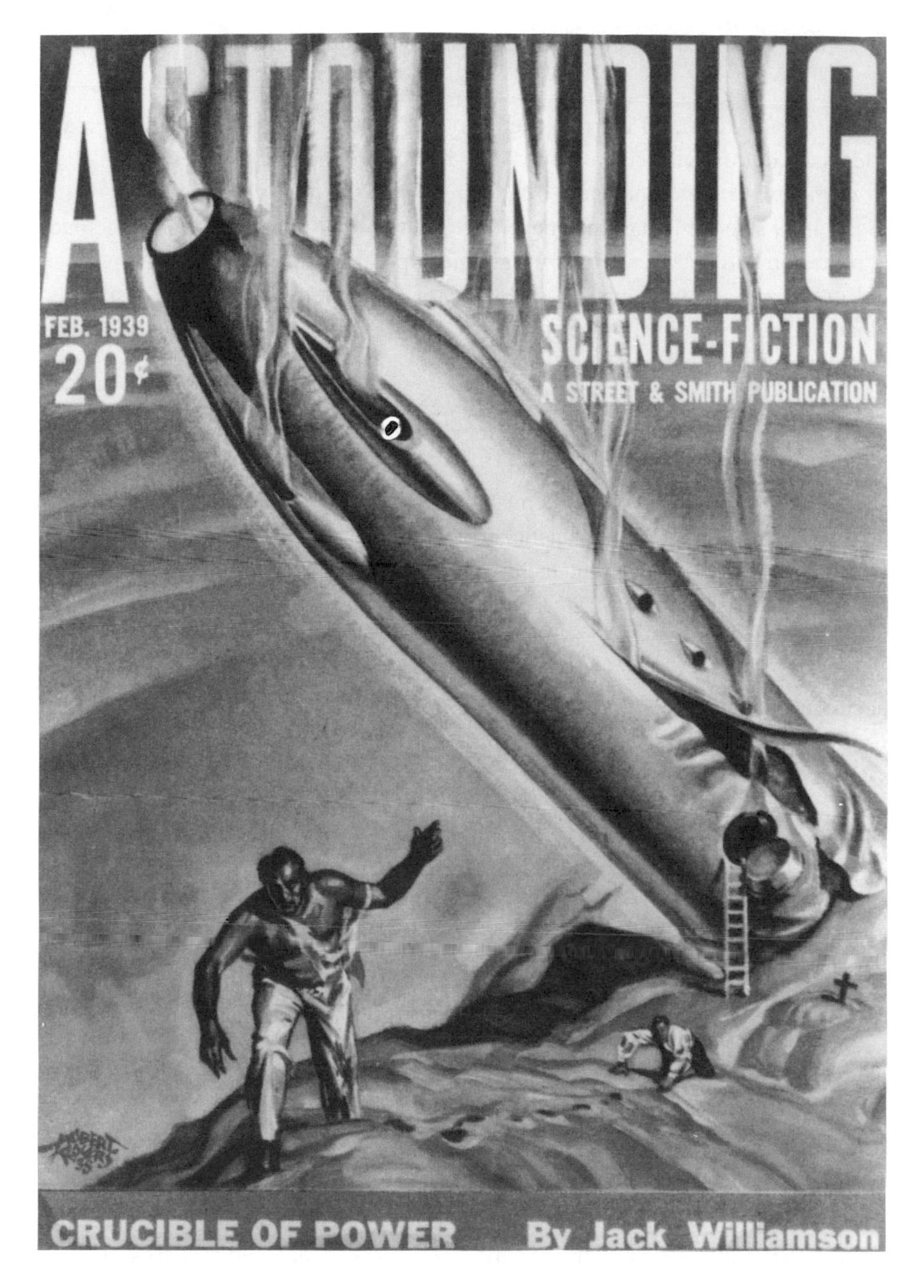

Members of the public remained skeptical about the possibility of space flight, relegating it to the realm of fantasy, the domain of pulp fiction magazines such as this 1939 issue of *Astounding Science Fiction.* Americans placed more faith in a future attended by atomic-powered airplanes and cancer cures than flights to the Moon, according to a 1949 Gallup poll. (©1939 by Street & Smith Publications; reprinted by permission of Dell Magazines, a division of Crosstown Publications)

followed. Producers of the new medium began offering a steady fare of space cadets to youthful viewers in June 1949 with the premiere of *Captain Video*. *Buck Rogers*, *Tom Corbett*, and *Space Patrol* premiered the following year.

By 1949 science fantasy had developed a loyal following and images of space exploration had become part of American popular culture. Much of the popular culture of space travel, however, was meant to be fanciful, not real. Visitors to Coney Island who rode the Trip to the Moon disembarked from a spaceship to be greeted by dancing Moon maidens and costumed midgets handing out samples of green cheese.[10] It is said that the producers of the wildly popular Captain Video confined their weekly budget for special effects to twenty-five dollars, producing such comical devices as the Captain's "opticon scillometer" and "cosmic ray vibrator."[11] E. E. Smith, a food chemist and part-time writer of science fiction, created the "space opera," intergalactic adventures featuring fleets of starships from which space cadets used exotic technologies to enforce Judeo-Christian ethics on villainous aliens. Smith's *Skylark of Space,* which he had begun writing in 1915, appeared in *Amazing Stories* in 1928.[12] Writers of science fiction and fantasy did not represent outer space as it actually existed any more than writers of Westerns accurately sought to portray the American frontier. Space was not meant to be real. Rather, it was meant to be entertaining. Science fiction writers used space, like writers of Westerns used the American frontier, as "a vast colorful backdrop against which any kind of story could be told," a philosophy that dominated space fiction for most of its history.[13]

As a consequence, the principles of space travel were not well understood among the public at large. In the case of other new technologies, such as airplanes and automobiles, people could witness actual operations firsthand. Although outrageous claims were made about the future of those technologies, those claims at least had their feet on the ground. With space travel, however, excepting the Europeans upon whom military rockets fell, most people had to experience the new technology vicariously. This gave the purveyors of popular culture considerable freedom in stretching technology to fit literary needs. Fantasy overwhelmed fact, and public understanding suffered as a result.

The golden age of science fiction, with its familiar aliens and intergalactic travel, did little to dispel disbelief among the public at large. Early science fiction, in the main, was not very scientific. The core of devoted readers remained clubby and small. Many people were exposed to science fiction, but few were so devoted as to acquaint themselves with the scientific facts involved. The 15 percent who responded affirmatively in 1949 to George Gallup's question about lunar expeditions undoubtedly exceeded the total population of science fiction devotees.

Faced with an extraordinary level of public skepticism, space boosters decided to deliver a more realistic message. David Lasser, one of the founders of the

American Interplanetary Society, published the first serious treatment of rocketry and space travel for the general U.S. public in 1931. Lasser explained how developments in the science of rocketry would make possible flights to the Moon, Venus, and Mars. Through his book, Lasser hoped that "the mists of misunderstanding, ignorance, and prejudice that surround the 'interplanetary rocket' question may be cleared up."[14] Both Lasser and his associate G. Edward Pendray sought to awaken public interest in the possibility of space flight.[15] In 1932 the *Brooklyn Daily Eagle* announced that Pendray would broadcast a talk on rocketry and space flight over W2XAB, the Columbia Broadcasting System's experimental television station. It must have been, said one historian, "the world's first TV space report."[16]

The proponents of space flight encountered extensive skepticism. In the 1936 volume of *Rockets Through Space,* another early English-language exposition on space travel, the British writer Philip E. Cleator complained, "Most people either do not believe that interplanetary travel is possible, or else they are utterly indifferent towards it. . . . There is no indication that the situation will be any different in the immediate future."[17]

Although none of the early works promoting space travel reached an exceptionally large audience, the themes they developed did. The principal theme was straightforward: human space flight was just around the corner. The obstacles to lunar and planetary exploration were not insurmountable, and the basic technologies had already been developed. Early publications provided readers with lessons on the dynamics of rocket flight and orbital mechanics as a way of proving this point. They also laid out an order of exploration that passed to the U.S. civilian space program practically unchanged.

Following the success of *Rockets: The Future of Travel Beyond the Stratosphere,* which had gone through a second edition in 1947, Willy Ley continued to produce books championing space travel. In 1949 he released the award-winning *Conquest of Space,* a collaborative effort with artist Chesley Bonestell.[18] In *Conquest,* as in the other books, Ley provided plain-spoken descriptions of the principles involved in orbiting the earth, building a space station, and exploring the Moon and other planets. *Conquest* opened with a description of a rocket launch from the White Sands Missile Range in New Mexico. The book's most distinguishing features were the illustrations by Bonestell, so remarkable that Viking Press gave him top billing on the title page. The pictures gave readers astonishingly realistic views of the earth as it might be seen by passengers in a rocket ship ascending into space. Thanks to telescopic images, Bonestell prepared equally realistic aerial views of the surface of the Moon, which looked remarkably like the photographs NASA would produce with real spacecraft more than a decade later. Having captured the attention of readers with these realistic views,

Bonestell offered illustrations of the surface of Mars and other planets based on astronomical information filtered through an active imagination.

Two years later Arthur Clarke published *The Exploration of Space,* another popular treatment of modern astronautics. This was Clarke's second book, *Interplanetary Flight* having appeared in 1951. Clarke was now thirty-three years old. Following his work on the BIS spaceship, he had served as a radar instructor in the Royal Air Force and earned a bachelor's degree in physics and mathematics. He had also distinguished himself by publishing a number of articles on electronics, including a 1945 proposal for a system of global communication satellites. Before the Second World War, space travel had been a hobby, spurred by a childhood interest in science fiction magazines. In 1951, he learned that he could support himself by writing about space travel, this being the first year his income as a writer exceeded the pay he received from his job as a scientific journal editor. Clarke had also dabbled in science fiction, producing a few short stories and a manuscript for a first novel on interplanetary flight. In *The Exploration of Space* he argued that the possibility of space flight "must now be regarded as a matter beyond all serious doubt":

> The conquest of space is going to be a very difficult, dangerous and expensive task. The difficulties must not, however, be exaggerated, for the steadily rising tide of technical knowledge has a way of obliterating obstacles so effectively that what seemed impossible to one generation becomes elementary to the next.[19]

In spite of his interest in electronics, which might have caused him to anticipate the potential for remotely controlled spacecraft, Clarke envisioned no suitable alternative to the widespread desire among space enthusiasts for human space flight. If his technical background suggested alternatives, his desire to entertain invalidated them. In his chapter on the spaceship, Clarke suggested that the exploration of space would occur in seven stages, all but the first requiring human crews. The stages, with a few technical changes, set forth the long-range plan that NASA adopted nearly one decade later:

> (1) Unmanned, instrument-carrying missiles will enter stable orbits round the Earth, and will travel to the Moon and planets.
> (2) Manned, single-step rockets will ascend to heights of several hundred miles, landing by wings or parachutes.
> (3) Multistage, manned rockets will enter circular orbits just outside the atmosphere and, after a number of revolutions, will return by rocket-braking and air resistance.
> (4) Experiments will be made to refuel these ships in free orbit, so that they can break away from the Earth, make a reconnaissance of the Moon, and return to the Earth orbit.

(5) The type of ship designed for a lunar landing will be flown up from Earth or assembled in free orbit, and after refuelling will descend on the Moon. The ship may then return direct to an orbit around the Earth, or it may make a rendezvous, *in an orbit round the Moon,* with tankers sent from Earth.
(6) While the exploration of the Moon is proceeding by the use of such ships and techniques, attempts will be make to refuel rockets for the journeys to Mars and Venus . . .
(7) Finally, landings will be made on Mars and Venus.

After humans reached the nearby planets, Clarke concluded, "the first era of interplanetary flight would be ended." The next era would be devoted to the problems of improving the efficiency of spaceships; building up bases on the Moon, Mars, and Venus; accumulating stores of fuel at useful places; and preparing for more demanding journeys to the outer planets and their moons.[20] Ultimately, Clarke believed, humans would touch intelligent life beyond the solar system:

> Even if we never reach the stars by our own efforts, in the millions of years that lie ahead it is almost certain that the stars will come to us. Isolationism is neither a practical policy on the national or the cosmic scale. And when the first contact with the outer universe is made, one would like to think that Mankind played an active and not merely a passive role—that we were the discoverers, not the discovered.[21]

In spite of the demonstrations of rocket power during World War II, the hopes of people such as Ley and Clarke continually struck a wall of public skepticism. Few believed that the U.S. government could be enticed to sponsor a massive program of lunar and planetary exploration. Serious plans for space travel had been envisioned but had not produced even the government authorization necessary to launch the first tiny Earth satellite.

This began to change in 1951. To broaden interest in their astronomy program, officials at the Hayden Planetarium, part of New York City's American Museum of Natural History, organized the first of three symposia on space travel. Although attended by only a few hundred persons, the products of the symposia eventually reached millions of Americans. Museum officials asked Willy Ley to coordinate a group of speakers for the first symposium. Ley's purpose in accepting the task was clearly promotional. "The time is now ripe to make the public realize that the problem of space travel is to be regarded as a serious branch of science and technology," Ley explained in his letter to potential speakers. "Invitations will be sent to institutions of learning, to professional societies and research groups, and also to the science editors of metropolitan newspapers and magazines (plus those out-of-town and foreign publications which have offices in New York)."[22] The first symposium was held on 12 October 1951. Ley, along with planetarium chair

Robert R. Coles and Robert P. Haviland, scheduled an appearance on the Nancy Craig television show for the afternoon following the symposium to report on their activities.

In his address to the symposium, Ley recounted the advances in rocketry that had made space travel possible. "Thirty years ago [in 1921] all serious thought about space travel consisted of a short book by Prof. Robert H. Goddard and a few articles in professional magazines." By 1931 the serious literature had grown in volume and scope and rocket enthusiasts had formed societies in Germany and in the United States. Small liquid-fueled rockets had reached altitudes of about three thousand feet. By 1941 rockets had risen 10 miles above the surface of the earth, and by 1951, two-stage rockets had attained a peak altitude of 135 miles. "The obvious question," Ley said, "is what will come next." He predicted that unmanned rockets would carry "all the way to the Moon," rockets would circle the earth in orbit, and nations would build manned space stations.[23] Prior to Ley's remarks, participants had been treated to an imaginary trip to Mars, part of the planetarium's current demonstration on "The Conquest of Space."[24] Ley was followed on the podium by Robert P. Haviland (a rocket expert with the General Electric Company), Fred L. Whipple (chair of the Department of Astronomy at Harvard University), Heinz Haber (a professor of space medicine), and Oscar Schachter (who spoke on the legal claims to outer space).

"It is obvious to all concerned that any project of such magnitude as the conquest of space can be successful only if it enjoys the full support of the public," Coles reiterated in opening the second symposium on 13 October 1952.[25] Ley coordinated the second symposium, inviting his old colleague Wernher von Braun to deliver one of seven formal papers. Von Braun and his German rocket team had moved to Huntsville, Alabama, where the forty-year-old von Braun oversaw the U.S. Army Ordnance Guided Missile Development Group as its technical director.[26] Anxious to summon forth public funding for space flight, von Braun called for a separate government program devoted solely to the exploration of space:

> Many a serious rocket engineer, while firmly believing in the ultimate possibility of manned flight into outer space, is confident that space flight will somehow be the automatic result of all the efforts presently concentrated on the development of guided missiles and supersonic airplanes. I do not share this optimism. . . . The ultimate conquest of space by man himself is a task of too great a magnitude ever to be a mere by-product of some other work.[27]

In calling for an independent space program, von Braun raised an issue with which speakers at the second and third symposia struggled. Many agreed with von Braun that space exploration should not be developed solely as a byproduct of military research or any other endeavor; they also understood the difficulty of justifying space travel on its own commercial or scientific merits, given its staggering cost. In introducing the speakers at the third symposium, museum official Joseph Chamberlain promised the audience, "You will hear a suggestion that we completely alter our stodgy and limited perspective concerning departure from earth and accept a new philosophy that may permit realization of this goal in our time."[28] Chamberlain had invited Arthur C. Clarke to coordinate the third and final symposium, held on 4 May 1954. Clarke had published five science fiction novels in the previous two years, including the classic *Childhood's End*, making him one of the most prolific and widely read advocates of space travel.[29] His symposium address on the history of the space flight idea was followed by a blunt assessment of the activities necessary to implement that ideal, by the commander of the Navy Bureau of Aeronautics, R. C. Truax, who laid out the obstacles confronting advocates of space flight like himself: "There is simply no overwhelming rational reason why we should try to set up a station in space, send a rocket to the moon, or take any other steps along the road towards interplanetary flight." The military utility was difficult to demonstrate, the commercial potential not worth the cost, and scientists were in general "a poverty-stricken lot." Eventually the United States might engage in space travel as an outgrowth of practical considerations, but "you and I," Truax noted, "would very likely not be alive to see even the beginnings." He wanted a space program soon and he wanted one simply for the excitement of it, because it was a great human adventure: "If the majority of the people of this country feel the same way, the arguments of immediate utility are unnecessary. One does not have to justify the manner in which he spends his own money. Ultimately every thing we do is done simply because we want to."

Government officials would be forced to support an ambitious space program if the public desired it. Members of the public would desire it if advocates excited their imaginations. Truax called on advocates of space flight to mount a crusade that would "fire the imaginations" of the uninformed and unconvinced. This work could not be carried out by a few informed rocket enthusiasts alone, he said, but "must be passed by newspapers and magazines, by commentators, by editors, by civic and fraternal organizations, by letter and by word of mouth from individual to individual.[30]

Members of the editorial staff from *Collier's* magazine had attended the first symposium on space travel at the Hayden Planetarium in October 1951. *Collier's* was a weekly magazine published in New York with a circulation of 3.1 million copies, making it one of the top ten magazines in the U.S. at a time when millions of Americans received information about current events through weekly and biweekly outlets such as *Life, Look, Collier's,* and the *Saturday Evening Post.*[31] Intrigued by statements emerging from the first symposium, *Collier's* managing editor Gordon Manning dispatched associate editor Cornelius Ryan to a less-publicized symposium on space medicine held in Albuquerque, New Mexico, early the following November. As described by Fred Whipple, who presented papers at both conferences, Ryan was initially skeptical about the possibility of space travel and artificial satellites. Following the formal presentations at the New Mexico conference, Whipple, Wernher von Braun, and Joseph Kaplin (a professor of upper atmospheric physics at UCLA) took Ryan aside for an evening of cocktails and dinner. An impassioned discussion ensued. As Whipple describes the meeting with Ryan,

> Whether or not he was truly skeptical, we persevered. Von Braun, not only a prophetic engineer and top-notch administrator, was also certainly one of the best salesmen of the twentieth century. Additionally, Kaplan carried the aura of wisdom and the expertise of the archetypal learned professor, while I had learned by then to sound very convincing. The three of us worked hard at proselytizing Ryan and finally by midnight he was sold on the space program.[32]

Ryan returned to New York and met with Manning. In preparation for what would eventually become an eight-part series of articles spanning two years, the editors initiated a series of discussions in New York with the leading advocates of space flight. The first issue to emerge from these discussions appeared on 22 March 1952 with a cover painted by Chesley Bonestell showing von Braun's design for a 265-foot-tall winged rocket dropping its second stage as it ascended past the forty-mile mark above the earth. Four of the speakers from the first Hayden symposium (Ley, Whipple, Haber, and Schachter) contributed articles, as did von Braun and Kaplan. The cover copy read: "Man Will Conquer Space Soon: Top Scientists Tell How in 15 Startling Pages."[33]

The eight-part series laid out what by then had become the conventionally accepted stages for the exploration of space. The 22 March issue opened with a lengthy article by von Braun describing plans for a large, Earth-orbiting space station, which he presented as the facility that would open up the new frontier. It could be built, he said, within ten to fifteen years. To introduce von Braun's article, Bonestell contributed what would become one of his most prophetic paintings—a two-page illustration depicting the 250-foot-wide, rotating space station, co-orbiting platforms, space "taxis," and a winged space shuttle. Ley explained an illustrated interior view of the space station, Haber discussed the problems of human survival in space, Kaplan gave a short lesson on the transition from atmosphere to space, Whipple described space-based astronomy, and Schachter discussed legal claims to the Moon and other celestial bodies.

To dispel public disbelief, advocates of space flight undertook a deliberate public relations campaign to convince Americans that space flight was real. Advocates presented their vision of the future in an eight-part series in *Collier's* that began with the 22 March 1952 issue. The cover was painted by the most influential space artist of the time, Chesley Bonestell. (By Chesley Bonestell with permission of the Estate and Bonestell Space Art)

The second set of articles, in the 18 October issue, described a journey from the space station to the Moon. "We will go to the moon in the next 25 years," the editors predicted after further discussions with their expert panel.[34] Von Braun contributed an article on the technical procedures for the voyage, and Ley provided a detailed description of the passenger vehicles. The following week, Whipple and von Braun described activities on the lunar surface, and Ley provided a short description of a lunar base. Stunning illustrations by Chesley Bonestell and Rolf Klep again graced the issues.

The next three issues, 28 February, 7 March, and 14 March 1953, were devoted to the problems of selecting, training, and protecting the people who would fly into space, confirming the assumption that humans would pilot the rocket ships of tomorrow. Only with the 27 June 1953 issue did the editors acknowledge the role of automation in space. Von Braun and Cornelius Ryan described what they called the "baby space station," a ground-controlled satellite that would collect scientific data. "What kind of scientific data do we hope to get?"[35] Von Braun and Ryan proposed collecting information on how three monkeys launched in the nose cone of their satellite would react during a one-way, sixty-day trip. Automated flight, in their view, existed for the purpose of determining the effects of space travel on living organisms, a prelude to human flight.

The ultimate aim of this early activity was presented in the 30 April 1954 issue of *Collier's*. Humans would explore Mars. Von Braun, who with Ryan had come to dominate the magazine's series, presented a plan for a two-and-a-half-year expedition to that mysterious, Earth-like planet. Fred Whipple introduced von Braun's article by posing the question that had fascinated astronomers and science writers for decades: Was there life on Mars? "There's only one way to find out for sure," Whipple concluded, "and that's to go there."[36] Chesley Bonestell, Fred Freeman, and Rolf Klep again provided vividly realistic illustrations depicting the proposed expedition.

As part of an aggressive campaign to promote the 22 March 1952 issue, the *Collier's* staff dispatched von Braun on a media speaking tour. Von Braun had emerged from the early sessions as the most adept exponent of space travel. Handsome and charismatic, with his slight German accent suggesting the archetypical rocket scientist, von Braun had an unusual talent for making space travel seem real. No single individual would attain his media status until astronomer Carl Sagan hit the cover of *Newsweek* in August 1977. Von Braun appeared on the "Camel News Caravan" with John Cameron Swayze and on Dave Garroway's "Today" show, promoting the theme that space travel would occur within the lifetimes of the viewing audience. Shortly after the *Collier's* series appeared, *Time* magazine characterized von Braun as "the major prophet and hero (or wild propagandist, some scientists suspect) of space travel."[37] *Life* magazine called him "the seer of space" and put him on the cover of their 18 November 1957 issue, devoted to the dawning space age.[38]

The most influential opportunity for public exposure of the space flight idea occurred in the spring of 1954 as the *Collier's* series ended. Walt Disney, who had already achieved considerable success with his animated cartoons and full-length motion pictures, had agreed to produce a one-hour weekly television program to begin on the American Broadcasting Company (ABC) television network in fall 1954. Disney was motivated by the desire to promote his Disneyland theme park, scheduled to open the following year. The television program was organized around the park's four themes: Adventureland, Frontierland, Fantasyland, and Tomorrowland. Of the four themes, Tomorrowland was the least developed. Disney asked one of his senior animators, Ward Kimball, to develop ideas for the television segment. Kimball, who had been following the *Collier's* magazine series, was very impressed that "there were these reputable scientists who actually believed that we were going out in space."[39] To assist with the show's story lines, Kimball called in Willy Ley, who in turn recruited von Braun and Heinz Haber. Ley, von Braun, and Haber, along with Kimball and Disney, appeared on the first show. It gave the space boosters access to an enormously large audience and a huge chunk of American popular culture.

Millions of Americans watched the first program, "Man in Space," which aired on Wednesday, 9 March 1955. (It was rebroadcast on 15 June and 7 September.) In opening the show, Disney noted that "one of man's oldest dreams has been the desire for space travel—to travel to other worlds. Until recently, this seemed to be an impossibility, but great new discoveries have brought us to the threshold of a new frontier—the frontier of interplanetary space."[40] Following an animated segment on the history of rocketry, Ley explained the development of the first multi-stage launch vehicles. Haber described the challenges of protecting space travelers from hazards such as meteorites and cosmic rays, illustrated by an amusing cartoon figure. Von Braun then presented the design for a large, four-stage rocket that could carry six humans into space and back. The show ended with an impressive animation depicting the first launch of the giant rocket ship from a small Pacific atoll and its return to Earth.

Von Braun and the others worked hard to convince viewers that space travel was real and that much of it could occur soon. "If we were to start today on an organized and well-supported space program," von Braun asserted, "I believe a practical passenger rocket could be built and tested within ten years."[41] His prediction was not far-fetched. The first flight of von Braun's three-stage Saturn 5 rocket with astronauts on board took place thirteen years later, on 21 December 1968.

Von Braun was more cautious in his predictions for a trip around the Moon. "Even though we now have the theoretical knowledge to make a trip to the Moon,"

The Disneyland theme park, which opened in 1955, featured a simulated space-station ride that passed over the surface of the United States from dawn to dusk. To promote his new theme park, Disney asked animator Ward Kimball to produce three hour-long programs for the popular "Disneyland" television show. Wernher von Braun and Willy Ley appeared on the shows to explain the mechanics of space flight and the need for an orbiting space station. (©Disney Enterprises, Inc.)

he said in the second Disney program on space flight, "it will be many years yet before our plans can fully materialize."[42] The second program, originally titled "Man and the Moon," aired on 28 December 1955. As before, von Braun emerged as the leading spokesperson for the space scientists and engineers, delivering a lengthy lecture on the construction of an Earth-orbiting space station and the preparation of the spaceship that would make the first voyage around the Moon. Von Braun and Kimball, who presented the major segments of the show (Ley and Haber did not appear), depicted the voyage to the Moon as one of the great adventures of humankind. "Such a trip has long been the dream of many men since history began," Kimball explained in introducing a humorous cartoon segment on lunar fantasies. Later in the program, four actors portrayed the first journey around the Moon, with dramatic close-ups of the lunar surface.

Von Braun appeared only briefly in the third program titled "Mars and Beyond," which aired in late 1957 while he was preoccupied with efforts to launch the first U.S. Earth satellite. He and Ernst Stuhlinger were photographed at work on plans for an interplanetary spacecraft powered by electromagnetic drive and did not speak. Most of the show consisted of animated segments and cartoons dealing with the development of life in the solar system and an imaginary trip to Mars.[43]

By then the series had accomplished the purpose sought by the space boosters. To millions of Americans it had portrayed space travel as something real, as no longer relegated to the realm of fantasy. The Disney series promoted this idea, as did the *Collier's* articles and other vehicles, simply by describing the voyages in an spellbinding, imaginative way.

In the summer of 1955, Disney opened his Anaheim, California, theme park, a monument to American mythology and popular culture. The original Tomorrowland sought to present a picture of the future as the Disney people envisioned it thirty-one years hence, in 1986.[44] Among its attractions, Tomorrowland offered the first visitors rides on an all-aluminum passenger train, pink and blue fiberglass boats, and an Autopia freeway. At the center of Tomorrowland, dominating the still-barren landscape of Disney's new park, rose an eighty-foot needle-nosed rocket ship, which Ley and von Braun had helped design.[45] The one-third scale model of an atomic rocket ship marked the entrance to Disney's Rocket to the Moon ride. As Disney himself explained, "After entering the Disneyland space port, visitors may experience the thrills that space travelers of the future will encounter when rocket trips to the Moon become a daily routine."[46]

The ride drew extensively upon the work undertaken for Disney's second television program on space. In a building behind the rocket, visitors received a fifteen-minute briefing on space travel, on film, before entering a circular chamber designed to simulate the ship's passenger cabin. Viewing screens on the floor and ceiling showed the earth recede and the Moon appear, and seats vibrated to simulate flight.[47] Disney wanted to make the ride as realistic as possible—a challenging task in 1955, when no film from space was available.[48] In only one respect did Disney delve into a bit of fantasy: guests on the ride looked down at the ruins of an ancient lunar base as the rocket traversed the back side of the Moon.[49]

The conquest of space as von Braun and Ley described it began with the construction of an Earth-orbiting space station. Consequently, the Disney people prepared Space Station X-1, an exhibit that opened with the park in mid-1955.[50] From a revolving platform, visitors watched the surface of the earth and the United States pass beneath them from dawn to dusk. As one Disneyland publication proclaimed, "In the future, according to scientists, space stations similar to the one at Disneyland will have full living quarters for several score of men and its own gravitational field."[51]

Disney and *Collier's* exposed millions of Americans to the possibility of space flight. They did so, moreover, in a manner that allowed people to visualize how space flight would actually appear. For years, lacking even primitive pictures from space, people had tried to imagine planetary landscapes and heavenly views.

This excited public curiosity, which provided a ready audience for artists who could depict cosmic images to an increasingly visual world. Artists had begun to

In addition to Space Station X-1, Disneyland featured a simulated rocket ride to the Moon. Wernher von Braun and Willy Ley helped design the eighty-foot-high model rocket ship that stood outside the entrance to the attraction. Through the Disneyland theme park and television series, space advocates helped convince millions of Americans that travel to the Moon and planets would occur soon. (©Disney Enterprises, Inc.)

portray landscapes on other worlds (independent of illustrations for works of fiction) in the late nineteenth century.[52] James Nasmyth furnished a set of pictures for the 1874 book on *The Moon* by James Carpenter.[53] Nasmyth prepared plaster models of lunar landscapes, which he photographed against black, starry backgrounds for most of his illustrations. The work of a number of space artists accompanied astronomical photographs and drawings in the 1923 book *The Splendour of the Heavens.*[54] Editors of various periodicals, including science fiction pulp magazines, commissioned astronomical art. The editors of *Astounding Science Fiction,* for example, displayed views of Mars and Saturn as seen from their moons in place of the more melodramatic illustrations promoting space fantasies.[55] Nearly sixty years after the event, Arthur Clarke still remembered "the splendid cover" of a 1928 edition of *Amazing Stories,* a painting of Jupiter with its atmospheric eddies and swirls.[56] Charles Bittinger illustrated a 1939 *National Geographic* article on recent developments in astronomy.[57] Like most other serious artists, he depicted astronomical wonders such as a terrestrial eclipse (an eclipse of the Sun covered by the Earth as seen from the Moon).

No artist had more impact on the emerging popular culture of space in America than Chesley Bonestell. Bonestell did for space what Albert Bierstadt and Thomas Moran accomplished for the American western frontier. Like Bierstadt and Moran, Bonestell's paintings took viewers to places they had never been before. Although the paintings were based on real sites, Bonestell used his imagination to exaggerate features in such a way as to create a sense of awe and splendor. He used light and shadow, as artists had done with the American West a century earlier, to portray space as a place of great spiritual beauty. Through his visual images, he stimulated the interest of a generation of Americans and showed how space travel would be accomplished. Many readers remembered the paintings of planets and spaceships more than the words in the articles that accompanied them.

Born in San Francisco in 1888, Bonestell was trained as an architect and designed numerous buildings in the San Francisco Bay area. In 1938, at the age of fifty, he changed careers, accepting a position as a special-effects artist at RKO studios in Hollywood. He became a highly paid specialist, painting backgrounds for such movie classics as *The Hunchback of Notre Dame* and *Citizen Kane.*

Since boyhood Bonestell had possessed an amateur's interest in astronomy. He occasionally prepared sketches of the Moon and planets in his spare time.[58] Using photomontage skills acquired from painting movie backgrounds, Bonestell prepared a series of paintings showing the planet Saturn as it would appear from five of its moons. He sold the paintings to *Life* magazine, which published them in its 29 May 1944 issue. The article contained what would become one of Bonestell's most famous illustrations—a stunning portrait of Saturn framed by snowy cliffs, as it might be seen from its giant moon Titan. Lit from behind, Saturn sat in a pastel blue sky, which, the editors explained, appeared blue instead of black because Titan was a moon with an atmosphere. In fact, as scientists later learned, the atmosphere of Titan is opaque. Any beings on the surface of this frozen moon would not be able to discern Saturn through the haze. That did not discouraged Bonestell nor other artists from preparing such fantastic scenes.[59]

In the tradition of astronomical art, Bonestell concentrated on landscapes rather than people or spaceships. He would occasionally add tiny figures to his landscapes, but these, as the editors of *Life* magazine explained, were "purely imaginary, put in to give scale."[60] Two years later, in 1946, *Life* published Bonestell's portrayal of a trip to the Moon. Again the paintings emphasized landscapes, particularly aerial views of Earth and the Moon. Only one of the paintings, an oblique perspective, featured the winged rocket ship that would carry the explorers on the voyage.[61]

Impressed by Bonestell's capacity to excite interest in space exploration, Willy Ley approached him to collaborate on the 1949 classic *The Conquest of Space.* Later, when *Collier's* magazine first contemplated its series on space exploration, its editors recruited Bonestell to meet with Ley and von Braun and others planning the stories.

Ley and von Braun provided Bonestell with the technical information necessary to prepare detailed paintings of spacecraft, and Bonestell in turn provided the space boosters with an imaginative visual representation of their ideas. Von Braun was especially influential. As Bonestell explained,

> Von Braun would send me sketches drawn on engineer's graph paper, which I converted into working drawings and then into perspective. The courses I had had at Columbia University [as an architecture student] enabled me to handle some very complicated problems, and my courses in structural engineering helped me to understand the mechanics of space machinery.[62]

A stunning painting of a V-2–type rocket ship sitting on the surface of the Moon with astronauts setting up experiments graced the jacket of *Conquest of Space.* Bonestell later collaborated with Ley and von Braun on *The Exploration of Mars.*[63] In addition to his usual landscapes, Bonestell contributed a series of 9 paintings depicting the expedition craft and party. Nearly all of Bonestell's paintings for the *Collier's* series contained a spacecraft, a space base, or other sign of the underlying exploration theme. Bonestell's art, along with paintings of rocket ships and space station cutaways by Rolf Klep and Fred Freeman prepared for the *Collier's* series, reappeared in two books based on the magazine articles.[64] At a time when the public had not yet accepted space travel, Bonestell's images showed it to be something real.[65]

Space art became increasingly technical under the influence of space boosters in the 1950s.[66] R. A. Smith, considered one of the pioneers of space hardware art, illustrated two of Arthur C. Clarke's early works on the exploration of space.[67] Jack Coggins illustrated two children's books that prominently displayed space travel technology.[68]

Public interest in the visual aspects of space encouraged American movie producers to depict space flight in realistic ways. In 1948 Robert Heinlein, a science fiction writer and unabashed promoter of space exploration, convinced movie producer George Pal to make a movie based on Heinlein's 1947 book *Rocketship Galileo,* a rather fanciful children's book.[69] Heinlein's screenplay depicted a trip to the Moon as it might actually occur. Pal recruited Chesley Bonestell to make the movie look accurate. The art work for the movie duplicated the cover of Bonestell and Ley's 1949 *Conquest of Space.* Bonestell chose for the movie destination the crater Harpalus, on the northern latitudes of the Moon, which allowed him to show the Earth near the horizon of his lunar landscape.[70] In one concession to dramatic imagery, he sculpted the crater walls to appear as if they had been carved by wind and rain. *Destination Moon,* released in 1950, provided large numbers of Americans with their first sense of what a lunar landing site might actually look like. It set new standards for the portrayal of space flight, returned a nice profit, and won the 1950 Academy Award for special effects.[71]

Encouraged by the popular reception to *Destination Moon,* Pal and Bonestell collaborated to produce *When Worlds Collide,* released in 1951. Bonestell again provided the dramatic artwork. Pal followed with *War of the Worlds* in 1953, for which Bonestell provided some art and technical advice. The special effects consumed six months of effort, whereas filming the actors took only forty days.[72] Buoyed by his success, Pal released *The Conquest of Space* in 1955.[73] The movie was based on the Bonestell-Ley book by the same name and on von Braun's earlier attempts to produce a Mars novel. Von Braun, Ley, and Bonestell all assisted with the movie, which depicts a voyage from von Braun's rotating space station to the surface of Mars.

In sharp contrast to the success of *Collier's* and Disney, this space booster failed. *Conquest of Space* was incredibly dull and a box office flop. Combined with huge cost overruns by the art and special effects department on the film *Forbidden Planet,* the failure of *Conquest* led Hollywood to abandon space realism for the next twelve years.[74] In 1957 the movie industry returned to established formulas in clunkers such as *Attack of the Crab Monsters* and *The Amazing Colossal Man.* Bonestell's Hollywood work dried up; Pal turned to science fantasy films such as *The Time Machine* (1959) and *Atlantis: The Lost Continent* (1960).

As the message of space exploration swept through American popular culture in the 1950s, public opinion began to change. People paid more attention to rocket technology and grew less skeptical of those promoting space exploration. By early 1955 the proportion of Americans who believed that "men in rockets will be able to reach the moon . . . in the next 50 years" had increased from 15 to 38 percent.[75] By 1957, following the launch of *Sputnik 1,* 41 percent agreed that humans would reach the Moon within twenty-five years, with the largest number correctly predicting that it would occur in "about 10 years."[76] All of this occurred prior to the creation of NASA and the decision to go to the Moon.

The public was not only better prepared to accept the reality of space travel by the mid-1950s but also prepared to accept the vision advanced by space boosters. Space travel was consistently depicted as humans exploring the last frontier; alternative space programs received scant attention in the organs of popular communication. Given the preponderance of attention devoted to space travel as human exploration, the public had difficulty imagining any other type of approach. This helped promote the belief, dominant within many sectors of the government, that space flight could not survive within political circles unless it emphasized human flight.

Top NASA officials embraced the paradigm of space exploration promoted by the space boosters, even before NASA was formed. Congress created NASA in large part through the transformation of the National Advisory Committee for Aeronautics (NACA), a forty-year-old collection of government laboratories whose employees spent most of their time conducting research on airplane flight. In early 1958, NACA was one of a number of government agencies vying for control of the new space program.[77] Even skeptics such as NACA director Hugh L. Dryden, who was not an enthusiastic booster of human space flight, helped further the space-flight dream. "The topic of our day is the new frontier, space," Dryden announced in an policy statement prepared for the annual meeting of the Institute of the Aeronautical Sciences in New York on 27 January 1958:[78]

> The escape of objects and man himself from the earth into space has long been the subject of science fiction writers and the comic strip artists. More recently, it has been a matter of interest to a growing number of serious-minded scientists. Now it has acquired a new sense of imminence and reality. Space travel has stirred the imagination of man to an extraordinary degree.

Dryden announced that NACA was prepared to play a leading role in the exploration of space, a position officially endorsed by the NACA Advisory Committee eleven days earlier.[79] "In my opinion," he said, "the goal of the program should be the development of manned satellites and the travel of man to the moon and nearby planets."[80]

Shortly after NASA came into being on 1 October 1958, engineers who had only months earlier worked for NACA established the Research Steering Committee on Manned Space Flight, commonly known as the Goett committee, to prepare a set of recommendations "as to what future missions steps should be." At the first meeting in May 1959, committee members agreed that they "should not get bogged down with justifying the need for man in space in each of the steps but outrightly assume that he is needed inasmuch as the ultimate objective of space exploration is manned travel to and from other planets."[81] Max Faget, who achieve fame as the designer of the blunt, conical-shaped spacecraft that took the first American astronauts into space and back, urged the committee to select a lunar expedition as its immediate goal, "although the end objective should be manned interplanetary travel." George Low, who would play a leading role as NASA's chief of manned space flight, concurred with the lunar objective, "because this approach will be easier to sell."[82]

Shortly after the creation of the National Aeronautics and Space Administration in 1958, agency officials adopted the long-range plan of lunar and planetary exploration promoted by space advocates such as Ley and von Braun. "Manned exploration of the moon and nearer planets must remain as the major goals," they wrote. This artist's rendition depicts the accomplishment of a key step in the completion of the spacefaring dream—construction of an environmental enclosure for a colony on Mars. (NASA)

At the May meeting, committee members adopted the exploration program promoted by space boosters. The official NASA program of space flight adopted every one of the steps proposed by Clarke in *The Exploration of Space:* suborbital tests of human spacecraft, orbital flights, a trip around the Moon, followed by a landing on it, reconnaissance of Mars and Venus by automated spacecraft, and human expeditions to those nearby orbs.[83] In only one respect did committee members depart from the traditional dream. Members could not agree on whether to build a large space station as a prerequisite to lunar and planetary exploration.[84] They agreed on the need for a small orbiting laboratory. The much larger space station, however, could wait. In the 1959 long-range plan, NASA officials announced that at least the first flights around the Moon could be conducted prior to the establishment of a "permanent near-earth space station."[85]

The paradigm of human exploration became part of NASA's organizational culture. People in NASA assumed that this was the way that space flight was done. During the 1984 debate over the proposed Earth-orbiting space station, one of

NASA's top engineers admitted that the civilian space agency would still want to build the facility "even if it could be proved that functionally everything conceived of today could be done by robots." NASA would build the station, he said, because "we think it is NASA's charter to essentially prepare for the exploration of space by man in the twenty-first century."[86]

The philosophy of space as human exploration dominated the thoughts of NASA policy planners in the decades that followed. In 1969, looking past the flights to the Moon, NASA officials proposed a course of action for the following twenty years. The NASA administrator who had overseen the crash program to get to the Moon, James E. Webb, had turned the agency over to his forty-seven-year-old deputy, Thomas Paine. Paine was a self-described "swashbuckler," an analogy he drew from his navy days. The son of a U.S. Navy commodore, Paine had served as a submarine officer in the Pacific during World War II. Though he left the navy for a career as an administrator and research engineer, the thought of naval traditions never left him. The Constitution permits Congress to grant letters of marque and reprisal, which is a device by which the government empowers private citizens to become buccaneers on behalf of the public good. As NASA administrator, Paine urged his officers to adopt that "swashbuckling, buccaneering, privateering kind of approach" to promoting the spacefaring dream.[87]

Paine proposed a bold post-Apollo space program nearly identical to the one set forth on the pages of *Collier's* magazine seventeen years earlier. His proposal (subsequently endorsed by the special vice-presidential Space Task Group) envisioned a very large space station, a winged space shuttle, large orbiting observatories, a lunar base, a space station in polar orbit around the Moon, a nuclear-powered transportation system for deep space travel, planetary probes and flybys, and a human expedition to Mars to be launched by 1983.[88] Though his proposals were not adopted, Paine's relentless proselytizing delighted advocates of the spacefaring dream. It also irritated White House officials so much that they removed him from his post as NASA administrator in 1970.

Paine reappeared in 1986 as chair of the National Commission on Space, having spent the interim as a captain of industry in the aerospace sector. The commission was created by Congress and the president to chart the nation's space goals for the twenty-first century. Its report opened with the famous 1952 Chesley Bonestell diorama portraying hopes for a permanent space station, accompanied by a winged space shuttle and a large space telescope. Underneath it a more recent illustration by space artist Robert McCall depicted the same facilities actually completed or approved: the current design for the *Freedom* space station, the existing NASA space shuttle, and the Hubble space telescope. Looking to the past, the commission members wrote that "while predicting the future can be hazardous, sometimes it can be done."[89]

In the report, commission members looked fifty years into the future: "We are confident that the next century will see pioneering men and women from many nations working and living throughout the inner Solar System. Space travel will be as safe and inexpensive for our grandchildren as jet travel is for us."[90] Members of the commission lavishly illustrated their report with paintings of lunar settlements, orbiting spaceports, transportation systems, robotic probes, space prospecting, and bases on Mars.

The dream continued. In 1987 NASA administrator James Fletcher established the Office of Exploration, set up to help win support for human expeditions to the Moon and Mars.[91] In his last year in office, President Ronald Reagan approved a national space policy that included a goal "to expand human presence and activity beyond Earth orbit into the solar system."[92] In July of 1989 a special NASA working group submitted a conceptual plan for a lunar outpost and a human expedition to Mars.[93] In a speech later that month from the steps of the National Air and Space Museum in Washington, D.C., President George Bush endorsed those goals. In addition to endorsing the lunar base and Mars expedition, Bush went beyond, becoming the first chief executive to suggest that humans would someday leave the solar system for nearby stars: "You who are the children of the new century—raise your eyes to the heavens and join us in a great dream—an American dream—a dream without end."[94] Bush said Americans "will follow the path of Pioneer 10," referring to the U.S. space probe that had already flown beyond the orbits of Neptune and Pluto: "We will travel to neighboring stars, to new worlds, to discover the unknown. And it will not happen in my lifetime, and probably not during the lives of my children, but a dream to be realized by future generations must begin with this generation."[95]

The idea of space exploration began as a fantastic, unattainable dream. Space boosters during the 1950s convinced the American public that the dream could become real. Government officials and committees set forth proposals that endorsed the primary initiatives. The basic vision of human space exploration never died. In many ways, it grew stronger with time. Winning the massive amounts of government support necessary to carry it out, however, was (and continues to be) very difficult. Space advocates need politicians to approve and fund the dream. Winning approval for major exploration steps like trips to the Moon required an additional step in imagination.

THE COLD WAR

> Control of space means control of the world.
>
> —Lyndon B. Johnson, 1958

> Chicken Little was right.
>
> —David Morrison, 1991

By themselves, the early efforts to promote the conquest of space were not sufficient to unleash the billions of dollars necessary to undertake the endeavor. The spirit of adventure and discovery to which much of the early promotional efforts appealed did not justify such a large commitment. Promotional efforts allowed the public to envision explorers making trips to the Moon and planets, but most politicians in the mid-1950s could not imagine appropriating the money to fund such expeditions. Members of an official U.S. science advisory committee repeated a familiar refrain when they admitted in 1958 that flights to the Moon and nearby planets would occur at some time in the future, but that "it would be foolish to predict today just when this moment will arrive."[1]

Having helped convince the American public that space travel was real, boosters faced an additional challenge: they had to conjure images that would promote the will to act. For this purpose space advocates found a ready supplement in public anxiety about the Cold War. Boosters not only promoted spaceflight through tales of adventure but also sought government support by scaring the American public. Public fears played a critical role in unleashing the billions of dollars necessary to begin the conquest of space. Enthusiasm for the grand vision drew force from nightmares about the cosmos as well as from pleasant dreams of space flight.

Public policies often emerge as the result of lengthy periods of preparation punctuated by precipitating events. From the late 1940s through the mid-1950s, advocates of space exploration prepared the American public for the conquest of space with elaborate visions of promise and fear. The precipitating events occurred on 4 October and 3 November 1957, with the launching of *Sputnik 1* and *Sputnik 2* by the Soviet Union. Both before and after these events, President Dwight D. Eisenhower and members of his administration sought to fashion a practical space effort that differed considerably from the grand vision of human exploration space advocates desired. The launch of the first Earth-orbiting satellites, combined with the previous promotion of space flight in the popular culture, overwhelmed the Eisenhower alternative and led subsequent political leaders to pursue much more ambitious goals.

When the Cold War ended, so did much of the rationale for the ambitious program of lunar and planetary exploration that fear had motivated. Space advocates, nonetheless, did not abandon the motivational use of terror. Rather than abandon their dreams, they sought out new dangers with which to promote interplanetary travel. Although not as powerful as the Cold War, the new nightmares followed the old formula, preparing public hopes and fears in anticipation of precipitating events that would eventually alter the public policy agenda.

During the Cold War not everyone shared the enthusiasm for the grand adventure of large space stations, lunar bases, sophisticated spacecraft, and voyages to nearby orbs. Not everyone accepted the prophecies of Wernher von Braun and other space pioneers. During the 1950s, a number of scientists and public officials put forth an alternative view. It was not as well presented as the adventurous vision, it never recruited a spokesperson as charismatic as von Braun, and the press tended to treat it as a dissent to the dominant vision rather than as a free-standing option. The alternative, nonetheless, enlisted one powerful advocate: Dwight D. Eisenhower, the first president to formulate public policy for the exploration of space.

Much of the attractiveness of Eisenhower's alternative arose from its low cost. During the 1950s, government officials identified a number of priorities on which to spend large sums of money. Expeditions into space were not among them. The United States undertook a $16-billion crash program to create a fleet of intercontinental and intermediate-range ballistic missiles in the hope of preventing nuclear armageddon.[2] It initiated a $26-billion program to complete the interstate highway system.[3] It intended to spend large sums of money to modernize the nation's schools.[4] Gigantic outlays for space adventures did not sit high on the list of national priorities.

In helping to prepare the articles on the conquest of space, the panel of experts advising the editors at *Collier's* magazine estimated the cost of the first major initiative in space. An Earth-orbiting space station and the fleet of large rockets necessary to support it would consume approximately $4 billion over ten years, they said.[5] When the *Collier's* panel issued its estimate in 1952, aggregate federal spending totaled only $68 billion. The first step toward exploration of the cosmos promised to consume the equivalent of 6 percent of all federal spending that year. Early estimates for an expedition to the Moon were even more staggering.[6] The editors at *Collier's* rightly pointed out that an ambitious space program would require an effort "equivalent to that which was undertaken in developing the atomic bomb."[7]

The division between advocates of the grand vision and those supporting a less-ambitious alternative surfaced during the second Hayden Planetarium Symposium on Space Travel held in New York City on 13 October 1952, which took place as the second and third issues of *Collier's* devoted to space travel reached their readers. In his remarks to symposium participants, Wernher von Braun displayed the *Collier's* articles as proof that full-scale exploration of the Moon "will be well in the realm of possibility" once work on an Earth-orbiting space station was complete.[8] Von Braun devoted most of his presentation to his primary specialty, the development of large rocket ships necessary to begin the conquest of space.

Dr. Milton W. Rosen, who immediately preceded von Braun on the podium, disagreed. Like von Braun, Rosen was a pragmatic rocket engineer. He was an active member of the American Rocket Society, the spacefaring group founded by science fiction fans twenty-two years earlier, which had changed its name in an effort to attract serious scientists and engineers. Following his college education, Rosen had accepted a job at the Naval Research Laboratory, a military research facility in southeastern Washington, D.C. Within the agglomeration of government agencies competing for a share of rocket and missile activities, the laboratory was somewhat conservative, its scientists and engineers devoted to the development of small rockets and satellite technology. When NASA was established, these people formed the nucleus of the satellite and science facility known as the Goddard Space Flight Center in Greenbelt, Maryland. Officials in the Eisenhower administration hardly could have found a existing government facility more devoted to their alternative vision of space exploration than the Naval Research Laboratory.

This was not the last time Rosen and von Braun would clash. In 1955, defense department officials gave the Naval Research Laboratory the responsibility for launching the first Earth-orbiting satellite, an assignment von Braun's army rocket team desperately wanted. At the time of the Hayden symposium, Rosen was the scientific officer in charge of the rocket program. The Viking rocket was the first stage of the launch vehicle that the Naval Research Laboratory would use in its attempt to launch its *Vanguard* satellite; Rosen would become technical director for Project Vanguard. When the rocket failed in a spectacular launchpad explosion in late 1957, von Braun grabbed the satellite assignment, and using his own Juno I rocket, launched the first U.S. Earth-orbiting satellite on 31 January 1958.[9]

At the symposium, Rosen explained the difficulties involved in launching even a small rocket toward space. Thousands of separate components had to work properly in order to avoid a failure. There were even greater obstacles to surmount in order to take humans into space. Reliability of components, cosmic and solar radiation, vehicle skin temperatures, rocket motor technology, and vehicle recovery all stood in the way of von Braun's dreams. Rocket scientists "have almost exhausted our store of basic knowledge," Rosen observed. "The engineer who has drawn the ingredients from the cupboard of basic research now finds that the cupboard is bare."[10] Rosen wished that the government would devote more resources to problems of space flight but admitted that this was not currently feasible. Too great a share of the nation's resources and scientific talent were devoted to military preparedness for the Cold War.

Given all of these limitations, what sort of a space program might the U.S. successfully undertake? "If we cannot build a space ship today," Rosen asked, "what can we do to advance the cause of space travel?" Rosen called for fundamental research on the problems of space flight to replenish the store of basic knowledge and urged his listeners to support a program of small Earth satellites. "Before we can attempt to transport human beings in a ship that orbits around the earth," he asserted, "we must produce a practical, reliable, unmanned satellite."[11]

Attuned to the newsworthy quality of controversy, the *New York Times* and *New York Herald Tribune* headlined the Rosen–von Braun debate. Both accompanied their stories with pictures of the two individuals posed alongside a twelve-foot-high scale model of von Braun's proposed three-stage rocket ship. "Two rocket experts argue 'Moon' plan," the *Times* headlined. "Army Expert Sees Platform 1,000 Mi. up in 15 Yrs," the *Herald Tribune* noted. "Navy Scientist Skeptical."[12] Both articles opened with von Braun's assertion that the United States within ten to fifteen years would construct a station one thousand miles out in space, serviced by large rockets, and both treated Rosen's comments as a dissent to von Braun's vision. The *Times* quoted Rosen's assertion that von Braun's designs "are based on a meager store of scientific knowledge and a large amount of speculation." The *Herald Tribune* featured Rosen's contention that the United States "would be throwing its money away" if it undertook "any one of the fantastic projects for a space ship that have been proposed in the last few years."[13]

Other outlets highlighted the dissenting view. After *Collier's* magazine claimed "Man Will Conquer Space Soon," writers at *Time* magazine responded with a cover story on the journey into space. "Separating facts and fancy about space travel is almost as difficult as a trip to the Moon," the authors observed. *Time* attempted to lay out the facts. A rocket ship capable of sending a small spaceship beyond the earth's gravitation well would have to be "as big as an ocean liner." Without scientific breakthroughs in areas such as rocket propulsion, the construction of large launch vehicles and space stations would be "a reckless leap into the blind future . . . a gigantic fiasco." This had not deterred the purveyors of public imagination, however. With little attention to the technical details, the writers said, toy shops, science fiction magazines, and television programs had "already zoomed confidently off into the vast ocean of space."[14]

The dissenting view was endorsed by groups of scientists waiting to conduct research in space. Such groups ridiculed the grandiose schemes of space boosters; some attacked von Braun personally. "He is the man who lost the war for Hitler," said one critic, who claimed that von Braun had drained the best brains and material away from the German war effort and now was trying to do the same to the United States as it sought to win the Cold War.[15] The manifesto for the dissenting point of view appeared in a thirteen-page pamphlet on outer space issued by President Eisenhower's science advisory committee, chaired by James R. Killian. Though trained as an engineer, Killian had developed a reputation as a person who could represent scientific points of view. He had spent his entire professional career in academia, rising from the modest position of assistant editor for a scientific journal published by the alumni association at the Massachusetts Institute of Technology to the presidency of that institution. During his years at MIT, he promoted a number of scientific projects, such as radar development and missile guidance systems, that found their way into military use. In the fall of 1957, following the launch of *Sputnik 1,* President Eisenhower asked Killian to come to the White House to serve as special assistant for science and technology. From that position, Killian issued the short *Introduction to Outer Space.*[16] Eisenhower found the manifesto "so informative and interesting" that he ordered the Government Printing Office to reproduce the statement and offer it to the American public for just fifteen cents per copy.[17]

The pamphlet set out a series of scientific questions that could be adequately addressed by a well-constructed program of satellites and automated probes. "Scientific questions come first," the statement argued, in measuring the value of sending rockets and satellites into space.[18] Earth satellites could study cosmic rays and solar radiation, assist with weather forecasts, transmit television broadcasts, and improve the clarity of astronomical observations. Automated probes could examine the origin of the Moon, search for life on Mars, and study the atmosphere of Venus.

The pamphlet placed little emphasis on the role of humans in space. Because humans were such adventurous creatures, the time would come when they would want to go into space and see the results of scientific efforts for themselves, but Killian and his colleagues were reluctant to predict when that might occur.

"Remotely-controlled scientific expeditions to the moon and nearby planets," they argued, "could absorb the energies of scientists for many decades."[19]

This perspective was repeated by subsequent science advisory groups. A special 1961 presidential transition committee cautioned incoming president John F. Kennedy that "a crash program aimed at placing a man into an orbit at the earliest possible time cannot be justified solely on scientific or technical grounds." The United States led the world in space research, the group argued, a position that an extensive manned space program could hinder "by diverting manpower, vehicles and funds."[20] A subsequent science advisory committee made a similar argument in 1970. The extraordinary cost of human space flight could not be justified solely on the grounds of science, technology, and practical applications, its members asserted. Instead, an excellent program of space science could be conducted for about half of the money then being expended on the civil space program.[21]

In spite of the care with which the dissenting view was presented, it never became part of the American culture in the same way the romantic vision did. This is not to say that the public accepted the grand vision unconditionally. Public opinion in the United States is notoriously ambivalent, a quality that affected civilian space as well as other government activities.[22] By the mid-1950s, most Americans believed that ventures such as a trip to the Moon would occur soon; at the same time most believed the government should not lay out large sums of money to accomplish the task. When asked in 1960 whether the United States should spend upwards of $40 billion "to send a man to the moon," 58 percent of the respondents to a Gallup poll responded no. Fifty-two percent of the same respondents nevertheless agreed that the venture would be accomplished within ten years.[23]

This pattern persisted throughout the formative decades of the U.S. space program. Public expectations remained high while the willingness to spend money remained low. Even by the mid-1980s, when interest in space stations and expeditions to Mars rebounded, the number of people who wanted to undertake these projects exceeded by a factor of two the number of people willing to increase the space budget to pay for them.[24]

As the dreams of space pioneers encountered the realism of American scientists, realism initially won. The efforts of space boosters to win financial support for their grand adventure had practically no effect on the Eisenhower administration. Eisenhower and his aides possessed an image of space exploration quite different from the one that dominated U.S. popular culture in the 1950s. Drawing on the dissenting view of American scientists, Eisenhower created what became the most visible alternative to the aims of space pioneers, and for a brief time, his vision defined U.S. space policy.

Eisenhower had spent most of his life in the U.S. Army. A 1915 graduate of the U.S. Military Academy at West Point, his skill at military planning so impressed his superiors that they promoted him over the heads of more than three hundred senior military officers to be supreme commander of all Allied forces in Europe. Though an career soldier, Eisenhower's simple Kansas upbringing left him suspicious of empire-builders in the Pentagon. He warned Americans in his farewell address that industrialists and military officials were acquiring "unwarranted influence" over the U.S. government, creating large, self-serving projects for which there was no justifiable need.[25]

As his biographer Stephen Ambrose has affirmed, Eisenhower came from a generation of military leaders for whom the Japanese surprise attack on Pearl Harbor was "burned into their souls."[26] He was absolutely determined to prevent a similar move by the Soviet Union on the United States or its Allies, and was far more interested in using space for this purpose than as a theater for some sort of exploratory opera. He characterized the promotional efforts of people who ignored this purpose in favor of space stations and flights to the Moon as "hysterical."[27]

From his wartime experience, Eisenhower knew that aerial reconnaissance, which revealed the movement of troops and munitions, was the key to preventing unforeseen attacks. The Soviet Union had raised the so-called Iron Curtain around itself and its European allies in part to prevent such reconnaissance. Existing reconnaissance technology, such as aircraft overflights and balloons, violated the airspace of countries behind the Iron Curtain. Soviet leaders viewed aircraft overflights as hostile incursions, a point underscored in 1960 when the Soviet Union shot down and tried U.S. spy plane pilot Gary Francis Powers.

Searching for a means to conduct effective aerial reconnaissance, Eisenhower approved the creation of the Technological Capabilities Panel whose 1955 report identified various scientific options for defense preparedness. The panel was dominated by scientists such as James Killian and Lee DuBridge, who, like Eisenhower, favored the alternative approach to space.[28] With regard to space activities, the panel proposed that the United States develop a small scientific satellite to be placed in orbit around the earth. In addition to its scientific value, the satellite would establish the principle that objects in orbit did not violate the territorial sovereignty of nations across which they flew. The Soviet Union, the panel reasoned, would be less likely to object to an internationally sponsored scientific satellite making the first orbital incursion than a satellite launched by the U.S. military. The scientific satellite thus would establish the important principle of free access, which in the future would allow the United States to fly military reconnaissance satellites across the Soviet Union without provoking protest or retaliation.[29]

The Soviet Union inadvertently established the freedom of space principle when it launched *Sputnik 1,* which crossed the continental United States shortly after it was launched. Because the Soviet's launch accomplished one of the primary goals of Eisenhower's space policy, some historians have speculated that the president was not as surprised by the Soviet achievement as the contemporary popular press portrayed.[30]

Eisenhower's vision for space relied to a great extent upon satellite technology. For Eisenhower and the members of his administration, satellites and automated probes accomplished most of what they wanted to do in space. Satellites reenforced the important freedom of space principle, gave the United States the aerial reconnaissance platforms it needed to detect military preparations and monitor arms agreements, and allowed scientists to conduct space-based research. They did so, moreover, at a fraction of the cost of manned missions, an important consideration in the budget-conscious Eisenhower administration.

Commensurate with the emphasis upon satellite technology, Eisenhower's space program deemphasized the role of human space flight. In 1958 Eisenhower approved an effort to develop a single-seat space capsule capable of placing an astronaut in orbit around the earth. Project Mercury, as it was known, was designed simply to test whether humans could perform any useful functions in the void.[31] Ignoring calls for more ambitious ventures, Eisenhower steadfastly refused to approve any "manned" program that went beyond the single-seat Mercury capsule. In late 1960, shortly before leaving office, Eisenhower specifically disapproved NASA's request to fund the keystones of its advanced human space-flight program—a three-person Apollo spacecraft and a powerful, liquid hydrogen–fueled rocket. In his final budget message to Congress, Eisenhower recommended that the United States undertake human space flights beyond Project Mercury only if "testing and experimentation" could establish "valid scientific reasons" for doing so, a not-so-subtle reference to the fact that space boosters wanted to use human exploration to boost national prestige.[32] In an accompanying report, his Science Advisory Committee dismissed the motives for advanced human space flight as "emotional compulsions."[33]

Eisenhower's alternative space program placed a great deal more emphasis upon space science than upon engineering feats. "Scientific questions come first," his advisory committee argued, when measuring "the value of launching satellites and sending rockets into space." Eisenhower was prepared to compete with the Soviet Union in scientific discoveries, in which he saw the United State holding a substantial lead. He wanted to weigh the value of research in space, moreover, against the benefits to be gained from investigations on Earth. "Many of the secrets of the universe will be fathomed in laboratories on earth," committee members wrote, and the national interest required that "our regular scientific programs go forward without loss of pace."[34] Eisenhower had no interest in entering into a race with the Soviet Union that depended upon large rocket boosters, a field in which scientific questions took a back seat to engineering capability and the

Soviets held a commanding lead. Interviewed after leaving the White House in 1962, he questioned the wisdom of President Kennedy's decision to engage the Soviet Union in a race to the Moon:

> Why the great hurry to get to the Moon and the planets? We have already demonstrated that in everything except the power of our booster rockets we are leading the world in scientific space exploration. From here on, I think we should proceed in an orderly, scientific way, building one accomplishment on another, rather than engaging in a mad effort to win a stunt race.[35]

Eisenhower did not want to set up a crash program to explore space on the scale of the Manhattan Project. He was already disturbed by the degree to which the arms race had bolstered what he called the "military-industrial complex" and did not want to give the aerospace industry another project on which to expand its base.[36] His desire for a balanced budget would not allow him to respond to every Cold War contingency, and he feared that a crash program would divert resources from more pressing needs. As he wrote after President Kennedy had approved Project Apollo, "This swollen program, costing more than the development of the atomic bomb, not only is contributing to an unbalanced budget; it also has diverted a disproportionate share of our brain-power and research facilities from other equally significant problems, including education and automation."[37]

NASA executives serving during the Eisenhower administration did not want to create separate organizations for human and robotic flight. Instead, they sought to merge NASA's "manned" and "unmanned" space activities. They created a single office of space flight at NASA headquarters for both human and robotic flight. They planned to create a single space projects center for both human and machine flight at what eventually became the Goddard Space Flight Center in Maryland.[38] That center was supposed to oversee both the recently approved Mercury human space-flight project and NASA's scientific satellite programs, but this intent quickly disappeared once President Kennedy approved the race to the Moon. As part of a full-scale 1961 reorganization to gear up for the lunar objective, NASA officials created a separate headquarters office for manned space flight.[39] Texas politicians, including Congressman Albert Thomas, head of the House appropriations subcommittee that oversaw NASA's budget, urged NASA to build an entirely new field center near Houston to oversee human flight.[40] Had Eisenhower's alternative prevailed, it is probable that human and robotic programs would have been merged within a single NASA center, dampening the schism that subsequently developed between "manned" and "unmanned" activities.

Eisenhower's alternative established a space program with far more emphasis upon satellites, robotics, and science than the boosters favored. It avoided the "boom and bust" cycle of crash programs and likely would have led to a closer relationship between human and automated space activities. Although the president normally plays a leading role in defining the scope and direction of the U.S. space program, in this case images in the public mind undercut Eisenhower's alternative, doing so because advocates of grander schemes adroitly played upon public fears about the Cold War.

Support for Eisenhower's space program within the U.S. government collapsed with the launch of *Sputnik 1* in early October 1957 and the launch of *Sputnik 2* one month later. *Sputnik 1* was the media event of the decade. Much of the sense of public security during the early years of the Cold War was based on the assumption that the Soviets were scientifically and technologically inferior. *Sputnik 1* shattered this assumption. The supposedly inferior Soviet Union succeeded in becoming the first nation to break free from earthly bonds. More frightening, the event suggested that the United States was no longer secure from a sudden nuclear attack. Early orbits of *Sputnik 1* crossed the United States, and readers of the *New York Times* had to be reassured that the satellite was not large enough to carry nuclear bombs that could be dropped on unsuspecting citizens.[41] In a frequently repeated theme, one writer for the *Times* observed that the space satellite fit the Soviet propaganda scheme "by implying that Soviet rockets can deliver heavy nuclear weapons."[42] In this respect, *Sputnik 2* was more significant than *Sputnik 1,* because the second satellite weighed 1,121 pounds, which *Time* magazine observed was "heavier than many types of nuclear warheads."[43]

President Eisenhower tried to downplay the significance of the event. On the day *Sputnik 1* was launched, he left the White House for a weekend of golf and rest at Gettysburg, Pennsylvania. The White House press secretary, James Hagerty, sought to pacify the press. In spite of reports to the contrary, Hagerty said, the feat had not caught the administration by surprise.[44] Defense Department officials discounted the military significance of the event. The launch was not evidence of Soviet superiority in missile development, nor could the satellite be used to bomb Americans while they slept.[45] Eisenhower himself joined the effort to calm the American public. "I see nothing at this moment . . . that is significant . . . as far as security is concerned," he announced at an October 9 press conference devoted to the subject.[46] In a televised address on American science and national security the evening of 7 November, he tried to reassure the nation with the fact that "the overall military strength of the free world is distinctly greater than that of the communist countries."[47]

The effort hardly accomplished its purpose. *Newsweek* called the Soviet satellite launch the "greatest technological triumph since the atomic bomb."[48] *Time* reproduced Washington senator Henry Jackson's complaint that it had been "a week of shame and danger."[49] News outlets compared the event to the splitting of the atom and the discovery of America by Columbus.[50] Outblitzed in the media, Eisenhower watched his popularity fall twenty-two points from its post-election high.[51]

In the wake of the *Sputnik* crises, the House Space Committee attacked Eisenhower's alternative as a "beginner" space program that failed to show "proper imagination and drive." Committee staff urged the administration to mobilize facilities throughout the nation in order to develop manned space stations, build large launch vehicles, and dispatch rockets to nearby planets.[52] Both the U.S. Air Force and U.S. Army drew up plans to put the first human into space. The director of the National Advisory Committee for Aeronautics, Hugh Dryden, announced that his agency was prepared to supervise "the travel of man to the moon and nearby planets."[53] Majority Leader Lyndon B. Johnson assembled a Senate preparedness investigating committee that held hearings from November to January and issued recommendations to strengthen the nation's missile and satellite programs.[54] Discontent with the Eisenhower space alternative reached near-hysterical proportions during this time.[55] Attempts by people in the Eisenhower administration to dispel space fears simply encouraged the belief that the president was inept and did not understand the nature of the challenge.

It is hard for people now separated from the events of the 1950s to appreciate how much the possibility of nuclear war dominated American popular culture. U.S. citizens had emerged from an exhausting world war during which their homeland was essentially secure from enemy attack to find that horrible weapons could now reach their shores. The Soviet Union learned how to fabricate atomic and thermonuclear weapons (using U.S. secrets, conservatives charged). Schools required children to practice civil defense drills and techniques for shielding themselves from nuclear blasts, a favorite method requiring students to dive beneath their desks at the sign of the first thermonuclear flash. Whole cities practiced evacuations, and warning sirens wailed weekly in civil-defense tests. Citizens constructed bomb shelters in their basements and back yards. The exercises may seem eccentric by modern standards, but they contributed significantly to public anxiety about an atomic attack during the 1950s.

For the most part, those promoting space exploration in the popular media did so for reasons of adventure and discovery. Adventure and discovery, however, could not elicit the billions of dollars required to mount an aggressive exploration program.[56] National security considerations could, particularly among those who believed space to be the "high ground" from which the Cold War would be decided. Space boosters increasingly tied their ambitions to popular fears about the nuclear age. The parsimony that motivated the Eisenhower space program seemed incomprehensible to Americans worried about the outcome of the Cold War. George Reedy, one of Lyndon Johnson's principal aides, predicted that

Eisenhower's inattentiveness to this issue "would blast the Republicans out of the water."[57]

The perception of space as the high ground of the nuclear age had begun to gain popular acceptance in the late 1940s. As part of their effort to promote space exploration, the editors at *Collier's* magazine published a story in their 7 September 1946 issue on the colonization of the Moon. In the article, American space booster and science fiction writer G. Edward Pendray issued a stern warning about the military significance of the undertaking:

> Its gravitational attraction is so small that rockets only a little faster than the German V-2s could bombard the earth from the moon. With the aid of suitable guiding devices, such rockets could hit any city on the globe with devastating effect. A return attack from the earth would require rockets many times more powerful to carry the same pay load of destruction; and they would, moreover, have to be launched under much more adverse conditions for hitting a small target, such as the moon colony. So far as sovereign power is concerned, therefore, *control of the moon in the interplanetary world of the atomic future could mean military control of our whole portion of the solar system.* Its dominance could include not only the earth but also Mars and Venus, the two other possibly habitable planets.[58]

At the end of his article, Pendray suggested that the Moon "may be the fortress of the next conqueror of the earth."[59] In making this claim, Pendray adopted a method of presentation repeated by space boosters for a decade thereafter: he offered an assertion without accompanying details. His argument on the military significance of the Moon is reprinted in its entirety. The absence of technical details allowed space boosters to maintain their assertion without proving it, which they could not have done since the assertion was in fact not true. The Moon provides a relatively poor platform for launching nuclear warheads toward the earth, as a comparative analysis of lift weights and trajectories will reveal.

To reenforce Pendray's claim, editors at *Collier's* showed their readers how such a nuclear strike could actually occur in a 23 October 1948 article titled "Rocket Blitz From the Moon." The article opened with an illustration of two V-2-shaped rockets rising out of lunar craters with a dome-shaped control center in the background. On the adjoining page, two large fireballs spread across an aerial view of New York City.[60] The nuclear blasts, drawn with stark realism by space artist Chesley Bonestell, were part of a larger literature portraying the effects of nuclear war on the United States. Once again *Collier's* assured readers that the Moon "could be the world's ideal military base."[61]

Promoters of space exploration repeatedly warned the American public of the military importance of space. In the 30 August 1947 issue of *Collier's,* science fiction writer Robert A. Heinlein teamed up with Captain Caleb Laning of the

As public interest in space flight grew, political support for major space spending remained thin. To win financial support, exploration advocates appealed to public fears about the Cold War, arguing that "control of space meant control of the world." To illustrate the point, *Collier's* ran a article describing how a hostile power could use the Moon as a platform to launch nuclear missiles against the earth. (By Chesley Bonestell with permission of the Estate and Bonestell Space Art)

U.S. Navy to explain how the absence of a space corps would leave the U.S. defenseless:

> Once developed, space travel can and will be the source of supreme military power over this planet—and over the entire solar system—for there is literally *no* way to strike back from ground, sea, or air, at a space ship, whereas the space ship armed with atomic weapons can wipe out anything on this globe.[62]

Just as V-2 rockets from Germany fell on London during World War II, space advocates warned that nuclear-tipped missiles from space could fall on New York during the Cold War. In this Bonestell rendition, one warhead has exploded near the Empire State Building, another in Queens. *Collier's* ran this painting alongside the illustration of rockets rising from the lunar surface in a vivid two-page introduction to the 1947 article "Rocket Blitz from the Moon." (By Chesley Bonestell with permission of the Estate and Bonestell Space Art)

The assertion reappeared frequently in the years that followed, commonly without elaboration. In explaining why the governments of Earth might be interested in constructing a space station, the author of a 1952 children's book on the conquest of space asserted that "from such a station it would be possible to launch giant rockets with explosives or atom bombs against almost any spot on earth."[63] Wernher von Braun advanced the same argument in his famous article in the 22 March 1952 issue of *Collier's,* calling for the construction of a space station:

> There will also be another possible use for the space station—and a most terrifying one. It can be converted into a terribly effective atomic bomb carrier. Small winged

> rocket missiles with atomic war heads could be launched from the station in such a manner that they would strike their targets at supersonic speeds. . . . In view of the stations ability to pass over all inhabited regions on earth, such atom-bombing techniques would offer the satellite's builders the most important tactical and strategic advantage in military history.[64]

In addition to its use as an "atomic bomb carrier," von Braun believed that his proposed space station would revolutionize military reconnaissance. He proposed that the space station be placed in a polar orbit, which would allow its occupants to act as global observers as the earth turned beneath them every twenty-four hours. "Troop maneuvers, planes being readied on the flight deck of an aircraft carrier, or bombers forming into groups over an airfield will be clearly discernible," von Braun stated. Using special optical instruments, he said, the view from orbit would be as clear as that from an observation plane flying close to the ground. "Because of the telescopic eyes and cameras of the space station, it will be almost impossible for any nation to hide warlike preparations for any length of time."[65]

In the introduction to their series on what they called the inevitable conquest of space, the editors at *Collier's* magazine warned of the consequences of falling behind: "The U.S. must immediately embark on a long-range development program to secure for the West 'space superiority.' If we do not, somebody else will. That somebody else very probably would be the Soviet Union." A space station under the control of the free world, the editors argued, "would be the greatest hope for peace the world has ever known." No nation could make undetected preparations for war, effectively tearing down the Iron Curtains of secrecy that communist leaders had erected around their nations. In the wrong hands a space station, the importance of which the editors likened to the development of the atomic bomb, would allow ruthless dictators to rule the world.[66]

Collier's insisted that "what you will read here is not science fiction."[67] In fact, the arguments were more fictional than real, a point underscored when Hollywood picked up the theme. The most critically acclaimed work of fiction to contain the message was the 1950 movie *Destination Moon,* which opens with an unsuccessful rocket launch and the news that the U.S. government has canceled support for the rocket development project. Seeking financing for a nuclear-powered rocket that can fly as far as the Moon, project leaders turn to private industry. At first industrial leaders are skeptical. Funds flow freely, however, once a retired military general explains:

> We're not the only ones planning to go there. The race is on, and we better win it, because there is absolutely no way to stop an attack from outer space. The first

> country that can use the moon for the launching of missiles will control the earth. That, gentlemen, is the most important military fact of our century.[68]

In retrospect, few of the early warnings about the military significance of space turned out to be true. Weapons technology allowed the construction of Earth-based missiles that could carry bombs to distant parts of the planet with far more speed and accuracy than rockets based on the Moon. Large missile-carrying submarines allowed military planners to conceal the location of nuclear weapons, a substantial advantage over the predictable location of orbiting platforms. Recognizing these facts, U.S. and Soviet leaders concluded by the mid-1960s that neither side could gain a military advantage by placing nuclear weapons in space and thus in 1967 signed a treaty agreeing not to do so.[69] Likewise, technology overtook the claims about stations in space. Scientists developed methods by which precise images could be transmitted remotely from satellites in space. As a consequence, automated spy satellites played a far more important reconnaissance role than astronauts peering through optical instruments.[70] It took a number of years for people in the U.S. government to realize this, but they eventually did. By the mid-1970s, the U.S. Air Force had abandoned its efforts to place sentries in space and accepted the virtues of satellite technology. Eisenhower's alternative space program, with its emphasis upon robotics, eventually came to dominate military space policy in the United States.[71]

During the 1950s, a number of groups tried to advance these realistic images of the military uses of space. At the time, few of their assertions took root. President Eisenhower's Science Advisory Committee, for example, added a special section on military applications to their famous 1958 brochure, correctly observing that the most important military uses of space would be provided by communication and reconnaissance satellites and predicting that satellites with telescopic cameras would be able to instantly transmit high-quality images back to Earth. As for the proposals to place bombs in space, committee members characterized these as "clumsy and ineffective ways" of delivering nuclear weapons:

> Take one example, the satellite as a bomb carrier. A satellite cannot simply drop a bomb. An object released from a satellite doesn't fall. So there is no special advantage in being over the target. Indeed, the only way to "drop" a bomb directly down from a satellite is to carry out aboard the satellite a rocket launching of the magnitude required for an intercontinental missile.

Even if the weapon were given a small push and allowed to spiral in gradually, that would mean "launching it from a moving platform halfway around the world." Schemes to drop bombs from space had "every disadvantage" compared to launching bombs from earth.[72]

Space boosters typically ignored arguments such as these as they continued to win converts through their claims about world supremacy. A number of factors allowed this to occur. First, public understanding about the reality of space travel was terribly thin. It was based as much on science fiction as on real experience, a situation that encouraged the application of misleading images to the new frontier. Second, it was easy for people who misunderstood orbital mechanics to draw incorrect analogies from the bombing campaigns of World War II. People who knew that bombs fell from airplanes were misled into viewing space stations and automated platforms as extensions of airborne bombers. Third, primitive technology also supported false claims. Images from first-generation spy satellites had to be retrieved through clumsy reentry schemes involving film drops, parachutes, and airplanes with skyhooks.[73] This awkward technology lent credibility to the notion that space stations would be a valuable reconnaissance tool.

The national media could have played an important role in educating the public about the realities of space. Unfortunately, they did not. They behaved irresponsibly, apparently sensing that hysterical assertions about threats from space were more newsworthy than calming assurances. Even the 1952 *Time* magazine cover story, the classic rebuttal to the *Collier's* series, did not challenge von Braun's warning about atom-tipped guided missiles spiraling downward through the atmosphere from orbiting space stations. Instead, the article characterized doubters as "timid military planners" and suggested that von Braun's schemes would sound even more practical if the public were allowed to see the military secrets that purportedly supported them.[74]

The willingness of the public to believe the worst drew much of its force from a larger concern with the way in which nuclear war might cause the world to end. This preoccupation lent credibility to the popular acceptance of space as the place from which both armageddon and salvation might arrive. Without the cultural obsession with nuclear war and final days, the more outrageous claims about the importance of conquering space would have not been so believable.

Since biblical times various people have advanced apocalyptic predictions, and public hysteria about impending doom has periodically swept across the face of civilization. Apocalyptic literature enjoyed wide popularity in both Jewish and Christian communities between 200 B.C. and A.D. 200. The Book of Revelation, composed some fifty years after the death of Christ, provided Christian believers with a rich though symbolic description of the final days, including predictions of war, famine, plague, earthquakes, and falling stars.[75] Establishing the exact time of the apocalypse has preoccupied many prophets and religious figures. A nineteenth-century New England farmer, William Miller, recruited some fifty thousand followers by deciphering one biblical chronology. His followers set the exact date of the end at 22 October 1844 and gathered together in their usual places of worship to pray and await the Lord.[76]

Public interest in the end of the world remained strong from the years just preceding World War II through the first decade of the atomic age. The thought that civilizations were preparing for another massive war joined with eventual understanding of the weapons developed for it encouraged public interest in various scenarios. In earlier times, religious leaders explained how the world might end; in the twentieth century, the public listened to scientists who looked to the sky.

In 1937 the Hayden Planetarium in New York presented a show that gave planetarium viewers a ring-side seat on the end of the world.[77] The show drew upon works of popular science that had appeared with increasing frequency after the turn of the century.[78] All of the threats presented by planetarium curators came from outer space. *Life* magazine featured the show with a 1937 article summarizing "Four Ways the World Might End," illustrated by Rockwell Kent, which offered illustrations of death by fire due to the explosion of the sun and a permanent ice age occasioned by its contraction. The latter, the authors incorrectly told their readers, would occur "within a few million years." The orbit of the Moon was unstable, they further explained, which would eventually cause it to come crashing toward the earth. A final scenario portrayed the breakup of the planet due to a celestial body passing too close to the globe. For this scenario, Kent portrayed a strange phenomena. The gravitational attraction of a very large body would for a split second snatch all human beings off the surface of the earth and propel them skyward, creating an astronomical equivalent of the ascension.[79]

Scientific descriptions of the end of the world entered popular culture through vehicles both popular and fanciful. As part of their eight-part series on "The World We Live In," editors at *Life* magazine gave their readership an illustrated view of the beginning and end of the earth. Counterbalancing colorful illustrations of the birth, the article featured a two-page painting by Chesley Bonestell portraying the dawn of the day on which the sun would explode.[80] The use of space travel to escape impending doom was a frequently used theme. Reporting on the Hayden Planetarium show in 1950, a writer for *Popular Mechanics* magazine announced that astronomers would be able to detect the impending catastrophe, giving humans "a few thousand years to get busy colonizing another solar system." The script for the Hayden Planetarium show ended with the promise that humans might evacuate the earth "and as a refugee on Venus find safety and peace,—if there is a Venus then."[81]

After 1945, with the explosion of the atom bombs at Hiroshima and Nagasaki, scientists added nuclear war to the list of doomsday scenarios. Scientists who had earlier sought to explain far-off ends of the earth found that they had helped create the means for an imminent conflagration. Writing for a 1947 issue of

Atlantic magazine, Albert Einstein predicted that "little civilization would survive" after a nuclear war.[82] In 1950 he extended his prediction by announcing in a speech broadcast over NBC television that "radioactive poisoning of the atmosphere and hence annihilation of any life on earth has been brought within the range of technical possibilities." By the early 1950s, atomic war was well established among writers of popular science as one of the ways in which the world could end.[83]

Throughout World War II, the skills of wartime photographers had exposed the American public to the face of war. Photographs of mass destruction, much of it caused by aerial bombardment, filled American magazines and newspapers.[84] With the advent of the atomic age, it became clear that such destruction could be visited upon the United States, only thousands of times worse. The popular press portrayed nuclear armageddon in much the same way they had reported World War II. Speculation about the effects of atomic bombardment increased following the news in September 1949 that the Soviet Union had exploded its first atom bomb.

An article in the 21 April 1953 issue of *Look* magazine was typical. The article opened with a drawing of a hydrogen bomb exploding in the sky above New York City, prepared by Chesley Bonestell. Authors explained the technical process for building a hydrogen bomb and presented a map showing how a bomb dropped over the Capitol in Washington, D.C., would knock down buildings and burn people as far away as Baltimore.[85] An earlier issue of *Collier's,* with a cover illustration by Bonestell, included an imaginative news story describing the effects of an atom bomb dropped over New York City.[86] *Collier's* later published an article describing "The Third World War," complete with another Bonestell illustration that showed Washington, D.C., in flames.[87]

None of these stories forecast the end of the world, a scenario apparently judged too depressing for periodicals that depended upon subscriptions purchased in advance. Instead, the stories promised deterrence through terror or safety through civil defense. Pessimistic scenarios found their way into fiction.[88] In Ray Bradbury's classic *The Martian Chronicles,* humans who have settled on Mars watch in horror as the earth explodes in an atomic conflagration.[89] Nevil Shute's chilling novel *On the Beach,* also a top-grossing movie, describes the breakdown of society in the aftermath of an atomic war as survivors wait for radioactive fallout to engulf them.[90] In one of the better mutant movies—*Them!*—giant ants charge out of an atomic test site in New Mexico to challenge civilization.[91]

By the mid-1950s, nuclear war preoccupied the American mind. Most Americans believed that another world war would occur within their lifetimes and that hydrogen bombs would be used in it. Many believed that all of humankind would be destroyed in an all-out nuclear exchange. In a series of public opinion polls throughout the 1950s, Americans consistently identified the threat of war as "the most important problem facing the entire country today."[92]

Science revealed terrible ways in which the planet might be destroyed. All of the doomsday scenarios, both astronomical and human in origin, fell from the sky. This led naturally to the conclusion that the future of the world would be determined by activities that took place above the surface of the earth. One of the most bizarre manifestations of this belief was the Unidentified Flying Object (UFO) phenomenon. Beginning in 1947, Americans in ever-increasing numbers began to report sightings of mysterious flying objects.[93] Media interest increased following the January 1950 issue of *True* magazine, which carried a story by Donald A. Keyhoe arguing that the objects were spacecraft under the control of intelligent beings from another planet.[94] In 1952 individuals began to report actual contacts, some personal and others via mental telepathy. Contactees told their stories on radio and television. Societies of believers arose, some of which attempted to spread the message that visitors had come to save humans from impending destruction.[95]

To those who believed in their extraterrestrial origin, flying saucers were like messengers from a higher civilization. Unlike angels and other religious figures who floated down from heaven in the past, flying saucers were not presumed to be the creation of the Deity. They were the product of advanced science, evidence of the length to which technology could carry civilization. Many viewed the saucers as a source of salvation from the treat of atomic war. In the classic 1951 science fiction film *The Day the Earth Stood Still,* a saucer lands in Washington, D.C., where its captain orders world leaders to control the nuclear arms race or face annihilation by a race of robot police:

> The Universe grows smaller every day—and the threat of aggression by any group anywhere can no longer be tolerated. There must be security for all—or no one is secure. . . . It is no concern of ours how you run your own planet, but if you threaten to extend your violence, this earth of yours will be reduced to a burned-out cinder.[96]

The saucer phenomenon fed on the latent distrust of government in American society. A central tenet among UFO partisans was the belief that the U.S. government was engaged in a vast coverup to shield the truth about flying saucers from the public at large. From this perspective, government officials had dropped flying saucers behind the cloak of government secrecy that shielded other technological developments in the nuclear age. UFO partisans alleged that government officials had obtained evidence as early as 1947 supporting the premise that flying saucers were real spacecraft from another planetary system.[97] Efforts by the U.S Air Force to explain UFO sightings, as with Project Blue Book, simply inflamed accusations that the government was hiding the truth.[98]

Reports of alien visitors did not originate in the modern era. In medieval times people spoke of encounters with demons and other creatures from the underworld.

Fear of Soviet domination of space, according to presidential science adviser James Killian, "created a crises of confidence that swept the country like a windblown forest fire." Some people wondered if aliens from outer space might rescue the people of Earth from nuclear war. In the classic 1951 science fiction film *The Day the Earth Stood Still*, a flying saucer lands in Washington, D.C., where its captain orders world leaders to abandon the nuclear arms race or face annihilation from above. (National Air and Space Museum)

Demons were even thought to engage in sexual unions with humans, a theme later repeated in reports of extraterrestrial encounters. The otherwise reliable nature of these reports, both medieval and modern, suggest that the phenomena may represent a form of mass hallucination, triggered in both eras by apocalyptic fears.[99] Once science overcame religious superstition, aliens from other civilizations replaced evil spirits as the visitors from beyond.

One striking piece of evidence supports this thesis. After an initial burst of interest, the rash of UFO sightings dropped off during the mid-1950s to an average of forty-six per month. This suddenly changed following the launch of *Sputnik 1,* when the number of reports rose sharply, with over six hundred sightings in the final three months of that year.[100] Baring the unlikely possibility that aliens actually stepped up observations of Earth, one is left with the plausible explanation that the launch of the Soviet satellite excited fears that caused people to detect more unknown objects in the sky.

From atom bombs to flying saucers, objects from space fell onto the public consciousness during the 1950s. Reassurances by political and scientific elites who told the public not to worry only fueled suspicions that the government was not disclosing all it knew. To the public at large, space technology seemed to be the source from which national salvation or doom would come. It was easy for Americans to believe that control of space would determine the future of the world.

Governmental officials, aware of the power of public opinion in a democracy, struggled to balance their perceptions with those of the citizenry. This proved especially challenging, because public opinion was based in part on a vision. Politicians who adopted the language and symbols of imagination gained an advantage in the quest for public support. By introducing the language and symbols of imagination into political discourse, politicians pushed official government policy toward those priorities contained in the image even though other officials wanted to go elsewhere.

Images that were promoted by space boosters and amplified by the news media became part of the political discourse, especially after the precipitating events of the *Sputnik* launches. By the late 1950s, public officials in increasing numbers adopted the language of popular anxiety about space and the Cold War. On 7 January 1958 Senate majority leader Lyndon Johnson stood before the Democratic Conference (the assembly of Democratic senators) and elevated space policy to the top of the Senate agenda. His words echoed the warnings of people such as Wernher von Braun and forums such as *Collier's* magazine:

> Control of space means control of the world, far more certainly, far more totally than any control that has ever or could ever be achieved by weapons, or troops of

> occupation. . . . Whoever gains that ultimate position gains control, total control, over the earth, for purposes of tyranny or for the service of freedom.[101]

Johnson insisted that this perspective was endorsed by "the appraisal of leaders in the field of science, respected men of unquestioned competence." He did not point out that most scientists greeted this viewpoint with considerable skepticism, nor did he observe that the "control of space" argument had been advanced by people more interested in adventure than scientific discovery.[102] Johnson gained no political advantage by emphasizing the skepticism that still transfixed the scientific community.

Instead, Johnson saw a political advantage in appealing to the public impression of space. He had grown up in the hill country of Texas, the son of a local politician of limited means. He was a rural progressive from the conservative South, a disciple of Franklin Roosevelt who had been elected to Congress at a time when conservatives were pulverizing the Roosevelt presidency. Johnson knew that conservative opposition to New Deal social programs was motivated by the same concerns that made politicians reluctant to initiate a big space program. As he told members of the Democratic conference, "Our decisions, more often than not, have been made within the framework of the government's annual budget. This control has, again and again, appeared and re-appeared as the prime limitation upon our scientific advancement."[103]

Johnson's remarks were directed at not only budget-conscious members of the Eisenhower administration but also fiscal conservatives within his own party who did not want the federal government to play a large role in domestic affairs. Johnson believed that conservative Democratic senators, particularly those from the South, could be motivated to vote for a large government presence in space as a matter of national security. This, he believed, would create the precedent he needed to press for government largess in areas such as health and education.[104] Johnson also understood how vulnerable Republicans had made themselves by failing to allay public anxiety.

The Cold War rhetoric was repeated by other Democratic senators who shared Johnson's views. In a Veterans' Day address in November 1957, Democratic senator Stuart Symington of Missouri announced that "the race for the conquest of space is today's major engagement in the technological war. We must win it, because the nation which dominates the air spaces [*sic*] will be in a position to dominate the world."[105] During his 1960 presidential campaign, Senator John F. Kennedy argued that the U.S. must win the race to conquer space: "Control of space will be decided in the next decade. If the Soviets control space they can control earth, as in past centuries the nation that controlled the seas dominated the continents. . . . We cannot run second in this vital race. To insure peace and freedom, we must be first."[106]

Public concern about the space race helped Kennedy maintain the most important illusion of his 1960 campaign, the nonexistent "missile gap" that Democrats charged the Eisenhower administration with creating.[107] Space spectaculars by the Soviet Union, accompanied by U.S. failures, helped perpetuate the illusion that the United States trailed the Soviets in the production of ballistic missiles, a charge Republicans failed to refute for fear of revealing the intelligence sources that provided the government with information on Soviet nuclear capability.

Officials within the bureaucracy who stood to gain from an expanding space program issued similar warnings. Brigadier General Homer Boushey, the deputy director for Air Force Research and Development, quoted Johnson's words in a speech before the Aero Club of Washington, D.C., on 28 January 1958. Asserting the military advantages of controlling space, Boushey announced that "he who controls the moon controls the earth." Boushey believed that the Moon would provide the United States with a platform for a powerful nuclear deterrent. If the Soviets attacked the United States, he explained, they would receive "sure and massive destruction" from the Moon some forty-eight hours later. "The moon," Boushey concluded, "represents the age-old military advantage of 'high ground.'" Boushey's remarks were reproduced in the 7 February issue of *U.S. News & World Report.*[108] Other military leaders issued similar warnings.[109]

Such statements exasperated Eisenhower and his advisers, who mightily tried to calm public fears. Interviewed for a 1958 issue of *Reader's Digest,* Chief of Naval Operations Arleigh Burke tried to explain that missiles launched from the earth offered a far more practical deterrent than orbital platforms.[110] The president's Science Advisory Committee summed up the arguments about the military significance of space: "Much has been written about space as a future theater of war, raising such suggestions as satellite bombers, military bases on the moon, and so on. For the most part, even the more sober proposals do not hold up well under close examination."[111]

These reassurances did little to alleviate the impression that the space race had turned the Eisenhower White House into what pundits labeled "the tomb of the well-known soldier."[112] National news outlets reporting the president's reassurances gave equal time to space cassandras preaching national doom. In a feature article in *Life* magazine shortly after the *Sputnik* launch, scientist George R. Price argued that "unless we depart utterly from our prevent behavior, it is reasonable to expect that by no later than 1975 the United States will be a member of the Union of Soviet Socialist Republics."[113] Even the president's science adviser, James Killian, acknowledged the difficulty of overcoming public perceptions:

> Sputnik I created a crises of confidence that swept the country like a windblown forest fire. Overnight there developed a widespread fear that the country lay at the

> mercy of the Russian military machine and that our government and its military arm had abruptly lost the power to defend the homeland itself, much less to maintain U.S. prestige and leadership in the international arena.[114]

Eisenhower was correct, but to little avail. In early 1961 President John F. Kennedy set aside the Eisenhower space program and embarked upon a massive buildup designed to make the United States first in space. In March Kennedy approved the allocation of funds necessary to accelerate development of the Saturn rocket program, and in May he established the goal of placing the first Americans on the Moon.[115] NASA's budget increased tenfold, from $524 million in the last full year of the Eisenhower administration (1960) to $5.3 billion in fiscal year 1965.

Even skeptics within NASA became advocates for the spacefaring dream. This outcome was by no means predestined. The groups that filed in to form the newly created civilian space agency represented a wide range of views, from romantic visionaries such as Wernher von Braun to practical conservatives such as Milton Rosen. The bulk of employees who formed NASA were neither romantic visionaries like von Braun nor advocates of Eisenhower's conservative satellite program. They were pragmatic American engineers, people such as spacecraft designer Max Faget. Faget worked for the National Advisory Committee for Aeronautics, which contributed all of its eight thousand employees to the newly formed space effort. The son of a renown public health physician, Faget became an engineer so that he could design airplanes. After a tour of duty as a submarine officer in the U.S. Navy, Faget went to work for the NACA laboratory at Langley, Virginia. He was assigned to what Langley executives called the Pilotless Aircraft Research Division, a euphemism for work on rockets and guided missiles. Faget and his colleagues tied models of guided missiles onto rockets and shot them out over the Atlantic at supersonic speeds to see how the models would fly. It was the closest thing to spacecraft testing that NACA did, but NACA engineers would not call it that. NACA employees did research on airplanes, a romantic undertaking in its own right. For most of the organization's existence, as historian James Hansen points out, *space* was a dirty word, and until the space race began, NACA employees were not much interested in what many regarded as that "Buck Rogers stuff."[116]

Pragmatic NACA employees might have created a balanced vision of space flight, halfway between the romantic dreams of the followers of people such as Wernher von Braun and the overly cautious Eisenhower alternative. They did not, and they failed to do so for a fairly simple reason. Like most Americans at the time, they found themselves caught up in the momentum of the Cold War, the battleground of which was space. If Americans lost that battle, many believed, the Russians would rule the world. Pragmatic American engineers jumped in to help beat the Russians in space, implementing the romantic vision that would dominate the American political scene for decades to come. Even Rosen joined the bandwagon, publishing a series of articles on the importance of human space flight. "As soon as the Sputnik went up," Faget explained, "all bets were off. It was pretty much a free for all."[117] Eisenhower's alternative disappeared under the force of Cold War necessities, and memory of it faded as Kennedy's vision filled the public mind.

As is often the case, the things people feared most did not occur. The fears were not real. Bombs did not fall from orbiting satellites nor rain down from the Moon. The Soviet Union did not conquer the world, even though it became the first nation to establish an Earth-orbiting space station in 1971. Public preoccupation with the threat of nuclear war began to wane in 1963.[118] The Cold War itself ended with the collapse of the Soviet Union in 1991. As concern about the Cold War weakened, so did the connection between the control of space and national security. This created a fundamental problem for people promoting the U.S. space program. The fears that furthered their romantic vision of space exploration no longer preoccupied the public mind. The United States found itself with a space program fashioned in response to an image of the world that never existed in fact and no longer existed in imagination. Without the Cold War, the civil space program became a cause in search of an explanation.

Social scientists for some time have studied the effects of failed beliefs.[119] True believers do not automatically abandon their cause when reality intrudes in discomforting ways. They rarely admit they were wrong or change their behavior, especially when they keep meeting other people who share their fantastic beliefs. Believers so sustained often seek new beliefs to validate old behavior and new interpretations to explain why prophecies fail to be fulfilled. Sometimes believers will deny that their prophecies in fact miscarried. Much of this applies to the behavior of people promoting the U.S. space program. Advocates of space exploration did not abandon their cause just because public interest in the Cold War waned. They did not retreat to less ambitious space efforts. Instead, they sought new rationales for continuing endeavors, and concocting fears remained part of that strategy. Some insisted that control of space still meant control of the earth, a position advanced by promoters of the Strategic Defense Initiative ("Stars Wars"), a space-based missile defense proposal that President Ronald Reagan endorsed in 1983. The position was also broached by military space commanders fighting the 1991 Gulf War, in which satellite technology played a decisive role.[120]

Because bombs did not fall from space, advocates of space exploration looked for other objects that did. Increasingly, natural objects such as comets and asteroids found their way into the politics of fear. In 1991, for example, Clark Chapman of the Planetary Science Institute and David Morrison of NASA's Ames Research Center announced that a person's chance of being killed as the result of an large asteroid strike was "more than six times greater than your chance of dying in a plane crash." This startling statistic, though exaggerated, contains an element of truth. On the average, 130 people die in commercial airline crashes in the United States each year. Based on a national population of 250 million, that produces an annual individual probability of one in two million. On the average, a major asteroid, one capable of killing everyone in the United States, strikes Earth every three hundred thousand years. (A more likely outcome is that one-quarter of the world's population would die.) The stretching of statistical credibility produces an average of 250 million deaths every three hundred thousand years, or an annual average of 833 fatalities, and that is more than the average annual number of fatalities produced by commercial airline crashes.

A major asteroid strike might not occur for 300,000 years (it might with equal probability occur tomorrow). When it does, however, it could devastate the earth. The massive impact could create a continent-wide firestorm. Nitric acid produced from the burning of the atmosphere might acidify lakes and soil and even poison the surface of the ocean. The temperature of the earth could alternatively plunge and rise, first as dust blocked out the sun and second as carbon dioxide pushed into the stratosphere produced a greenhouse effect. "Agriculture and commerce would probably cease," Chapman and Morrison warned, "and most people in the world would die."[121]

Entertaining the view that doom and salvation come from above, writers of science fiction have for some time contemplated the consequences of large objects striking the earth. In 1951 producer George Pal used the possibility of a celestial flyby to portray the end of Earth in the movie *When Worlds Collide.* In the film, which won an oscar for its special effects, two wandering planets intersect Earth's orbit. Before the second planet collides with the earth, a small group of humans escape to the first on a giant rocket ship, where they begin civilization anew.[122] In *Lucifer's Hammer,* residents of California are bombed back into a new dark age by the effects of a massive comet smacking the earth. Arthur C. Clarke showed how Earthlings might deflect approaching asteroids in the 1993 novel *Hammer of God.* Such works of imagination combine the best traditions of science fiction with the thrilling fascination of disaster stories.[123]

Public interest in massive collisions was not generated by fiction alone. In 1980 Luis and Walter Alvarez made the startling suggestion that a large asteroid strike was responsible for the death of the dinosaurs. Cross-sections of rock formations show evidence of a catastrophic event at the end of the Cretaceous era sixty-five million years ago; analysis suggests that an asteroid in the ten- to fifteen-kilometer range was responsible. This was not a freak event, moreover.

Astronomers and geologists believe that really big asteroids fall every ten to thirty million years, whereas smaller objects with the potential for global transformation (at least one kilometer in size) hit Earth every three hundred thousand years.[124]

Three hundred thousand years is a long time. Remote possibilities can motivate governmental spending, nonetheless, when real events publicize the prophesies of doom. Such an event occurred in the summer of 1994. A comet perhaps ten kilometers across, named Shoemaker-Levy, split into a string of fragments that struck the planet Jupiter in mid-July. The largest fragment left visible shock waves in the Jovian cloud cover as large as the earth. Space advocates reminded the public that much smaller strikes could have devastating consequences on the Earth. In 1908, for example, a small asteroid about sixty meters across entered the atmosphere above Tunguska, Siberia, where it exploded and flattened trees over an area twice as large as the city of New York.[125]

Space advocates used the possibility of asteroid strikes to resurrect a variety of space proposals. At the least, boosters argued, the swarm of medium- to large-sized comets and asteroids, thousands in number, whose paths intersect that of the earth should be catalogued, their orbits calculated. The asteroid belt should be explored, they added. As one advocate group observed, asteroids are the nearest extraterrestrial neighbors beyond the Moon and "logical sites to develop the techniques of human deep-space exploration." Many offered techniques for deflecting the orbits of comets and asteroids. Leading candidates included mass drivers, perhaps using fuel manufactured from the objects themselves, or nuclear explosions in a standoff mode. If asteroids could be deflected, others observed, they could also be moved into safe orbits around the earth and mined for propellants, metals, or gold. Ultimately, some said, the presence of so many objects intersecting the earth's path would force the human race to disperse itself onto other habitable spheres. Any advanced civilization, astronomer Carl Sagan warned, must become spacefaring in order to protect itself from cosmic bombardment. The eventual choice, Sagan concluded, "is spaceflight or extinction."[126]

In 1990 the House Committee on Science, Space, and Technology requested that NASA conduct two workshops as a starting point for further tax-financed support. Workshop participants in turn urged Congress to immediately fund a census of all large, Earth-crossing asteroids. The participants took the name of the proposed census from an award-winning science fiction story: "We call this proposed survey program the Spaceguard Survey (borrowing the name from the similar project suggested by science-fiction author Arthur C. Clarke nearly 20 years ago in his novel *Rendezvous with Rama*."[127] In August 1994, shortly after Shoemaker-Levy struck Jupiter, the U.S. House of Representatives passed legislation directing NASA, in conjunction with the U.S. Department of Defense and other countries, to "identify and catalogue within 10 years the orbital characteris-

Cold War fears helped motivate the large outlay of funds necessary to take the first steps into space. When the Cold War ended, so did much of the motivation for an ambitious space program. Seeking other nightmares with which to scare up political support, space advocates warned of the need to protect the Earth from comets and asteroids falling from space. Any advanced civilization, astronomer Carl Sagan argued, must go into space in order to survive. This 1995 cover of *Astronomy* magazine showed the beginning of the end again for New York City. (Painting by Bob Eggleton)

tics of all comets and asteroids that are greater than 1 kilometer in diameter and are in an orbit around the sun that crosses the orbit of the Earth."[128]

Space enthusiasts looked to asteroid dangers as a basis for completing old fantasies. In reality, any space program to emerge from asteroid fears would probably not resemble the dreams of spacefaring pioneers. The asteroid census requested by the House will be completed entirely with ground-based telescopes on Earth. Who knows what sort of interception and deflection devices might be used? The physics of deflection favor interception as far from Earth as possible, as the energy required to move an object increases dramatically as it closes in. Precursor missions, well in advance of deflection attempts, would be needed, in as much as different materials respond to perturbations in different ways. This could easily favor, as participants at one NASA workshop observed, "low-cost missions using small, lightweight spacecraft," substantially different in practice from the large human expeditions envisioned in the spacefaring dream.[129]

Warnings about asteroids and comets striking the Earth mobilized a response feeble by comparison to space efforts incited by the Cold War. The Cold War really scared people, and asteroids do not. The former motivated otherwise pragmatic individuals to implement elements of an extraordinary dream. It set the employees of the National Aeronautics and Space Administration on a mission from which they did not deviate, even when the cause of that mission disappeared.

APOLLO .... THE AURA OF COMPETENCE

If we can send a man to the moon, why can't we clean up Chesapeake bay?

—Tom Horton, 1984

Shortly after taking office, President John F. Kennedy approved the crash program to put Americans on the Moon.[1] By shifting the national space effort into high gear, he established the modern U.S. space program with its emphasis on large-scale engineering, big science, and human exploration. The budget of the National Aeronautics and Space Administration increased tenfold.[2] Eisenhower's approach to space, with its stress upon satellite technology and scientific achievements, disappeared.

President Kennedy's commitment was strong but not unalterable. During the 1960s, a number of efforts were made to release the United States from the self-created mandate to be first in space, particularly the goal "before this decade is out, of landing a man on the Moon and returning him safely to earth."[3] Even Kennedy himself questioned the wisdom of the decision he had made.

The most serious effort to derail Project Apollo occurred in 1965. Budget Director Charles Schultze informed President Lyndon Johnson that fall that cost overruns in the space program were eating up the funds that Johnson needed to run his War on Poverty and prosecute the war in Vietnam. Schultze urged Johnson to cut the NASA budget by a total of $600 million, a decision that would have effectively abandoned the Moon race and deferred the lunar landing "into the 1970s."[4]

To maintain interest in an increasingly expensive space program, advocates of exploration diversified their rationale. They sought new grounds on which to sustain the commitment that Kennedy had made. One of the most important was the creation of an aura of competence around the fledgling space program.

A generation of Americans grew up during the 1960s watching NASA astronauts fly into space, beginning with fifteen-minute suborbital trajectories and culminating in an eight-day trip to the Moon. Television coverage of real space adventures was long and intense; there were moments of danger and seasons of accomplishment. In the public mind, the civilian space agency established a reputation as a government organization that could take on difficult tasks and get them done. Project Apollo was widely viewed as triumph of effective government as well as a demonstration of technology.[5]

NASA's ability to fashion a sense of competence helped boost public confidence in the capability of government in general. Although it was not the sole contributor, the new space program joined other forces to produce a record level of public confidence in government. Responding to an opinion survey in 1964, 76 percent of Americans polled expressed confidence in the ability of government "to do what is right" most or all of the time.[6] An all-time high in the history of polling, this level of public trust allowed the government to move ahead with other large initiatives during the 1960s, including the War on Poverty and civil rights reform, and helped maintain faith in the value of government-sponsored science and technology.

Refusing to follow the advice of those who wanted to defer the Moon landing, Johnson did not delay Project Apollo. He provided NASA with sufficient funds to complete this national goal. Many factors motivated Johnson to do so. He felt a personal commitment to the program he had helped create, aerospace contracts had been distributed to states whose representatives helped form his congressional majority, and most important, politicians like Johnson recognized the symbolic importance of projecting a sense of competence over the government as a whole.

In the beginning, the aura of competence surrounding the early space-flight program owed as much to perception as reality. Without question, NASA's early space-flight record could have supported accusations of incompetence as easily as images of success. During its formative years, NASA completed all of its human space flight missions without enduring a major accident or a nationally televised fatality. This was a great accomplishment at the time, given the well-known tendency of rockets to explode or fly off course.[7] Even so, the new agency frequently suffered technical difficulties and launch delays. There was plenty of ammunition for an assault on NASA's capabilities had the media wanted to launch one.

Rocket boosters often misperformed, and even if a launch went well its mission might flounder in space. During the first year of operation (1958), all four of NASA's major launch attempts misfired. The situation was little better in 1961, when nine of twenty-four major launch attempts failed.[8] Fortunately, none had

In 1958, concurrent with the founding of the civilian space agency, President Dwight Eisenhower directed NASA employees to proceed with the first cautious step toward human space flight. Project Mercury provided NASA and its contractors with the funds to develop a single-seat space capsule capable of flying into space and returning safely to Earth. The rudimentary spacecraft, ready for launch, sat on top of a modified Atlas intercontinental ballistic missile. (NASA)

people on board. Beginning in 1961, NASA launched the first in its series of *Ranger* probes, a robotic mission designed to produce the first closeup pictures of the Moon. NASA made six consecutive attempts to fly the *Ranger* spacecraft between 1961 and 1964; all six failed. Congress launched an investigation, but the incidents hardly tarnished NASA's reputation. NASA recovered and carried out the mission with the remaining three spacecraft.[9] During the second attempt to push an American astronaut to the edge of space in 1961, the *Mercury* space capsule sank into the Atlantic Ocean before navy helicopter pilots could retrieve it, nearly drowning astronaut Gus Grissom.[10] Some suspected that Grissom had panicked and prematurely blown the capsule's side hatch into the water, but the press portrayed "little Gus" as a national hero.[11] John Glenn's 1962 orbital flight was repeatedly delayed by bad weather, including a nationally televised 27 January scrub just twenty minutes before lift-off. The press moaned but did not openly complain.[12] When the *Vanguard* rocket had exploded on its launch stand during the first attempt to orbit a U.S. satellite in 1957, the press had roasted the Naval Research Laboratory, the directing organization.[13] NASA failures escaped similar treatment.

Although neither the public nor the media seemed chagrined by NASA mishaps, members of the Kennedy administration were. Kennedy's aides frequently discussed the political ramifications of potential space-flight disasters as the new administration prepared to take over the White House. The issue surfaced during the 1960 presidential campaign and persisted through the spring of 1961 as officials in the Kennedy administration pondered whether to commit the United States to a space race with the Soviet Union. Aides considered how they might distance the new president from Project Mercury, the agency's high-profile human space-flight initiative, should an accident prevail. A special transition team openly urged Kennedy to disassociate himself from the project or risk blame for its failure.[14]

Kennedy's advisors had good reason to urge caution about the high-profile space effort. Blunders plagued virtually every aspect of the project during this period. In the summer of 1960, while Kennedy was campaigning for the presidency, the first test of the Mercury-Atlas system ended in calamity when the rocket exploded mysteriously one minute after lift-off. Although no astronaut was on board, the accident delayed the flight program by six months. On election day 1960, the capsule's escape system misfired. Two weeks later, the Mercury-Redstone combination lifted four inches off the launch pad and shut down. In March of 1961 the escape system failed again, and in April a Mercury-Atlas test flight had to be destroyed by the range safety officer when the autopilot sent the rocket astray.[15] The aerospace trade journal *Missiles and Rockets* warned its readers during the summer of 1960 that Project Mercury "could easily end in flaming tragedy."[16] A space-flight disaster with an astronaut on board would have undercut public confidence in an already besieged administration and derailed discussions leading up to the lunar commitment. Understandably, Kennedy looked grim as he watched Alan Shepard's nationally televised fifteen-minute suborbital flight from the White House on 5 May 1961.[17]

Shepard's successful mission contributed significantly to a public mood of euphoria and pleasure with NASA's space effort.[18] After a string of malfunctions, the program seemed to be off the ground at last. Step by step, NASA rocket engineers identified the weak points in their technical program and corrected them. Five more Mercury astronauts followed Shepard into space, with John Glenn becoming the first American to orbit the earth on 20 February 1962.

All of the flights with astronauts on board were nationally televised, and thousands of people made the pilgrimage to the beaches at Cape Canaveral to watch the rockets blast into space. One hundred thirty-five million Americans watched John Glenn's orbital flight on national television.[19] The flight stopped traffic in New York's Grand Central Station, as thousands of commuters paused to watch the lift-off on a sixteen-foot television screen.[20] It was wonderful theater, contrasting sharply with the shroud of secrecy surrounding the Soviet space program.

In the beginning, the civil space program was marked by exploding rockets and crashing spacecraft. During the first test of the Mercury-Atlas system, an automated test with no astronaut on board, the rocket exploded one minute after lift-off. Flight engineers reassembled the *Mercury* capsule from debris collected from the ocean floor. Advisers warned president-elect John F. Kennedy to disassociate himself from the project or risk blame for its failure. (NASA)

With such intense national interest, the press did not need to create controversies in order to maintain viewer attention. Television commentators frequently downplayed dangers that would have captivated public attention had they been properly explained. Instead, press coverage reflected the calm "can-do" attitude that NASA sought to portray. During John Glenn's orbital flight, instruments suggested that the heat shield on his *Friendship 7* spacecraft had come loose. In a desperate measure, flight controllers told Glenn to reenter the atmosphere with his retrorockets still strapped on. Only after Glenn had returned safely to earth did the press emphasize what the consequences might have been. Had the signal been accurate (it was not), Glenn would have faced a fiery death as his unprotected space capsule reentered the atmosphere.[21] Time after time, a government agency whose space program had begun with exploding rockets put its reputation on the line and carried out another mission, each more complex or daring than the last. That, at least, was the appearance NASA transmitted during the 1960s. Errors and malfunctions were forgotten as a feeling of competence grew around the U.S. space program.

Performance alone did not maintain the sense of competence that came to envelop the nation's space program during the 1960s. Improving performance certainly supported the image of success, but it was not sufficient to explain the reverence in which the space-flight program was held. NASA did not operate an error-free space effort during this time, and, in fact, the actual record of human and instrumented flight could have supported an alternative interpretation had opinion leaders wanted to go that way. Why then did the impression of competence prevail?

One of the most important factors that helped create the aura of competence was the personalization of the American space effort through the astronaut corps. By promoting the astronaut corps, agency advocates and media leaders were able to reduce complex technical issues to personal values such as bravery and patriotism. The first group of astronauts were remade into the embodiment of American values in such a way that few wanted them to fail. Accusations of failure would have called into question the special values that defined the American experience during the mid-twentieth century.

The seven Mercury astronauts were presented to the public at a NASA press conference in Washington, D.C., on 9 April 1959, just six months after the creation of the space agency. To the surprise and ultimate consternation of some NASA leaders, they immediately became national celebrities and the leading symbols of the fledgling space endeavor.[22] Both NASA officials and the press contrived to present the seven astronauts, whose public images were as carefully controlled as those of movie idols or rock music stars, as embodiments of the leading virtues of American culture in the 1950s.[23]

The press was fascinated by the apparent willingness of the astronauts to risk their lives for the good of a national cause. Tom Wolfe captured the method of this imagery some twenty years later in *The Right Stuff.* The astronauts were not brave in a stupid, unknowing way. Any fool could recklessly throw his or her life away:

> No, the idea here . . . seemed to be that a man should have the ability to go up in a hurling piece of machinery and put his hide on the line and then have the moxie, the reflexes, the experience, the coolness, to pull it back in the last yawning moment—and then to go up again *the next day.*[24]

Remembering that first press conference, Wolfe cut to the essential question the reporters had circled around. Were the astronauts afraid they were going to die?

> They had volunteered to sit on top of rockets—which *always blew up!* They were brave lads who had volunteered for a suicide mission! . . . And all the questions about wives and children and faith and God and motivation and the Flag . . . they were really questions about widows and orphans . . . and how a warrior talks himself into going on a mission in which he is bound to die.[25]

On 9 April 1959 NASA officials presented the seven Mercury astronauts to the Washington press corps. The apparent willingness of these men to risk their lives on shaky spacecraft and flaming rockets enthralled the American public. The astronauts became instant heroes and the most visible representatives of the expectations placed on the newly created civil space program. (NASA)

In introducing the Mercury Seven to their readers, the editors of *Life* magazine ran a two-page photograph that showed the astronauts at a long table smiling broadly with their hands raised toward the sky. The editors explained the hazards, "yet when asked for a show of hands by those who thought they would come back alive, the answer came unhesitatingly, unanimous." John Glenn, who would become the first American to orbit Earth, lifted both hands above his head.[26]

The bravery of the astronauts touched emotions deeply seated in the American experience at that time. Young and courageous, each sat alone in the single-seat Mercury capsule, like the "lone eagle" Charles Lindbergh crossing the Atlantic Ocean thirty-two years earlier. Facing personal danger, they fit the myth of frontier law enforcers, whose grit filled the substance of Hollywood matinees and television screens.[27] As military test pilots, the astronauts recalled the sacrifices required to produce the Allied victory in World War II at a time when military service was still held in high regard. Their personal exploits even evoked the substance of one of America's most popular sporting events: in exploring the appeal of the astronauts, Wolfe drew on his interviews with Junior Johnson, a legendary transporter of moonshine whiskey whose degree of physical courage propelled him to the top ranks of professional stock car drivers.[28]

Test pilots and race car drivers were thought to be a hard-living, hard-drinking lot.[29] In private, a number of the astronauts behaved this way, a fact revealed by Wolfe. At the time of their presentation, however, the astronauts were hardly cast in this light. To a public clamoring for personal details, the Mercury Seven were presented by the press as the personification of the clean-cut, all-American boys whose mythical lives popularized family-oriented television programs during the 1950s and 1960s.[30] They were portrayed as brave, God-fearing, patriotic individuals with loving wives and children. Addressing a joint session of Congress after his orbits around the world, astronaut John Glenn announced to the wildly cheering crowd, "I still get a real hard-to-define feeling down inside when the flag goes by" and got away with it.[31]

One year after their presentation to the American public, the Mercury Seven signed a contract by which they gave *Life* magazine the exclusive right to their personal, first-hand stories.[32] Astronaut stories ran in twenty-eight issues of the weekly magazine between 1959 and 1963.[33] Although *Life* reporters followed the astronauts at home and on the road and witnessed occasional indiscretions, such tidbits did not find their way into the mainstream press.[34] One of the reporters for *Life* later explained:

> There was no explicit editorial direction, but the deal Life made with NASA and the seven individuals created a strong bias toward the "Boy Scout" image, because all pieces under the astronauts' bylines had to be approved by them as individuals, as a group, and by Shorty Powers [NASA's public affairs officer for the astronauts] and whomever happened to be in charge at the moment in Washington.

The astronauts, the reporter continued, were the "main architects" of the image,[35] using their status as national heroes to enhance their influence in a flight program dominated by rocket scientists and engineers. NASA as well was anxious to perpetuate the mythology of the astronauts, although government officials were nervous about unrestricted access.[36] NASA used its astronauts to promote the space program, parading them through the White House and across Capitol Hill. Reporters cooperated because it made great copy and permitted them to tag along like technology groupies on a great American tour.

The press knew the astronauts endured the same personal difficulties as any other cross-section of well-educated, middle-class Americans. In fact, the astronauts probably endured more, given the pressures of the flight schedule and the temptations afforded celebrities. Said one of the writers for *Life:*

> I knew, of course, about some very shaky marriages, some womanizing, some drinking and never reported it. The guys wouldn't have let me, and neither would NASA. It was common knowledge that several marriages hung together only because the men were afraid NASA would disapprove of divorce and take them off flights.[37]

Life introduced their readers to the astronauts' "brave wives and bright children." Wrote the staff: "If the U.S. was getting a bargain in its calm, brave astronauts—and it was—it could also take pride in the wives they had waiting back home."[38] When NASA made its first flight assignments, *Life* ran a two-page picture of three astronauts gathered with their wives and children on the beach at Cape Canaveral watching a test flight of the Redstone-Mercury system that would carry the first two into space. Drawing on the conjunction of the common and the exceptional, the staff wrote: "In shorts and summer hats, carrying cameras and field glasses, the group looked like sightseers whose next stop might be Cypress Gardens or Marineland. But the men were not vacationers. They were Astronauts from Project Mercury, the prime candidates for a violent, historic event."[39] In 1962 *Life* ran a poignant story about astronaut Scott Carpenter, returning with his father and four children to the Colorado mountains that he had loved as boy. The family climbed rocks and explored caves, a personal sanctuary from the difficulties of the world.[40] It was a moving portrayal of a moment most fathers hope to share with their progeny. The story gave no indication that Scott Carpenter's marriage with his wife Rene was headed for divorce.[41]

The astronauts appeared at a time when NASA desperately needed to inspire public trust in its ability to carry out the nation's space goals. Rockets might explode, but the astronauts shined. They seemed to embody the personal qualities in which Americans of that era wanted to believe: bravery, youth, honesty, love

President Kennedy looked noticeably concerned as he watched the first flight of an American astronaut on a suborbital trajectory into space. He need not have worried. The Redstone rocket performed flawlessly, and the U.S. Navy recovered astronaut Alan Shepard and the *Mercury* space capsule only eleven minutes after its water landing. Thus began a twenty-five-year string of U.S. space missions in which NASA always brought its astronauts back alive. (John F. Kennedy Library)

of God and country, and family devotion. How could anyone distrust a government agency represented by such people? The trust the public placed in the astronauts spread through NASA and to the government as a whole. As one of the *Life* reporters summarized:

> Life treated the men and their families with kid gloves. So did most of the rest of the press. These guys were heroes, most of them were very smooth, canny operators with all of the press. They felt that they had to live up to a public image of good clean all-American guys, and NASA knocked itself out to preserve that image.[42]

Of equal importance in building the sense of competence was the glorification of science and technology. The early space-flight program commanded public attention at a time when the American public still worshiped science and technology. For much of the twentieth century, progress had seemed to sprout from invention and scientific discovery. Moreover, science and technology was viewed by many as the means by which the United States would win the Cold War. In the words of David Halberstam, the "best and the brightest" applied the energy of intelligence toward the solution of national problems.[43] The space program epitomized this trend and reminded Americans of the achievements that could be realized through the peaceful uses of science and technology, an image not overlooked by its promoters.

The history of the Seattle World's Fair well illustrates the growing public interest in science and technology. In the mid-1950s, civic leaders in Seattle sought participation for what they called the Century 21 Exposition. At the time, Seattle was a remote city noted for its bad weather and airframe industry. Shortly after the launch of *Sputnik 1* and *Sputnik 2* in 1957, civic leaders, who were struggling to identify a theme that would attract financial support for the fair, linked arms with officials in science agencies who wanted "to awaken the U.S. public to the significance of the general scientific effort and the importance of supporting it."[44] The latter wanted to rectify the slight to scientific developments that had taken place in the U.S. pavilion at the 1958 Brussels World's Fair. That exhibit had featured hamburgers and wide-screen movies, whereas the Soviet Union had displayed Earth satellites and spacefaring dogs.[45] Together with Washington senator Warren G. Magnuson, scientists and civic leaders won government funding for a museum-sized pavilion that sought to communicate "the innate joy of science," including exploration of the universe.[46] "No one knew whether a single huge presentation devoted solely to science would interest the public," its designers wrote.[47] They did not need to worry. By the time the Seattle World's Fair opened in April 1962, the U.S. space program was in full throttle and the public was clamoring for information about science and technology.

The exhibit treated science as a process of discovery. Although a number of exhibits within the pavilion's five large buildings dealt with space science, including a large planetarium, the science center deliberately avoided the display of space technology.[48] In apparent compensation, NASA constructed a separate building around the corner from the science pavilion. In it, the agency presented exhibits on rocketry, satellites, technology, tracking and communication, and human space flight.[49] The highlight of the NASA exhibit was John Glenn's *Friendship 7* space capsule, returned to Earth and eventually on to Seattle as part of its world tour.[50] Space exploration was also featured in the Washington state pavilion and the Ford Motor Company's simulated journey into space.

The following February Glenn's capsule was installed in Washington, D.C., in a small tin hanger on the south side of the national Mall. It joined a modest but enormously popular exhibit of aircraft and space memorabilia in what was known as the National Air Museum. This exhibit in turn formed the core of the National Air and Space Museum, the construction of which Congress authorized in 1966.[51] In Huntsville, Alabama, home of NASA's Marshall Space Flight Center, local leaders won support for what became the U.S. Space and Rocket Center and U.S. Space Camp.[52]

During the early 1960s, the public possessed an apparently insatiable appetite for information about space science and technology. Sensing this interest, television networks and print journalists went to elaborate lengths to inform the public about the details of space flight. Newspapers provided elaborate accounts of rocket technology, orbital dynamics, life support, guidance and control, and reentry mechanics. They explained communication blackouts, space medicine, rendezvous and docking, and a host of other details regarding space exploration.[53] Words in print followed the tradition established by writers of popular science two decades earlier, whose books led readers step by step through the details of space flight without much interpretation or philosophizing.[54] Even Madison Avenue joined the bandwagon, using public interest in space technology to market products such as Tang, a powdered orange drink supplied to the astronauts for their voyages into space.[55]

Communication experts wondered whether space flight could be presented effectively in the new medium of television. In the beginning, extended space flight did not seem to lend itself to visual presentation. Although the launches were spectacular, the voyages offered little visual excitement. Television cameras did not go into space for real-time coverage until 1968.[56] Except for lift-off and splashdown, the astronauts and spacecraft were out of sight. Moreover, launch preparations were monstrously slow, snippets of action hardly punctuating hours of inactivity. Media critics wondered how television producers could fill the great void of air time in a medium that required constant action and moving pictures. There was plenty of air time to fill. When John Glenn became the first American to orbit the earth, the three television networks scrubbed the regular Tuesday programming and devoted their entire day-time programming to the flight.[57] Adding the scrubbed launch on 27 January and the parades and ceremonies that followed, total television coverage of the story averaged twenty-nine hours per network.[58]

How did the networks fill all of that time? They rarely sought to explain the NASA bureaucracy, except to compliment the NASA-industry team. They hardly ever spoke about the politics of national space policy, even though controversies remained. Very little time was devoted to space propaganda or to selling future missions. Instead, when not presenting personal interest stories dealing with the astronauts and their families or public reactions to the flights, the media treated the American public to a huge, decade-long lesson on space science and technology. Given the state of television technology at the time, this was done with great creativity. Mock-ups were built so that newscasters could explain spacecraft maneuvers, scientific experiments were acted out in makeshift laboratories,[59] networks commissioned animation sequences to demonstrate rocket staging and reentry, and newscasters took cameras to industrial plants in which space capsules were being built.

Television easily adapted to space flight, and in fact enjoyed a particular advantage over print journalism. Newspapers and magazines had to describe space technology with diagrams and words, whereas newscasters could show viewers the machinery with their hands. Newscasters such as Jules Bergman became media celebrities by explaining space technology using mock-ups or other props. During the flight of Scott Carpenter, the second American to orbit the earth, Bergman placed himself in one of the biomedical harnesses that the astronauts wore to monitor their vital signs. The harness revealed that Bergman suffered as much stress during the twelve hours of television coverage as Carpenter did in orbit.[60] One media critic observed that "the moment the rocket goes out of sight is precisely when television can let its imagination soar."[61]

Much of the coverage of early space ventures placed newscasters and journalists in the position of professors explaining the wonders of science and technology to attentive students. Humorist Art Buchwald spoofed the tendency of newscasters to lay science lessons over periods of inactivity with his parody of the press covering an early morning launch from Cape Canaveral:

> "The sun has just come up, David, and it's quite a sight to see."
>
> "Could you describe it to us?"
>
> "Well, from where I'm standing, it's round and looks like a great big fiery ball. Scientists have informed me it's 85 million miles from the earth and it's very hot. As you can see, it's rising from the east. . . . I have been told that without the sun the earth might not sustain life."[62]

Science and technology can be frightening. Doomsday scenarios of atomic warfare terrified Americans during the 1950s, as did images of exploding nuclear reactors in subsequent decades. The national space program, however, provided an example of science under control. It helped to maintain the faith that science meant progress and discovery.

The emphasis upon science and technology served to displace other, more pejorative, images of government. It displaced the language of exposé, cost overruns, and management errors. Those accusations fell on the space program, especially in later years, but in the early days they were drowned out by the chorus of confidence. Science lessons displaced the language of partisan wrangling and policy disputes. Some people complained about the funds lavished on space, but those opinions were background noise set against the wider symphony of general praise.[63] Rarely did the media trounce on the space program as an example of the horrors of Big Government. Project Apollo was one of the largest undertakings ever set upon by the federal bureaucracy, but the engineers and scientists who ran NASA were cast as citizen soldiers in the Cold War, not empire-building desk squatters with lifetime sinecures. To a generation of Americans in the early 1960s, the space program was a national joy ride. It was a demonstration of personal character, of the power of science and technology under control, and of the ability of Americans to complete a task simply because it was hard.

Drawing a slogan from the circus, the *Herald Tribune* called the space program not " 'The Greatest Show on Earth,' but . . . The Greatest Show, period."[64] Even critics gushed over the early space program. Editorial writers at the *New York Times,* which had scoffed at space flight when Robert Goddard first proposed a rocket shot to the Moon, praised President Kennedy's decision to conduct a space program "second to none" and called the flight of John Glenn "one of our finest hours."[65] Editorial writers at the *Washington Post,* often sour on the value of human space exploration, compared Glenn's flight to Columbus's discovery of America. Responding to the *Gemini 3* flight a few years later, *Post* editorial writers exclaimed that "this is wholesome competition in a race in which the ultimate winner is mankind itself."[66]

Much of the worthiness of the national space effort arose from the difficulty of the undertaking. Human flights to the Moon and robotic probes to the planets seemed incredibly perplexing to a public barely accustomed to rocketry. They even appeared difficult to NASA engineers.[67] The space race thus provided a national self-examination, a trial of the ability of Americans and their government to surmount great obstacles, just as the mobilization for World War II had tested the American system two decades earlier.

The political language supporting the early space program emphasized this perspective. Politicians understood the degree to which public impressions about the challenges of the space program fostered confidence in government, and their understanding can be traced through their own words. Speaking at Rice University in September 1962, President Kennedy provided his most comprehensive explanation of the reasons for exploring space. "We choose to go to the moon in this decade and do the other things," he said, "not because they are easy, but because they are hard, because that goal will serve to organize and measure the best of our energies and skills."[68] Kennedy had made a similar comment in proposing the Moon race on 25 May 1961. Defending the choice, Kennedy told Congress that no other objective in space "will be so difficult or expensive to accomplish."[69]

Several early space documents reveal the degree to which members of the Kennedy administration sought to use the hard challenges of space to build trust in American society and its government. The persons among whom they wished to build that trust were not Americans, however, but people in nations uncommitted in the Cold War. Writing in response to President Kennedy's request for an assessment of alternatives, Vice President Lyndon Johnson assured Kennedy that "other nations will tend to align themselves with the country which they believe will be the world leader—the winner in the long run," and space spectaculars were increasingly being used by people in those nations as an indicator of world leadership. In a May 1961 memorandum, Defense Secretary Robert McNamara and NASA administrator James Webb argued that the United States "needs to make a positive decision to pursue space projects aimed at enhancing national prestige." Achievements in space, they explained, "symbolize the technological power and organizing capacity of a nation." National prestige was exactly the arena, they continued, in which the United States lagged behind.[70] Kennedy understood this argument and rarely strove to justify the lunar objective on scientific grounds.[71]

As a consequence of its symbolic value, Kennedy realized much of the advantage of the lunar mission simply by setting the goal. The willingness of the United States to challenge the Soviet Union in such an undertaking impressed foreign leaders. The Moon race signaled the willingness of the U.S. government to contest the Soviet Union at all levels of military and technological competition, of which space was merely one. In many ways, the actual mobilization for the voyage was anticlimactic to the decision to go, which helps explain why Kennedy became reluctant to complete the voyage once that advantage had been gained.

Realizing almost immediately that Project Apollo would empty the federal treasury, Kennedy quietly sought a way to curtail the endeavor. He could not back down alone without calling into question governmental capability and Cold War commitments. Instead, he began to explore the possibility of reshaping the program from one of competition into one that fostered international cooperation.[72] In June 1961 Kennedy made a direct proposal to Soviet premier Nikita Khrushchev at the summit meeting in Vienna for a joint expedition to the Moon.[73] He repeated the call in 1963 in a speech before the United Nations:

> Why, therefore, should man's first flight to the moon be a matter of national competition? Why should the United States and the Soviet Union, in preparing for such expeditions, become involved in immense duplications of research, construction, and expenditure? Surely we should explore whether the scientists and astronauts of our two countries—indeed of all the world—cannot work together in the conquest of space.

Kennedy closed his speech by urging, "Let us do the big things together."

Kennedy's cooperative vision was steadfastly resisted by the Soviet Union. In public, Soviet leaders were noncommittal, dismissing the proposal as premature, but in private, they viewed the offer as a ploy to gain advantage in the Cold War. Cooperation in space would open the Soviet society to U.S. scrutiny and steal attention from one of the few arenas of Soviet success. At the Vienna summit, Khrushchev linked space cooperation to other Cold War issues, insisting that Kennedy first withdraw his military forces from bases along the Soviet border.[74]

Kennedy's cooperation initiative had a second, less-intended consequence. It undercut domestic political support for the race to the Moon, thereby providing an opportunity for congressional critics to propose large reductions in the budget for Project Apollo. Buoyed by the thought that the Moon race was no longer the Cold War priority it had once been, opponents moved to cut funds for what one called "a manned junket to the moon." The House of Representatives began the assault in 1963 by removing $600 million from Kennedy's $5.7-billion NASA budget request. The administration objected on the grounds that the cuts would interfere with the deadline for reaching the Moon. Members of Congress questioned White House support for a deadline that had apparently lost its priority. In the Senate, Arkansas senator J. William Fulbright moved to cut 10 percent more from the NASA appropriation. The president's allies prevailed on that vote, but assaults on Apollo's budget followed almost yearly thereafter.[75]

As interest dwindled in the symbolism of space spectaculars for people overseas, preoccupation increased in the value of domestic consumption. Space accomplishments helped promote confidence in government at home. This became more and more important as the counterculture movement with its distrust of national authority set in. Such distrust threatened to plunge the country into political turmoil and a rejection of middle-class values. President Johnson in particular understood that he could not abandon the Moon race without undermining public confidence. Political leaders such as Johnson increasingly spoke about the inspirational effects of the lunar voyage within the United States.

Advocates of space exploration needed all the help they could get. By 1965 the assault against civil space expenditures from opponents within the government was fully underway. Space expenditures, running at more than $5 billion per year, dwarfed other priorities, such as the $1.8-billion War on Poverty and the $2 billion set aside to improve elementary and secondary education.[76] When asked in early 1966 to identify ways to cut the growth in federal spending, Budget Director Charles Schultze, a pragmatic economist with a strong commitment to social reform, recommended that President Johnson defer the lunar landing until the 1970s and abandon post-Apollo space efforts, a recommendation that produced the second largest sum of savings in Schultze's litany of cuts. Failing to win

Johnson's approval in 1966, Schultze tried again the following year. He warned that NASA might fail to meet the lunar goal even if Johnson restored the funds: "It would be better to abandon this goal now in the name of competing national priorities, than to give it up unwillingly a year from now because of technical problems."[77]

Johnson agonized over the prospective cuts. He did not want to abandon the commitment to put Americans on the Moon by the end of the decade. At the same time, he knew that he had to endorse substantial cuts in federal spending in order to win congressional support for the 10 percent tax surcharge needed to finance the war in Vietnam. Johnson spared Project Apollo but decimated NASA's spending plans for exploration efforts beyond. Responding to the protests of NASA administrator James Webb, Johnson replied that he personally did not want "to take one dime from my budget for space appropriations" but agreed to do so in order to satisfy the budget cutters supporting his tax bill.[78] Exhausted by the budget battles, Webb resigned the following year.

In resisting efforts to undermine the lunar goal, Johnson acknowledged the degree to which space flights contributed to domestic confidence. "Somehow the problems which yesterday seemed large and ominous and insoluble, today appear much less foreboding," he announced after the completion of the *Gemini 5* flight in August 1965. Americans, he proclaimed, did not need to fear problems on Earth when they had accomplished so much in space.[79]

In his last month as president, Johnson welcomed the crew of *Apollo 8* to the White House. He spoke of the progress in the space program he had helped to launch a decade earlier. Using words that recalled the difficulty of the endeavor, Johnson said, "If there is an ultimate truth to be learned from this historic flight, it may be this: There are few social or scientific or political problems which cannot be solved by men, if they truly want to solve them together."[80]

Through the 1960s, the space program provided successive examples of a government program that worked well. This in turn inspired one of the greatest feats of imagination in American history. NASA's accomplishments allowed members of a society founded on the mistrust of government to believe that their political institutions could organize an incredibly complex endeavor and accomplish it successfully. As Lyndon Johnson suspected it would, this trust allowed the federal government to proceed with other ambitious initiatives. Some of them, such as the Vietnam War and the War on Poverty, failed. The space program, however, continued to shine.

Throughout the late 1960s and early 1970s, the sense of competence that transformed the space program into an object of national pride coexisted alongside a general crankiness toward technology. As the national space program progressed, public fascination with science declined. Americans who had supported technology races hardly a decade earlier remembered that science was as capable of producing disasters as progress. Though NASA retained its aura of competence, and space exploration remained a technology of optimism, political leaders found little support among the public at large for ambitious new initiatives.

Exploration advocates still dreamed of space stations, space shuttles, orbital observatories, lunar bases, robotic tours, and expeditions to Mars. Such desires required a level of government funding equal in value to that provided during the height of the Apollo moon effort.[81] To begin these initiatives, NASA officials and their allies had to ensure that the hardware and technical work force assembled for the flights to the Moon did not vanish once the landings began. In 1965, as spending for space exploration peaked at $5.5 billion, NASA officials established the Saturn-Apollo Applications Office at NASA Headquarters and began to press for new missions that would maintain the production of Saturn 5 and Saturn 1B rockets and related spacecraft. The office sought approval for additional scientific and technological missions "in earth orbit, lunar orbit, and on the lunar surface . . . through use of Saturn/Apollo hardware." The main proposals to emerge from this near-term endeavor were an orbital workshop, later named *Skylab,* and a large telescope to be operated by astronauts working in space.[82]

NASA supporters put forward these proposals during a major sea change in public attitudes toward technology. Up until that time, technology had remained a force that in the mind of the public seemed to produce more good than evil. Broad-scale attacks on technology such as the "ban the bomb" movement had remained on the fringe of American political life.[83] Other antitechnology protests possessed a mystical quality that placed them well outside the mainstream of American public opinion. The October 1967 March on the Pentagon illustrated the madcap quality of the latter. The march was an early event in the protests against the War in Vietnam. As a gesture of defiance against American military technology, protesters planned to encircle the Pentagon with a sufficient number of demonstrators to perform a rite of exorcism that would tear the building from its foundations and cause it levitate above the ground. The government, which had no apparent objection to the levitation, would not permit the demonstrators to form the human chain around the structure that was essential to the rite on the grounds that the chain would block employee access. Arrests inevitably followed.[84] To the participants, the exorcism was no more absurd than the public's faith in the saving power of technology.

As the sixties progressed, attacks on technology became more respectable, moving away from the protest fringe. In her 1962 book *Silent Spring,* Rachel Carson questioned the value of the chemical revolution that had transformed agriculture and industry, warning Americans that they were poisoning their environment.[85] In 1968, in *The Population Bomb,* Paul Ehrlich raised the specter of environmental collapse in a world plagued by uncontrolled population growth and dwindling resources, an image that became more plausible as pessimism about technology spread.[86] Works of fiction amplified the theme. Joseph Heller and Kurt Vonnegut, in two 1960s classics, portrayed the application of military technology as acts of insanity, a popular counterculture theme.[87]

Throughout the 1950s, aerospace technology was presented to the American public as one of the principal means by which the Cold War would be won. Thousands of middle-class Americans went to work for the aerospace industry, serving as civilian soldiers while providing themselves a degree of job security unprecedented in previous times. No event did more to transform American attitudes toward aerospace technology than the constant bombardment of images from Vietnam. An extremely powerful military-industrial complex made possible by scientific and technical elites visually failed, day after day on the broadcast news, a conclusion unavoidable after the January 1968 Tet Offensive. With all the technology it could muster, the U.S. military could not contain a poorly equipped guerrilla band assembled by a Third World country with a subsistence economy.

A sharp downturn in aerospace employment accompanied the failures in Vietnam. Workers who had come to rely upon government aerospace support for their economic security found themselves the objects of layoffs, cutbacks, and scorn. The sense of disillusionment arising from these developments undercut trust in government, even among its strongest supporters, and disillusionment made possible that which was unthinkable one decade earlier: the abandonment of new technologies.[88]

Defeat of government support for the Supersonic Transport (SST) marked the political acceptability of the belief that some technologies were not worth pursuing. The congressional votes in March 1971 to discontinue government funding for the SST were a slap in the face to those who believed that new technologies always bred progress. Congress had supported the SST throughout the 1960s, ever since President Kennedy proposed that the government assist American industry with this "logical next development" in aviation technology.[89] But media attacks and citizen protests mushroomed as the sixties progressed. In 1967 William A. Shurcliff, a Harvard physicist, formed a citizen's organization with the relatively innocuous task of protesting the assault on tranquility generated by sonic booms from a fleet of supersonic aircraft. By 1970, this relatively modest objection had been joined by premonitions of ecological disaster. Exhaust from the high-flying aircraft, opponents claimed, would cause global warming and destroy the protective ozone layer in the upper atmosphere.[90] Economic arguments also played a role, but the final defeat of the SST was due largely to the fact that it became a symbol of misguided progress among a public increasingly suspicious of new technologies.

Throughout this transformation of American culture, the civilian space program provided the one best example of technology both benign and effective. The only major event tarnishing the reputation of the civilian space program during this period was the fire at launch complex 34 on 27 January 1967 that claimed the lives of astronauts Virgil Grissom, Edward White, and Roger Chaffee during a test of the Apollo spacecraft.[91] Rather than reduce support for the civilian space effort, however, the tragedy actually served to strengthen public resolve, as opinion polls reveal.[92]

Public support for NASA space flights reached a peaked toward the end of 1968, coinciding with the famous Christmas voyage in which astronauts Frank Borman, James Lovell, and William Anders became the first humans to leave the gravitational well of Earth and travel to the Moon. On Christmas Eve, from their orbit around the Moon, the astronauts sent back the first closeup television pictures of the lunar surface. The pictures conformed remarkably to the images anticipated by space artist Chesley Bonestell some twenty years earlier. As the lunar surface passed below them, the crew read the story of Creation from the first ten verses of the Book of Genesis. "Merry Christmas and God bless all of you," they closed, "all of you on the good earth."[93]

Coverage of the flight was uniformly favorable. "Not since Christopher Columbus's first voyage to the 'new world' have men embarked upon a journey comparable to that begun by Apollo 8," gushed editorial writers for the normally dour *New York Times.* The voyage "vividly juxtaposed science as the instrument broadening man's reach into the universe with science as the source of weapons that may destroy humanity." Journalists and cartoonists portrayed the *Apollo 8* capsule as a Christmas star.[94] The voyage of *Apollo 8* provided a moment of hope and serenity in what had been an otherwise ugly year. The American people had endured riots, the assassinations of Robert Kennedy and Martin Luther King Jr., the Tet Offensive in Vietnam, and a turbulent contest for the presidency. Said editors for the *Los Angeles Times,* "The Apollo 8 flight, therefore, comes as a welcome talisman of future good fortune—a kind of reassurance that we are still a nation capable of great enterprises."[95]

During the 1960s space travel escaped nearly all the apocalyptic portrayals that overwhelmed other images of technology. Misused military technology, for example, provided the basis for the classic 1964 satire on official stupidity and nuclear war, *Dr. Strangelove,* as well as the surprise ending for the popular 1968 film *Planet of the Apes,* in which Charlton Heston discovers that the simian planet on which he has landed his spacecraft is really a post-apocalyptic earth. Environmental destruction motivated the 1971 release *Silent Running,* in which the last hope of species rescued from a sterile Earth resides with keepers of large, terrarium-like structures in space. Stupid government officials order keepers to destroy the greenhouses and all they protect.[96]

The optimism accorded space technology was carried to spiritual levels with the 1968 release of Stanley Kubrick's *2001: A Space Odyssey.* The film stunned audiences with the remarkably quality of its special effects, setting new standards for the cinematic portrayal of space travel. The screenplay, by Arthur C. Clarke, suggests that space technology, both Earth-based and extraterrestrial, has the capacity to radically transform human evolution in positive ways. After a battle with the onboard computer (the HAL 9000 provides the one example of technology gone awry), astronaut Dave Bowman travels through a strange monolith to an unexplained destination in a corner of the universe. There he is reborn as a superbeing, a star child who is transported back to the solar system to gaze down at planet Earth.[97]

Throughout the 1960s, space travel continued to be associated with new beginnings and social justice, themes emphasized in the most popular television shows to deal with space travel during that decade. In *Lost in Space,* which began its three-year run in 1965, the Robinson family is selected from more than two million volunteers to leave the critically overpopulated Earth. Their sabotaged spacecraft flies off course, allowing them to meet a variety of alien characters in what becomes an extraterrestrial morality play. A similar impulse guided the highly influential *Star Trek* series, which began its initial three-year run the following year. The crew of the starship *Enterprise* sets out to explore worlds where no humans have gone before. Creator Gene Roddenberry used *Star Trek* to address the social issues of the times: superpower conflict, fascism, civil rights, and interracial sexuality. These two series eschewed the more terrifying aspects of space exploration examined in alien monster movies such as *The Thing* (1951) or the bizarre plot twists exploited in the *Twilight Zone* (1959–64).[98]

One of the few widely viewed examples of a dangerous space technology during this period appeared with Michael Crichton's *Andromeda Strain,* published as a novel in 1969 and released as a film the following year. In the story, a *Scoop* satellite searching for extraterrestrial life brings a virulent organism back to Earth, where the organism threatens to break into the general population after killing all but two residents of a small southwestern town. The story excited fears that space exploration would bring some unknown retribution on the earth, a possibility that prompted NASA to quarantine the first astronauts to return from the Moon.[99] Occasionally, as with the 1979 release of *Alien,* producers and writers terrified the public with the misuse of space technology. In *Alien,* the crew of the space freighter *Nostromo* is sent to retrieve twenty tons of mineral ore and, without the crew's consent, a vicious alien that proceeds to kill every human on board except Sigourney Weaver.[100] In general, however, space travel in popular culture continued to be associated with rebirth and hope, themes repeated in later releases such as *Close Encounters of the Third Kind* (1977), *E.T.: The Extraterrestrial* (1982), and *Cocoon* (1985).[101]

The aura of competence created by the newly created civil space program and its astronaut corps helped boost public trust in government to all-time highs. Politicians could not undercut the commitment to journey to the Moon without subverting the trust that supported government in general. On 16 July 1969 the Apollo-Saturn space vehicle carrying the first astronauts to the lunar surface lifted off from the Kennedy Space Center. (NASA)

NASA's own flight program helped maintain the image of space technology as benign. Astronauts from Earth landed on the surface of the Moon on 20 July 1969. The landing was far more difficult than the public knew. Steve Bales, a twenty-six-year-old flight controller at NASA's Mission Control in Houston, told astronauts Neil Armstrong and Buzz Aldrin to ignore an alarm from the spacecraft computer during the final descent. To compound that problem, Armstrong and Aldrin left themselves less than thirty seconds of fuel as they searched for a smooth place to land. Had the fuel tanks run dry, or the alarm been real, the onboard computers would have fired the ascent engines in an attempt to separate the upper half of the landing module and the astronauts from their moving platform, a very dangerous maneuver.[102]

NASA officials prevailed again when one of the oxygen tanks supporting the flight of *Apollo 13* exploded halfway to the Moon, depriving the crew of the breathing air and electricity they needed to survive. NASA engineers pieced together a plan for the astronauts to squeeze into the powered-down lunar module and ride it like an interplanetary lifeboat back to Earth. As the rescue surged from crises to crises, NASA made the return trip look almost routine.[103]

Even as other big government initiatives failed, the U.S. space program remained one of the few examples of a large government program that worked. From the summer of 1969 to the end of 1972, successive teams of American astronauts explored the surface of the Moon. Even as Americans grew disenchanted with other endeavors, space travel remained one of the few forms of technology about which commentators remained optimistic. Politicians who did not want to fund extravagant space missions nonetheless would not allow the human space-flight program to die. (NASA)

Repeatedly during this period the press praised the capabilities of the men and women who managed the nation's space effort. In spite of the obvious negligence that necessitated it, the rescue of *Apollo 13* was widely hailed as NASA's "finest hour," an impression that grew with time. By the time director Ron Howard made the big-budget *Apollo 13* film more than twenty years later, NASA's astronauts and flight controllers had become legends in American popular culture.[104]

NASA's impression of competence lasted until 1986, well after other icons had fallen from public favor. It ended with the first live television broadcast of a space flight fatality, the 28 January loss of the seven astronauts on the space shuttle *Challenger.* With the accident, NASA's image moved from that of an agency that could do no wrong to a bureaucracy that could do little right. Forty-seven percent of those responding to a Media General/Associated Press public opinion poll indicated that their confidence in NASA had been shaken by the event, and only one-third of them said after two years that it been restored.[105] A succession of errors amplified the impression of an agency in decline—the myopic Hubble space telescope, an antenna problem with the *Galileo* Jupiter probe, the mysterious loss of the billion-dollar *Mars Observer* probe, and more exploding rockets, fortunately with only instruments on board.

In actuality, NASA conducted a less accident prone flight program in the 1980s than it had during the 1960s when the sense of competence arose. NASA endured far more launch and mission failures during the 1960s, when the space program was less routine and more experimental.[106] As history shows, however, facts do not always play the dominant role in the creation of images about the U.S. space program.

In the early 1970s, after the first landing on the Moon, President Richard Nixon examined the various options for the future of the U.S. space program. He confronted the most important decision affecting the U.S. space effort since President Kennedy had challenged Americans to race to the Moon one decade earlier. Even though public confidence in the goodness of space travel remained high, public support for an extensive government effort remained low. Only once since the mid-1960s had the proportion of people who wanted to "do more" in space exceeded the number who wanted to "do less." By 1973, 59 percent of the American public responding to a Gallup public opinion poll wanted the government to cut spending on space, compared to only 7 percent who wanted to spend more.[107] At the same time, public confidence in the overall space effort remained high. The public did not want to spend money on space exploration, but they were proud of the nation's accomplishments.

Various factions within the White House fought over the future of the space program. Many opposed any large new initiative. Problems with inflation had forced White House officials to reduce spending and set a $3-billion cap on NASA appropriations, far less than the budget levels attained during the Apollo years. Much of the interest in Cold War competition with the Soviet Union had disappeared, as well as the sense that the United States had to engaged in a continuing space race.[108]

On the other hand, the U.S. space program remained one of the few shining examples of a big government program that worked. Further cuts, warned Caspar Weinberger, then deputy director of the president's Office of Management and Budget, would send a signal that "our best years are behind us." In an August 1971 note to the president, Weinberger mused that "America should be able to afford something besides increased welfare, programs to repair our cities, or Appalachian relief and the like."[109]

Nixon read the memo and scribbled a note in the margin: "I agree with Cap." Nixon did not want to be the president who shut down such a source of national pride, especially so soon after the defeat of the SST. His need to maintain political support in aerospace states such as California and Texas contributed to his decision to maintain the human space flight effort, but so did his sense that NASA oversaw one of the few remaining technologies of optimism at that time.[110] Nixon agreed to move ahead with one new initiative, a modest Earth-to-orbit space transportation system that preoccupied NASA's human space-flight activities for the next twenty-five years.[111]

The aura of competence surrounding the NASA space-flight program helped maintain political support for these endeavors. Project Apollo was motivated by Cold War nationalism and nuclear paranoia, justifications that quickly disappeared. By fulfilling the project, Americans proved that they possessed the skill, technology, and wealth to complete voyages to other spheres. The sense of accomplishment thus engendered helped maintain support for the space program in general, even as national leaders debated exactly where that program should go.

MYSTERIES OF LIFE

> If we find the answer to why it is that we and the universe exist, it would be the ultimate triumph—for then we would know the mind of God.
>
> —Stephen Hawking, 1993

Space exploration received government support during its infancy largely as a result of nationalistic concerns. The desire to rekindle national pride, to grasp military advantages from the cosmos, and to build confidence in the U.S. system of government created governmental majorities for large space expenditures. Space exploration promised to do all these things. Without these promises, it is doubtful that the civil space effort would have progressed much beyond the modest research program advanced by the Eisenhower administration.

To its most devoted advocates, however, space exploration promises far more: to continue the quest begun centuries earlier to supplant religious dogma with science as a means of understanding the universe. Space exploration addresses the great mysteries of life. How did the universe begin? Where did the solar system come from? Are humans alone, or is the universe teeming with life? How will it all end? Since Galileo Galilei employed a pair of converted spectacles to observe the moons of Jupiter, advocates of this new way of understanding have argued that natural observation will provide answers to questions such as these.

For many in the exploration business, this quest has a spiritual quality, promising answers to cosmological questions that have intrigued humans throughout history and have inspired great myths and religions.[1] By probing the mechanics of the universe, humans can find answers to questions that have encouraged spiritual introspection since thinking began. It will allow humans, in the words of scientist Stephen Hawking, to know the mind of God.[2]

To a great extent, the images that portray space exploration constitute a form of cultural anticipation. Not only do such images frame questions, they also anticipate answers. Scientists may urge caution, but popular culture advances conclusions. The answers people expect to find in the cosmos already exist in their own minds, created by works of fiction and cultural traditions. The universe itself places a heavy burden on these expectations. Initial reports from the void, sent back by automated satellites and planetary probes, reveal a solar system that confounds traditional beliefs: it does not conform to the images that dominate popular culture. This has not defeated the advocates of space exploration, however. Rather than revise their views, they simply look harder.

No expectation has had more influence on support for space exploration than the belief that humans are not alone. Those who promote exploration possess an almost universal faith in the doctrine that life exists on other spheres. Even the most skeptical scientists have difficulty concluding that God or whatever natural force created life did so on a single planet in the entire universe.

For much of modern history, exploration advocates advanced the view that life, either primitive or advanced, could be found within the solar system, most probably on Venus and Mars. This belief drew sustenance from the stimulating but factually incorrect theory of solar-system evolution widely known as the nebular hypothesis, advanced by the French mathematician and astronomer Pierre Simon de Laplace. Laplace postulated that the planets arose from a vast rotating nebula extending into space beyond the orbit of the farthest planet. The mechanics of rotation forced the nebula to separate into rings, which in turn coalesced into planets, beginning with the spheres most distant from the sun and ending with the planet Mercury.[3]

The proposition that natural forces created the planets in stages, accompanied by growing acceptance of the theory of biological evolution, suggested to the popular mind that the planets constituted a sort of evolutionary time machine. On Mars, people could view Earth as it would become millions of years hence, once its oceans dried up and its atmosphere grew thin. Any life that arose on Mars would be ancient and any civilizations far older than those on Earth. The planet Venus, by contrast, was viewed as an embryonic planet still shrouded in primeval haze. As late as 1934 a writer for *Nature* magazine reported that "the opinion is quite generally held by astronomers that Venus is not as far advanced in evolution as the earth, while Mars is a much older world."[4]

Improvements in telescopic imagery during the nineteenth century allowed astronomers both professional and amateur to make detailed observations of Earth's neighbors in space. These improvements were accompanied by a rash of reports purporting to present evidence of life on nearby spheres. Some were perfect hoaxes, such as the wholly fictional report by the *New York Sun* in 1835 that astronomers had sighted animals on the Moon. Others were more serious. Twelve years earlier a Munich astronomer, Franz von Paula Gruithuisen, announced that he had discovered a walled city on the lunar surface. What Gruithuisen observed was a strange formation of mountain chains near the crater Schroter, amplified by an active imagination. The announcement, according to space historian Willy Ley, "created a stir which can be traced through the whole literature of that time."[5]

Notions about an extraterrestrial presence on the Moon persisted even into the twentieth century. Frustrated by the absence of visible evidence, promoters of popular culture suggested that fortifications or other structures might be placed where they could not be seen. In their three-part television series on space exploration, animators working for Walt Disney made such a suggestion in portraying an imaginary trip around the Moon. The four-person crew makes a startling discovery as they cross the terminator on the back side of Earth's nearest neighbor, the side previously hidden from human view. The crew reports: "Captain, I'm getting a high geiger count at 33 degrees." . . . "Contour mapper shows a very unusual formation at about 15 degrees 7 latitude and meridian 210." "Get some flares in that area quick."[6] As flares light up the lunar surface, the crew looks down on the ruins of a lunar colony of unknown origin on the back side of the Moon. Arthur C. Clarke and Stanley Kubrick continued this tradition in the 1968 classic *2001: A Space Odyssey,* a film in which lunar explorers discover a monolith of extraterrestrial origin buried beneath the surface of the Moon by a spacefaring civilization millions of years ago.[7]

Detailed observations of the Moon, even in the nineteenth century, revealed a stark and lonely world with little evidence of life either local or extraterrestrial. Rather than discourage speculation, these findings simply moved speculation to new realms. Astronomers watching the planet Venus observed a faint luminosity in the atmosphere of its night sky. This led to suggestions that its inhabitants might be holding giant festivals, or burning large stretches of jungle to produce new farm land, or even attempting to communicate with the earth. The physical similarities between Venus and Earth led the popular science writer Camille Flammarion to conclude that its inhabitants should be "but little different from those which people our planet. As to imagining it desert or sterile, this is a hypothesis which could not arise in the brain of any naturalist."[8]

The notion of Mars as an ancient planet held special appeal and helped create the Martian myth, which dominated the popular culture of space throughout the twentieth century. The myth drew support from reports that Mars harbored conditions suitable for the development of life (an atmosphere and liquid water) that had been present at least as long as they had been producing life on Earth. This hope gave rise to a number of popular accounts promoting the notion that most certainly vegetation and possibly advanced life-forms could be found on Mars.

The widely disseminated findings of Giovanni Schiaparelli and Percival Lowell, who saw markings on the Martian surface that looked like canals, encouraged such views.

Schiaparelli, who observed the markings in detail during the close approach of 1877, called them *canali,* or "channels." Schiaparelli was a well-trained astronomer, with more than twenty years of experience in the field. As director of the Milan Observatory, he had already established his reputation by observing asteroids, comets, and meteor showers. He reserved his judgment on the meaning of the markings on Mars in spite of the fact that the *canali* and surrounding areas seemed to change with the expansion and contraction of the Martian north pole. It was not necessary to attribute the markings to the work of intelligent beings, he wrote in 1893, as they could as easily be natural features on the surface of the planet, like sea channels on the surface of Earth.[9]

In English, however, *canali* became canals, a translation that had tremendous popular significance. The greatest works of engineering on Earth during the nineteenth and early twentieth centuries were canals, including the Suez Canal, completed in 1869, and the Panama Canal, begun in 1904. The ability to organize the construction of canals was, in the nineteenth century, the signature of an advanced civilization. To the informed public, Schiaparelli's findings required little interpretation. If canals criss-crossed the surface of Mars, the planet must harbor a highly intelligent civilization.

This vision excited the imagination of Percival Lowell. Born into a famous Boston Yankee family in 1855, Lowell had a fortune sufficient to pursue whatever fancy intrigued him without the inconvenience of formal training in the subject. After graduating from Harvard in 1876, he devoted himself to literature and travel, producing a series of books on the Orient. As an amateur astronomer, Lowell was aware of the findings on Mars. He returned to the United States with the intent of building his own observatory and making his own investigations during the planet's close approach of 1894. Friends at Harvard suggested a site on a high mesa near Flagstaff, Arizona. With technical support from Harvard astronomers, supplemented by his family's ample resources, Lowell completed the observatory by the spring of 1894. He gazed with amazement at the polar ice caps and located by his own account nearly two hundred canals.[10] Between 1895 and 1908 Lowell published three books that helped create the popular image of Mars.[11]

Lowell concluded that the canals he mapped were the product of an ancient civilization struggling to survive on a dying world, a world far older than Earth. "His continents are all smoothed down; his oceans have all dried up," he reported in 1895.[12] Life on Mars confronted a growing scarcity of the fluid essential to existence, he asserted, the only available water being that which came from the semiannual melting of the caps of snow at the Martian poles. Lowell detected a pat-

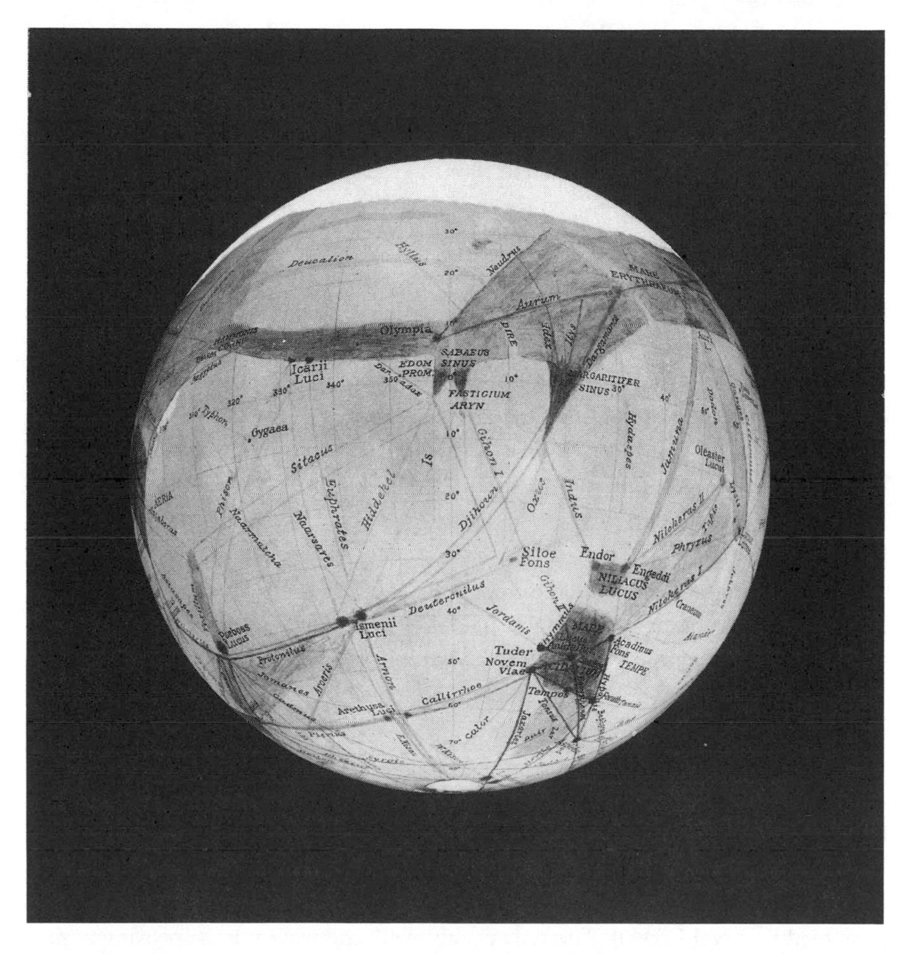

Unattainable expectations plagued the civil space effort as it matured. The vision of space exploration that so excited public audiences was based on a series of expectations that were hard to satisfy. Percival Lowell created one of the most enduring expectations with his turn-of-the-century drawings of canals on Mars. Lowell insisted that the canals he viewed were the work of an ancient civilization struggling to maintain its existence on a dying world. (Lowell Observatory photograph)

tern to the canals—straight lines criss-crossing the surface of the planet meeting in what he perceived to be oases of vegetation and life. The pattern led Lowell to conclude that the canals must carry water between the poles and the more habitable portions of the planet. The regularity of the lines and the knowledge that water does not run uphill led Lowell to the unalterable conclusion that intelligent beings must be pumping water back and forth.

Because canals crossed the entire planet, Lowell also deduced that Martians had abolished war: only a necessarily intelligent and nonbellicose race could "act as a unit throughout its globe." The pressures of survival on a dying world would further sharpen mental capabilities. "To find, therefore, upon Mars highly intelligent life is what the planet's state would lead one to expect."[13] Sadly, Lowell concluded, by the time Earthlings developed the technology required to reach Mars, the civilization might have died away as the last of its water disappeared.[14]

Scientists cautioned the public against embracing Lowell's conclusions. In a book-length review of *Mars and Its Canals,* the famous naturalist Alfred Russel Wallace correctly observed that the markings Lowell saw might be due to craters and cracks on the planet's surface. Russell faulted Lowell for not considering scientific findings to the effect that the Martian atmosphere was exceptionally thin and appraising the effect of a thin atmosphere upon on the retention of heat. The mean surface temperature on Mars, Wallace correctly predicted, would be well below the freezing point of water and thus hardly capable of supporting a canal-building civilization.[15]

Cranky scientists did little to dim popular interest in life on the red planet, as Martian tales swept the public fancy. In 1897 British novelist H. G. Wells began to serialize *The War of the Worlds.* If the Martians of Lowell's imagination did not make war on themselves, they were certainly capable of making it on other planets. "To carry warfare sunward is indeed their only escape from the destruction that generation after generation creeps upon them," Wells explained.[16] As a student at the Normal School of Science in London, Wells had studied under the famous biologist T. H. Huxley, an outspoken supporter of Darwin's evolutionary theory. Wells wanted to examine how Darwin's theory might operate on a totally separate world.[17] In seeking to explore the views of Charles Darwin, however, Wells promoted Percival Lowell. The remarkably well-enduring tale, appearing at precisely the same time as Lowell's books, exposed thousands of readers to the notion of Mars as the abode of a superior but dying civilization.

The story's special appeal, as Wells himself recognized, lay in its power to suppose that creatures on a nearby planet might evolve to the point that they could destroy human civilization on Earth.[18] The idea that life on Earth might be primitive both technologically and biologically in comparison to other planets assaulted the views of religious authorities who opposed Darwin's theory. A superior race of Martians bent on conquering the earth posed a inalterable challenge to the belief that God planned to make Earth the seat of intelligent life and English-speaking people rulers of the world. If Earthlings could be snuffed out by an extraterrestrial invader, as easily as humans had destroyed species or primitive cultures on Earth, then one could hardly view homo sapiens as the ultimate objective of God's plan. In his third book on Mars, Lowell admitted that the existence of Martians would force humans to revise their cosmology: "Their presence certainly ousts us from any unique or self-centered position in the solar system, but so with the world did the Copernican system [alter] the Ptolemaic, and the world survived this deposing change."[19]

Over and over again, fictional Martians provided humans with a counterpoint to earthly life-forms.[20] Beginning with the *Princess of Mars* in 1912, Edgar Rice Burroughs introduced a generation of readers to the exotic flora and fauna of this nearby planet and one of most beautiful and scantily clad heroines in modern literature. Ten more books followed as Burroughs reenforced the prevailing notion of a dying planet on which survivors of a once-mighty civilization struggled against the perils of a thinning atmosphere and warring tribes.[21] On 30 October 1938 Orson Welles made radio history with his newscast-style presentation of *War of the Worlds,* proving once again the appeal of the Martian invasion tale.[22] In 1950 Hollywood released *Rocketship X-M,* a competitor to *Destination Moon,* in which space explorers encounter a dying Mars inhabited by a race of blind, mutated survivors. As the rocket crew approaches Earth (insufficient fuel dooms their return), they warn the people below of their startling discovery: an all-out atomic war blew an advanced Martian civilization back into the stone age.[23] In 1953 Martians attacked Earthlings again as the movie version of *War of the Worlds* played to large audiences.[24]

By the time Ray Bradbury published his *Martian Chronicles,* Martians had disappeared, at least in their physical form. The Martians of Burroughs's imagination had evolved to the point where they disappear into each other's minds. Bradbury's Martians had moved to another dimension, a theme presented in many works of fiction that explore the consequences of evolution.[25] Settlers from Earth arriving on Mars encounter the ruins of an ancient civilization and an invisible race that uses telepathic powers to repel early explorers.[26]

In his third and final program on space exploration, broadcast in 1957, Walt Disney reenforced prevailing conceptions about Mars. Much of the program, titled "Mars and Beyond," dealt with the search for life in the solar system. Disney's animators prepared drawings of exotic plants and animal life on the Martian surface. Out of respect to the Martian myth, the program ended with pictures of flying saucers skimming through the planet's atmosphere:

> When earthman finally walks upon the sands of Mars, what will confront him in this mysterious new world? Will any of his conceptions of strange and exotic Martian life prove to be true? Will he find the remains of a long-dead civilization? Or will the more conservative opinions of present-day science be borne out with the discovery of a cold and barren planet, where only a low form of vegetable life struggles to survive?

The apparent discovery of canals on Mars spurred speculation about exotic life-forms, and even serious astronomers allowed that vegetation had developed there. Following the release of *War of the Worlds*, H. G. Wells published a magazine article in which he described "feather-covered men nine or ten feet tall." (Ordway Collection/U.S. Space and Rocket Center)

The first interplanetary expeditions, the series writers concluded, would give humans the opportunity to understand "the miracle of life as it exists in all its countless forms throughout an infinite creation."[27]

For much of the twentieth century, the Martian myth dominated the popular culture of space, among both the public at large and serious scientists. Bruce Murray, a professor of planetary science who helped NASA conduct the *Viking* probes to Mars, observed in 1971 that Mars had grabbed hold of human emotions in such a way as to distort both popular opinion and scientific desire: "We want Mars to be like the Earth. There is a very deep-seated desire to find another place where we can make another start, that somehow could be habitable."[28]

Belief in the plurality of worlds drew strength not only from works describing conditions in outer space but also from the whole history of exploration on the earth. The hope that life could be found elsewhere in the universe found support in the memory of earthly expeditions. For centuries earthly explorers had amazed the public with tales of strange terrestrial creatures, leading the public to associate ventures into the unknown with new life-forms. People promoting space exploration drew upon this expectation. It is doubtful that the general public would have paid as much attention to space exploration had its early proponents insisted on an dead and uninhabited universe.

Much of the anticipation of new life-forms flowed from what Daniel Boorstin has termed the effort to catalogue the whole creation.[29] Since antiquity, natural scientists have sought to provide the public with descriptions of the flora and fauna that inhabit the earth. In A.D. 77, Pliny completed his *Natural History,* an encyclopedia that included within its numerous books descriptions of the known animals, insects, and plants, including legendary creatures and popular folklore. One of the greatest literary achievements of antiquity, the work remained an authoritative source for fifteen hundred years, reminding educated readers that foreign lands teemed with exotic beasts. Generously illustrated bestiaries, natural histories that described real and imaginary animals, appeared in medieval times. To the delight of readers, such books described the habits of exotic beasts, often inaccurately. Elephants had no desire to reproduce, a twelfth-century Latin bestiary reported, and other books repeated stories about an Arctic goose whose young did not hatch from eggs but grew up in shells hanging from pieces of wood.[30]

In earlier times most people rarely ventured more than a slight distance from their homes. As a result, readers had to take on faith the portraits of exotic creatures observed by travelers to distance lands. Occasional zoos and traveling circuses provided some empirical evidence,[31] but most people lacked practical guides and could hardly distinguish between a flying fish (a real creature) and mythical creations such as the Griffin, a cross between a lion and an eagle.[32] Ancient and medieval catalogues overflowed with such beasts. The fourteenth and fifteenth centuries, when, as one observer noted, "no travellers' tales seem too gross for belief," produced some of the most impressive.[33] One of the most popular catalogues was *Mandeville's Travels.* Its author supposedly set out from England in 1322 and traveled to the Near East, India, and China. He described species of wondrous variety, including animals with human faces, lambs that grew out of vegetables, and a race whose faces appeared on their torsos.[34]

As biologists discredited animal myths from the earth, explorers continued to reveal new species.[35] Reports of astounding variety continued as Europeans explored the Western world. Journeys into the Americas produced reports of the amazing species to be found there, including the grizzly bear and pronghorn antelope.[36] Expeditions often employed naturalists for the purpose of cataloguing the strange life-forms to be found in foreign lands. One of the most influential was the report from the naturalist aboard the HMS *Beagle.* Conducting an expeditionary voyage between 1831 and 1836, the crew of the *Beagle* traveled along the coasts of South America and across the South Pacific. The naturalist was Charles Darwin, twenty-two years old when the *Beagle* set sail. Darwin reported the results of the expedition in a series of books published between 1839 and 1846. He described species as wondrous as those to be found in medieval mythology: giant tortoises on the Galapagos Islands and herds of lizards that swam out to sea.[37] On the coast of South America at Tierra del Fuego, he described a race of humans so savage that he remarked that they might as well have been creatures from another world.[38]

The discovery of other spheres in space, accompanied by the rush of reports describing exotic creatures across the earth, strengthened the belief in a plurality of worlds. The acceptance of extraterrestrial life was well established in the minds of respectable scientists and other educated people by the end of the nineteenth century.[39] Attempts to portray Earth as the sole source of life, or even as the asylum of intelligent beings, were feeble and doomed. The concept of humans as lonely travelers on the only habitable planet in the entire universe was never able to capture imaginations in the same way that the idea of extraterrestrial life did. For those attentive to the intentions of the Deity, the existence of life on other worlds now seemed to be part of God's plan.

Belief in Earth as the sole source of living creatures required believers to adopt views of the cosmos that could not maintain a large following in the modern world. Aristotle's argument for a single world, for example, depended upon a boneheaded theory of physics. In Aristotle's view, heavy elements in the universe coalesced at a single spot to form the earth, whereas lighter elements such as fire rose toward the heavens to form stars.[40] The observations of Copernicus and Galileo spoiled that argument, because heavy elements had obviously coalesced above the earth to form other planets. Saint Thomas Aquinas maintained his belief in a single world on the grounds that to do otherwise would require the believer to accept the idea that evolution proceeded on the basis of chance.[41] One perfect world created intentionally, Aquinas argued, was more in accord with God's purpose than many worlds, some randomly more perfect than others. Darwin's theory of evolution pretty well extinguished that point of view. To the extent that God was involved at all, he had apparently created a wide variety of species, leaving some to thrive and many to die. The chance fortune of environmental conditions determined the winners and losers according to the laws of natural selection. So much for the idea of a single perfect world.

That left the single world theory sitting on a one-legged stool: the doctrine of catastrophe. The only doctrine that promised to harmonize natural science with religious hope, it maintains the uniqueness of Earth by suggesting that life arose according to observable natural processes that are nonetheless extraordinarily rare events. According to one such theory, the planets of the solar system were ripped from the mass of the Sun when another star happened to pass close by.[42] Such a freakish occurrence would make the solar system and the earth nearly one of a kind.

Catastrophic theories have enjoyed a good deal of popularity. They also have attracted an unconventional following. Between 1950 and 1955 Immanuel Velikovsky published three books that attempted to reconcile astronomy with biblical accounts of antiquity. The books captivated public attention while irritating serious scientists. Velikovsky argued that life on Earth developed in response to great upheavals recorded in ancient texts as the miracles. Based on his analysis of ancient texts, Velikovsky suggested that a comet brushed Earth around 1500 B.C., parting the Red Sea and reversing the planet's rotation. The comet returned some fifty years later, according to his interpretation, when it again interrupted Earth's spin and accounted for such terrestrial phenomena as the collapse of the great walls of Jericho during Joshua's siege.[43]

The eccentric quality of catastrophic theories repels serious scientists. Since the time of Copernicus, scientists have sought to dismantle views of the universe organized around the centrality of man. Centrality theories hold that God created humans on Earth in order to populate it with beings in his own image and, according to some, set the universe to rotate around the earth.[44] Through natural observations, astronomers removed the earth from the center of the universe, placing it instead on a remote spiral arm of one of a hundred billion inconspicuous galaxies. Biologists, in the meantime, established the evolution of life. The processes that formed the solar system and kindled life on Earth seemed not rare but common. This suggested to the best minds that life must exist on other planets and in other solar systems.

If life existed elsewhere, how had it evolved? Imagination joined science as people sought to describe the inhabitants of the celestial realm. A variety of works both serious and fanciful described the plurality of worlds. Bernard de Fontenelle, a French writer of popular science and history, composed a widely read seventeenth-century tract. *Entretiens sur la Pluralité des Mondes* (Conversations on the Plurality of Worlds) appeared in 1686 and was a huge popular success. Its bestiary of planets described luminous birds that lit the night sky on Mars and sluggish creatures on Saturn that moved slowly to protect themselves from that planet's extreme cold.[45] Following the publication of *War of the Worlds,* H. G. Wells wrote an article for *Cosmopolitan* magazine in which he showed how Martian creatures might adapt themselves to local conditions, captivating readers with descriptions of a local ruling class nine to ten feet tall, feather-covered, with tentacles in place of hands.[46] In 1951 Kenneth Heuer described the types of beings that other planets in the solar system might produce, "a conclusion to which the mind is almost necessarily led." Living pillars with a sodium-based chemistry graced Jupiter, and on Venus the first explorers might find thinking and talking trees.[47] In 1966 astronomer Carl Sagan speculated on the life-forms that might arise on other worlds, suggesting that "organisms in the form of ballasted gas bags" might swim like plankton-eating whales through the thick Jovian atmosphere.[48] These interplanetary bestiaries departed little from medieval tomes, with fact indistinguishable from fancy. A 150-pound bat could glide through the air in a world that retained a dense atmosphere, and animals with six legs could hop across a low-gravity world.[49]

Writers of science fiction contributed even stranger life-forms to such bestiaries of space.[50] In the 1951 version of *The Thing,* scientists discover a carrot-shaped monster frozen in a spacecraft in the Arctic ice shield.[51] In *Alien,* a 1979 release, Sigourney Weaver battles a predatory creature utterly hostile to human life.[52] George Lucas produced the ultimate science fiction bestiary for his movie *Star Wars.* Drawing on the tradition of the frontier tavern in the Hollywood western, Lucas created the Mos Eisley Cantina, an intergalactic bar occupied by a grubby assortment of alien low-life.[53] Four of the five top-grossing motion pictures of all time as of 1991 dealt with life-forms from other planets, led by the children's tale, *E.T.: The Extraterrestrial.*[54]

For much of history, people have wanted to believe in the earth as the center of life. "Though the belief that our world was the material center of the Universe has long been dead," astronomer Henry Norris Russell observed in 1943, "the supposition that it was . . . unique in being the abode of creatures who could study the Universe has lingered long."[55] The supposition gave meaning to life by elevating the role of homo sapiens in the cosmos. It supported the view that humans were the result of a deliberate process designed to create beings who could contemplate their existence and imagine God. Now science was suggesting that evolution occurred elsewhere. If just a single speck of vegetation was found on a planet such as Mars, British astronomer H. Spencer Jones observed in 1940, it would follow that "life does not occur as the result of a special act of creation or because of some unique accident, but that is the result of the occurrence of definite processes." On any planet capable of supporting it, life will gain a foothold. "Given the suitable conditions," he concluded, "these processes will inevitably lead to the development of life."[56]

The new cosmology did not banish God from the universe. Rather, it encouraged insights into the Creator's intent.[57] If it was God's plan to create life, then he must have done it elsewhere. "God exists," insisted one turn of the century astronomer, summing up the implications of life on Mars, "and He did not create habitable spheres with no object." Habitable spheres were created for the purpose of being inhabited.[58] Writing some fifty years later, Harvard astronomer Harlow Shapley suggested that the idea of a plurality of worlds "inspires respect and deep reverence." If theologians found it difficult to accept the idea that the God of humanity was also the God of gravitation and hydrogen atoms, at least theologians might be willing to consider the reasonableness of extending to other thinking beings "the same intellectual or spiritual rating" they give to us.[59] So powerful was the expectation of extraterrestrial life that even religious figures acknowledged its creation as God's will.[60]

The search for living beings elsewhere in the universe excited spiritual debate, if for no other reason than the possibility that the God of humans might be busy overseeing other planets.[61] By advocating the plurality of life, scientists addressed one of the great mysteries of life. They could no more avoid the religious implications of this message than the followers of Charles Darwin could avoid the need to debate the church.

The actual search for life on other spheres began with the planet Mars. For decades the general public had been bombarded with images of life-forms on that nearby planet. Although serious scientists discounted the possibility of finding intelligent life, they held high hopes for the discovery of lower forms. For years following the Schiaparelli-Lowell discoveries, the popular press assured the public that some form of life would be found by the first explorers on Mars. Fuzzy images from Earth-based telescopes showed the seasonal advance and retreat of the polar caps on Mars and captured pictures of dark areas that seemed to vary in shape and color from season to season. Displaying a sequence of images in a 1944 issue of *Life* magazine, writers for that popular outlet announced that "it is logical to conclude that the vast regions on Mars that change from green to brown in seasonal cycles are covered by vegetation."[62] Confirmation of even dry mosses or lichen, the first observable evidence that the life process had begun elsewhere, would be a stunning discovery.[63] Mars provided the first practical test of the plurality of worlds thesis. The results, to say the least, were disappointing.

On 4 November 1964 NASA launched a 575-pound space probe toward the planet Mars. The speculations of nearly a century rode on the *Mariner 4* spacecraft as it glided along a curving trajectory toward the planet of so many dreams. The spacecraft arrived on 14 July 1965. Scientists at NASA's Jet Propulsion Laboratory cheered as sensors on *Mariner 4* located the planet and began taking the first of twenty-two pictures. Digit by digit, the spacecraft's radio transmitted the coded data back to Earth. The third picture revealed a surface feature, a circle of sorts some twelve miles across. By the seventh photograph, the features had become clear. Mars was covered with craters. The closeup pictures showed a surface pocked with ancient craters. There were no oceans, no artificial canals, and no oases of vegetation. Mars looked lifeless. It looked, as the horrified scientists were forced to acknowledge, like the Moon.[64]

"Mars is dead," announced *U.S. News and World Report.* "There are no cities, oceans, mountains or even continents visible on Mars."[65] Inspecting the photographs, President Lyndon Johnson observed that "life as we know it with its humanity is more unique than many have thought."[66] The extreme lunarlike fea-

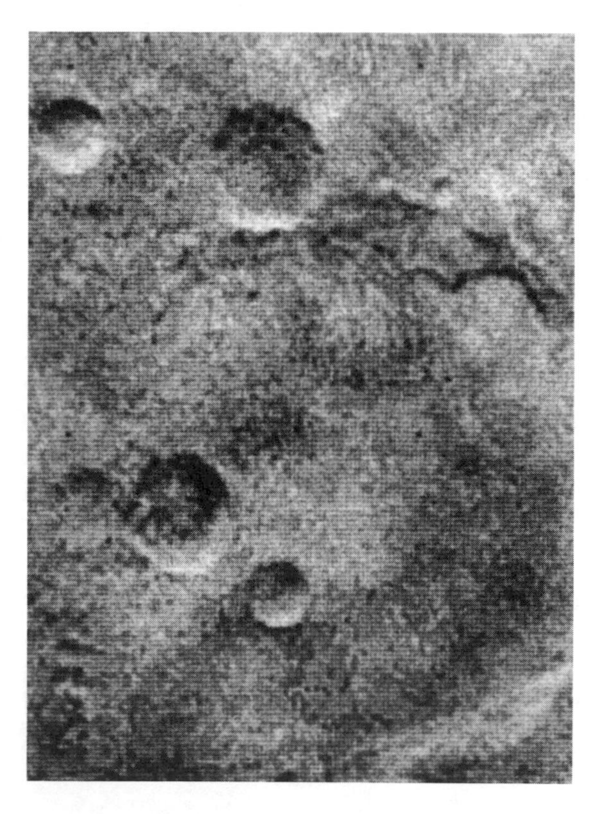

The real Mars dashed expectations. In the summer of 1965, the *Mariner 4* space probe produced the first closeup pictures of the red planet. This is the first picture in which viewers could distinguish the strange surface features. There were no canals, no signs of life. Mars was pocked with craters. To the disappointment of exploration advocates everywhere, Mars looked like the Moon. (NASA)

tures shocked even the scientific teams overseeing the mission. Many had hoped to see evidence of weathering that might suggest the presence of free water so essential to life on Earth. As *Mariner 4* sped past the planet, it beamed radio signals back to Earth through the Martian atmosphere, measuring its density. The results were just as depressing as the photographs. The atmosphere proved remarkably thin, hardly capable of protecting embryonic surface life from the constant bombardment of cosmic rays and ultraviolet radiation.[67]

Scientists tried to stem the general disappointment. Astronomer Carl Sagan compared the photographs from *Mariner 4,* taken from six thousand miles above the Martian surface, with similar satellite photographs of Earth that also showed no signs of life. Astronomer Clyde Tombaugh announced that photographs of a large crater on Mars contained markings coinciding with the position of a short canal mapped more than a half century earlier by Percival Lowell.[68]

NASA tried again in 1969, as *Mariner 6* and *Mariner 7* zipped by Mars. "All is not lost for the astronomical romantics," the *Washington Post* reported. "The famed canals of Mars showed up clearly."[69] The announcement prompted Arizona senator Barry Goldwater at a special congressional hearing to request the pictures that showed the long-sought after canals. NASA officials sheepishly explained that no such pictures existed. What the pictures in fact revealed were discolored areas on the Martian surface, light and dark splotches that resembled broad canals "under poor seeing conditions."[70] In spite of the discouraging results, NASA executives hoped that scientific interest in Martian exploration would not wane. One Mars expert hired to interpret the photographs for NBC television, however, was not so sure. "If no one told you, you wouldn't know if this was Mars or the moon," he complained.[71]

The search for life got a small boost when *Mariner 9* swung into orbit around Mars on 13 November 1971. The planet was shrouded under a global dust storm when the spacecraft arrived, and scientists had to wait several months before the atmosphere cleared enough to permit detailed photography. Appearing first above the dust clouds were four tall volcanos, the largest more than twice the elevation of Mount Everest. As the dust finally settled, scientists watched in amazement as the spacecraft transmitted images of canyons and chasms that looked like nothing less than dry riverbeds.[72] If water had carved these features, life might lurk in protected pockets of the planet's soil.

To search for life on the surface of Mars, NASA dispatched two *Viking* spacecraft, which arrived in their Martian orbits in the summer of 1976. Each carried a 272-pound lander (plus fuel) that descended to the surface of the planet and searched for life. The landers carried a variety of instruments, including a slow-working camera and a biological package to test the soil for living microorganisms. Excited scientists let their imaginations soar. "I keep having this recurring fantasy," Carl Sagan reported. "We'll wake up some morning and see on the photographs footprints all around Viking that were made during the night, but we'll never get to see the creature that made them because it is nocturnal." Sagan wanted a night light put on *Viking*. He also joked about putting out bait.[73]

Once again, the probes disappointed expectations. The cameras took pictures of a cold, dry, rock-strewn surface. The biology package produced chemical reactions like those one would expect in the presence of living organisms, but efforts to detect organic molecules in the Martian soil failed.[74] Mars was "deader than Elvis," wrote *Washington Post* staff writer Kathy Sawyer.[75] Concluded one commentator,

> The exploration of the planet Mars in the 1960s and 1970s, culminating in the landings of two instrumented Viking spacecraft in 1976, ended a long dream of Western civilization. Since these explorations obtained a negative answer to their most interesting question, that of Martian life, they are doubtless regarded by many as a failure.[76]

By making Mars a duller place, the missions extinguished a long-held dream. The Mars of Lowell and Wells was gone, at least for a while. Scientists looked to other frontiers.

Where else was life to be found? Probes to the planet Venus, in size Earth's sister world, revealed a hellish sphere, hardly the primeval place of carboniferous swamps pictured by early visionaries. With its thick atmosphere, mainly carbon dioxide, Venus has a greenhouse effect run amuck. The surface temperature is nearly nine hundred degrees fahrenheit, and the clouds contain sulfuric acid. What were potentially the most hospitable places in the solar system showed no signs at all of advanced life-forms.

This did not rule out intelligent life beyond the solar system, however. If life develops through a natural process, as most scientists believe, then beings might be found on suitable planets turning in orbits around other medium-sized yellow stars. Writers produced formulas using probabilities to calculate the number of civilizations with whom Earthlings might possibly communicate. The formula developed by astronomer Frank Drake was typical. The number of "communicative civilizations" in the Milky Way, he suggested, was equal to the product of

1. the number of stars in the galaxy;
2. the fraction of stars with planetary systems;
3. the number of planets per system that are suitable for life;
4. the fraction of suitable planets where life might actually develop;
5. the fraction of living planets that produce intelligent life forms;

6. the fraction of life forms that are willing and able to communicate with other civilizations;
7. the average life expectancy of a technological civilization expressed as a fraction of life of the planet.[77]

All of the numbers in the equation are estimates. Some of the numbers are terribly small. The probability of a technological civilization arising might be but one in one hundred, and the chance that it lives to a ripe old age might be equally scanty. Still, the number from which the calculation begins is astonishingly large. Astronomers believe that the number of stars in the Milky Way exceeds two hundred billion.[78] Even small probabilities multiplied by large numbers produce impressive results. Millions of civilizations with national space programs could have developed within the Milky Way alone. The galaxy could be teeming with intelligent civilizations at this very time.[79]

The potential existence of long-lived, technologically advanced civilizations is a fascinating possibility, one that has motivated authors of science fiction for some time. Between 1951 and 1953, Issac Asimov issued the foundation trilogy, three sweeping books assembled around a plot line originally developed for a set of short stories written for *Astounding Science Fiction.*[80] One of the most influential science fiction writers of the twentieth century, Asimov began publishing fantastic stories while a student at Columbia University. He supported himself as a professor of biochemistry until the income from the foundation series and other works allowed him to write full time. The foundation trilogy describes a galactic empire in which a spacefaring civilization has spread itself over twenty-five million inhabitable planets. After twelve thousand years of relative peace, the civilization is collapsing. Efforts based on the science of psychology to shorten the period of impending barbarity guide the plot.

Galactic empires, human or otherwise, have provided the backdrop for some of the most popular science fiction stories of all time. In the original *Star Trek* television series, the forces of the Federation do battle with the Klingon Empire, a weakly disguised effort to replicate the nuclear superpowers of the 1960s in an intergalactic setting.[81] In the famous *Star Wars* cinema series, Princess Leia Organa represents the resistance in a battle against the evil military leader of the Galactic Empire, Darth Vader.

In Asimov's foundation trilogy, humans travel between distant corners of the galaxy with as much ease as Americans of the twentieth century flew between cities on Earth.[82] What writers of fiction imagined, however, rocket scientists could not produce. Not content to wait for the development of spaceships of the interstellar kind, scientists sought to communicate with extraterrestrial civilizations in more immediate ways. In 1960 astronomer Frank Drake used the eighty-five-foot-

If advanced life-forms could not be found on Mars, perhaps they walked on more distant worlds. Scientists and fantasy writers alike encouraged the belief that humans were not alone. In 1982 movie producer Steven Spielberg introduced one of the most endearing aliens of all time in the top-grossing film *E.T.: The Extraterrestrial.* (©1982 by Universal City Studios, Inc.; courtesy of MCA Publishing Rights, a Division of MCA, Inc.; all rights reserved)

wide dish of the National Radio Astronomy Observatory to listen for cosmic calling cards. Drake hypothesized that any advanced technological civilization would produce radio and television waves and might even send out radio signals announcing itself to other civilizations. After reviewing the possibilities, Drake tuned the West Virginia radio observatory to 1,420 megahertz and pointed it at two nearby stars.

To his astonishment, Drake received a very strong signal of intelligent origin from the apparent vicinity of Epsilon Eridani, a star some ten light years from the Earth. For a moment, it appeared that Drake had found the proverbial needle in the cosmic haystack. As the first person to discover extraterrestrial life, Drake would have been as famous as Albert Einstein. Who else might be transmitting signals at that wavelength? Upon further investigation, Drake discovered that he had inadvertently tuned in to an Earth-based signal emitted as part of a secret military experiment.

This did not discourage scientists in their effort to contact extraterrestrial civilizations. In the early 1970s, scientists advising NASA on its *Pioneer* spacecraft probes urged the inclusion of a small plaque containing a symbolic message from Earth. The spacecraft were designed to fly by Jupiter and Saturn, a journey whose momentum would carry the probes out of the solar system. NASA agreed to attach an engraved gold-anodized aluminum plate to the spacecraft antenna support struts.

The plate caused an immediate controversy. In addition to locating the planet from which it had come, it included figures of a man and a woman as they would appear unclothed. The vision of male genitalia and female breasts being carried on a government-sponsored space probe to an alien civilization outraged some Americans. "Isn't it bad enough that we must tolerate the bombardment of pornography through the media of film and smut magazines?" one letter writer to the *Los Angeles Times* declared. Now "our own space agency officials have found it necessary to spread this filth even beyond our own solar system." Responding a few days later, another letter writer suggested that NASA might "visually bleep out the reproductive organs of the man and the woman. Next to them should have been a picture of a stork carrying a little bundle. Then if we really want our celestial neighbors to know how far we have progressed intellectually, we should have included pictures of Santa Claus, the Easter Bunny, and the Tooth Fairy."[83]

Scientists and their government allies sought public funding for the search to discover extraterrestrial life. During the early 1970s NASA proposed Project Cyclops, a powerful ground-based listening system. Some fifteen hundred antennas, each one hundred meters in diameter, would have been linked by computers into one enormous listening device.[84] Estimated to cost $20 billion, the project was never approved. In 1974 Drake and Sagan reversed the normal process and sent

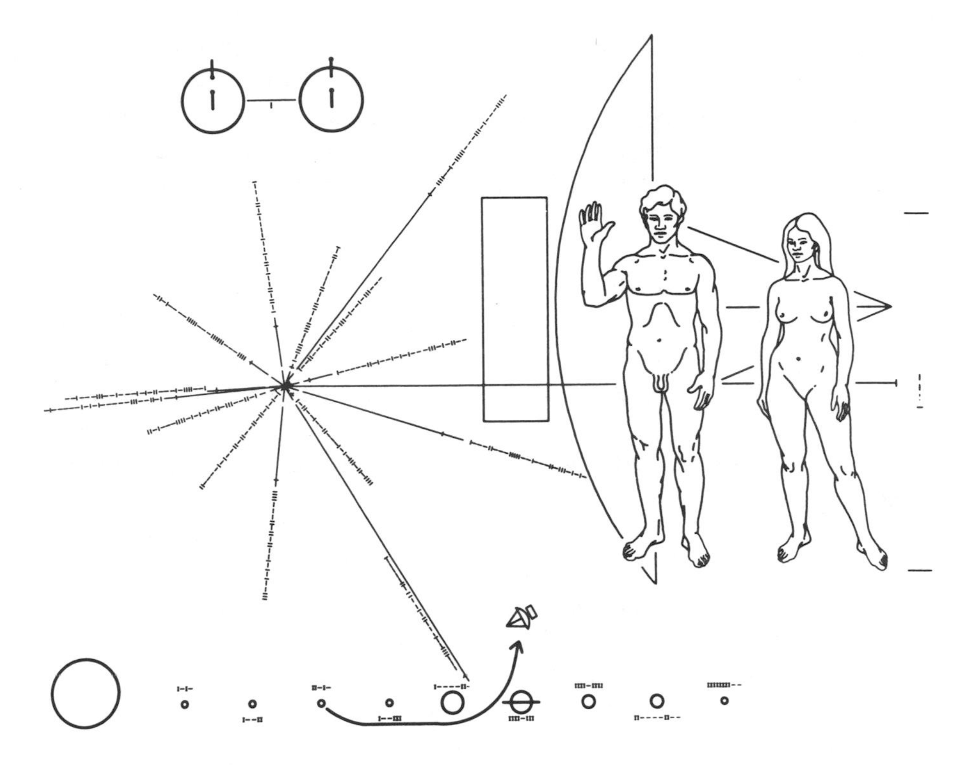

If humans could not travel to habitable planets around distant stars, perhaps they could send messages that would. In 1972, NASA launched *Pioneer 10* on a trajectory that took it past Jupiter and out of the solar system. The space probe carried a gold-anodized aluminum plate designed to show scientists from other planets who might intercept it millions of years hence where the probe came from, when it was launched, and who sent it on its way. Earth scientists also sent out radio signals and listened for broadcasts from distant worlds. (NASA)

out an electronic greeting card, aiming the Arecibo Radio Telescope in Puerto Rico at the globular star cluster M-13 in the constellation Hercules and beaming a message. By Sagan's estimate, the chances of hitting a technological civilization in the more than one-hundred-thousand-star cluster was fifty-fifty. At the speed of light, however, the message would not arrive for twenty-four thousand years.[85]

In 1975 NASA administrator James Fletcher delivered an address to the National Academy of Engineering in Washington, D.C., in which he asserted that the universe was teeming with life. Our own galaxy, he said, "must be full of voices, calling from star to star in a myriad of tongues." He criticized the desire of pragmatists to structure the space program in such a way as to produce only direct and immediate benefits to humans on Earth. Through their preoccupation with what Fletcher called the "now syndrome," people were losing sight of the revolutionary breakthroughs that more adventurous minds might provide. "It is hard to imagine anything more important than making contact with another intelligent race," he observed. "It could be the most significant achievement of this millennium, perhaps the key to our survival as a species."[86] Fletcher's words were rooted in his own religious experience. A lay minister in the Church of Jesus Christ of Latter-Day Saints, Fletcher subscribed to the theological doctrine that God had created a plurality of worlds populated with intelligent beings. His position as NASA administrator allowed him to search for empirical verification of these beliefs.[87]

NASA continued to promote the Search for Extraterrestrial Intelligence (SETI) throughout the 1970s and 1980s.[88] "Life as we know it here on Earth appears to be the result of universal laws of chemistry and physics," NASA officials observed in defense of the effort. "Elsewhere among the 400 billion stars in the Milky Way galaxy or in one of the hundred billion other galaxies, the same processes may have produced beings who stare at the heavens and wonder about other occupants of their universe."[89] In spite of NASA's enthusiasm for the effort, the project received a skeptical reception in the U.S. Congress. Politicians viewed it as an extravagance the country could ill afford. One likened it to "buying fur coats for your cows while your children were freezing."[90] In 1978 Wisconsin senator William Proxmire granted SETI his Golden Fleece Award, which he periodically bestowed on projects he felt misused the public treasury.[91] Members of Congress complained about efforts to locate "little green men" and sought to eliminate SETI's funding.[92]

In the face of mounting criticism, NASA officials refocused the project and gave it a new name. The Microwave Observing Project would survey the heavens for microwave radio signals, the most likely means of communication between intelligent civilizations. NASA officials estimated that the project would cost a paltry $135 million over ten years, about one-third the cost of a single space shuttle flight. Counting on its success, supporters called it "the biggest bargain in history."[93] Congress was unimpressed. In 1993, after a brief startup, lawmakers terminated the project. Senator Richard Bryan, the Nevada Democrat who led the cancellation fight, argued that the chances of finding intelligent life were too remote to justify the expenditure. "The SETI program has found nothing," he observed. "All the decades of SETI research have found no confirmable signs of extraterrestrial life."[94]

Even serious scientists had grown skeptical. If the universe was teeming with various life-forms, where were they? A lush, wet planet like Earth would hardly go unnoticed in a universe full of spacefaring civilizations. A single civilization, moving out from its home planet at the rate of one one-hundredth the speed of light could colonize the whole galaxy in less than ten million years, a short period of time in astronomical terms.[95]

Flying-saucer buffs believe that Earth has already been visited, but the absence of detectable evidence such as radio transmissions seems to discredit this. The other possibilities are frankly disturbing. A favorite theory, advanced frequently during the height of the Cold War, suggests that technological civilizations might be remarkably short-lived. What takes billions of years to evolve might extinguish itself after only a few decades.[96] Carl Sagan adroitly pointed out the effect of limiting Drake's formula to technical civilizations that last so short a time. The number of technical civilizations existing at any time would quickly approach one. Observed Sagan, "There might be no one for us to talk with but ourselves."[97] If a society that masters space technology simultaneously masters atomic destruction, it may be taking the final step on the evolutionary scale.

Others began to question the plurality of worlds thesis. Perhaps the evolution of life, contrary to popular thought, really is a cosmic accident. The slightest changes in the characteristics of the Earth would leave it a much less desirable place to live.[98] Even the placement of the Moon, with its ability to create modest tidal pools, might be critical to the evolution of life. "If we were typical," said one observer trying to discredit the plurality of worlds doctrine, "we should not exist."[99] The process of evolution would have begun on other planets and created advanced civilizations in our cosmic neighborhood long ago, and colonists from other worlds would have reached Earth before human history began. This did not occur. Based on the available evidence, all life on Earth has a common origin and developed without any subsequent outside interference. The fact that extraterrestrial settlers have not arrived casts doubt on the vision of a galaxy ringed with civilizations.

By 1990 the national space program had begun to push the edge of the imagination envelope. Conventional, well-established images such as life on Mars had been worn down. To learn the true nature of the universe, scientists had to venture to the outer limits of credibility, looking for black holes, planets around other solar systems, evidence of hyperspace, and echoes of the Big Bang. Only on the outer limits were investigators likely to find reality.

Understandably, this created political problems for the NASA space program. The need to push the imagination envelope occurred at a time when excessive deficits prompted cuts in questionable spending. Some science programs survived; others disappeared. Congress terminated the renamed High Resolution Microwave Survey because it fell over the edge of political credibility, being too easily associated with little green men and flying-saucer cults. "The opponents of the program have frequently poked fun at it," observed Maryland senator Barbara

Mikulski in a futile effort to save the project, "and one can understand why. . . . Have we all not seen those pictures? 'Extraterrestrial alien with Bush at Camp David'; 'Extraterrestrial alien with Clinton at Martha's Vineyard.'" The answer to the mysteries of life, Mikulski said as she parodied her opponents, had already been "on the front page of the *National Enquirer.*"[100]

Searching for echoes from the Big Bang is hardly less exotic than searching for radio signals from extraterrestrials. The former project, however, survived, in part because the images on which it drew seemed less bizarre. The $230 million Cosmic Background Explorer (COBE) sought to examine the way in which the universe began, a matter close to the heart of theology.

Scientists believe that the entire universe was created some ten to fifteen billion years ago from a single point of unity, smaller than the dot at the end of this sentence. During its initial stages of expansion, the material making up the present universe was intensely hot. Any hot object emits radiation and continues to do so as it cools. That is why a soldier with special goggles can see tanks and jeeps at night even after the vehicles have been parked and shut down. They continue to emit background radiation. As strange as it may sound, the background radiation from the period of the Big Bang is still around, and its frequency can be calculated with mathematical precision. Scientists can detect the radiation using antennas on Earth, but to obtain measurements with sufficient accuracy to test the Big Bang theory, they need to send their instruments into space.[101]

NASA launched the two-and-one-half-ton COBE satellite into Earth orbit on 18 November 1989.[102] Scientists were ecstatic as the results poured in. "They have found the Holy Grail of cosmology," exclaimed physicist Michael Turner.[103] In effect, COBE took a snapshot of creation a few hundred thousand years after the Big Bang. The space probe detected background radiation at precisely the predicted frequency. Moreover, it resolved one of the most significant problems with the Big Bang theory. The galaxies could not have formed from a uniformly expanding universe, because gravity would act equally on all available matter. From the ground, with imperfect instruments, the background radiation seemed just that—uniform in every direction. With a clearer view, COBE detected ripples or blotches in the background radiation, exactly what was needed for matter to condense in irregular ways.[104] "It is the discovery of the century, if not all time," said Cambridge professor Stephen Hawking.[105] "It's like looking at God," declared astrophysicist George Smoot.[106]

The COBE program and its findings were a success because they fit so well with theological images already in people's minds. All theologies contain creation stories.[107] The story found in Genesis tells of a creation beginning with a great burst of light. God then separates the light from the darkness. In the first phase of creation after the Big Bang, as scientists describe it, the universe was all light—an opaque fog so thick that light scattered as soon as it formed. COBE took a picture of the universe at the point at which the universe became transparent, when the darkness divided from the light. In Genesis, "God said, 'Let there be light' . . . God then separated the light from the darkness. . . . Thus evening came, and morning followed—the first day."[108] Said veteran space journalist Kathy Sawyer: "The announcement [of COBE's results] had such impact because the scientists' themes are in apparent harmony with the biblical version of creation recorded centuries ago by scribes who had no inkling of relativity, particle physics or other elements of modern cosmology."[109]

In 1895 H. G. Wells wrote *The Time Machine,* in which he sent a character from Victorian England spinning through the future almost to the end of the world.[110] Time travel is a favorite theme in science fiction, but not as fantastic as writers might think.[111] In promoting the Hubble space telescope, NASA officials pointed out that it like all other telescopes acted as a cosmological time machine: "Because light travels across the universe at a finite speed (186,000 miles/sec.), the deeper astronomers look into space, the farther back in time they look."[112] Because of the unprecedented clarity an orbital platform offers, the space telescope could peer back close to the beginning of time. "From its vantage point outside the Earth's murky atmosphere, the HST will be able to probe a distance of 14 billion light-years—offering views of galaxies so distant that they will appear as they were when the universe was formed."[113]

The space telescope can look back in time; it cannot transport objects through it. For years the possibility of moving backward and forward in time has been the province of science fiction writers and scientific crackpots. Time travel raises a seemingly insurmountable paradox, well represented in the motion picture *Back to the Future.* In that film, Michael J. Fox travels back in time and meets his mother as a young girl. To his horror, Fox's mother falls in love with him and spurns his father. Unless he can reunite his parents, Fox will not be born.[114] In physical terms, he will create another universe in which he does not exist.

This is not as farfetched as it sounds. The leading theoretical candidate for the unification of all known laws of nature allows for time travel and alternative universes. That is the theory of hyperspace or, in its most advanced form, superstring theory. "Everything we see around us," says popular science writer Michio Kaku, "from the trees and mountains to the stars themselves, are nothing but vibrations in hyperspace.[115] The theory suggests that the universe may operate through more than the four familiar dimensions that people commonly perceive (height, width, depth, and time). The laws of nature may fit together into one unified theory when more dimensions are added. The fact that we cannot perceive these dimensions does not alter the possibility that they exist.

The added dimensions allow for the presence of "wormholes," tunnels that create shortcuts through space and time.[116] Even more impressive is the possibility that wormholes may provide pathways to different universes. Physicists have suggested that ours is only one of a vast number of parallel universes connected to one another by a web of wormholes. Travel through such features is theoretically possible.[117] Fantastic children's stories in which young people step through looking glasses or wardrobes into other worlds may turn out to be the inspiration for twenty-first-century physics.[118]

Already the Hubble space telescope has photographed the signature of a black hole.[119] Because no light emerges from them, black holes cannot be photographed directly, but the vortex of material around a black hole leaves a distinctive signature, visible from a telescope above the atmosphere of Earth. The laws of mathematics suggest that matter entering a black hole might reemerge in a parallel universe.[120] Watching matter being yanked into a potential black hole, said the director of the Space Telescope Institute in Baltimore, "was like having a telephone line to God."[121]

Costly, complicated projects with armies of government-supported employees are required to test theories such as these. In 1993, after appropriating hundreds of millions of dollars to start the project, Congress voted to terminate the Superconducting Supercollider. This "window on creation" was designed to accelerate two beams of protons and send them speeding in opposite directions within a circular tube fifty miles in diameter. As the beans approached the speed of light, they would collide, releasing subatomic particles. Among other possibilities, the supercollider might have revealed the secrets of superstrings, leading the world into twenty-first-century physics.

The science was simply too esoteric for most politicians. Struggling to defend the project, congressional supporters issued excuses about photon-beam cancer therapy and trains capable of traveling at three hundred miles per hour. Supporters were unable to articulate a persuasive vision of direct benefits for the simple reason that the consequences of the project were fantastic. "The costs are immediate, real, uncontrolled, and escalating," observed Congressman Sherwood Boehlert of New York. "The benefits are distant, theoretical, and limited."[122]

During the 1950s, advocates of space exploration worked hard to promote their dreams. They convinced the public that space travel was something desirable and real, not just the fantasy of a small group of believers. Drawing on cultural expectations and familiar theology, advocates created images capable of eliciting government funding, images that helped propel the civil space program through moon shots, space telescopes, and planetary probes. The reality of space science, however, quickly outdistanced the vision that made it acceptable. Within forty years, space science was pushing against fantasy once more.

In an effort to revive public interest, and to satisfy their own curiosity, space scientists returned to a familiar theme: the search for extraterrestrial life. Following the demise of SETI and the disappointing biology experiments on the surface of Mars, the search for extraterrestrial life using public funds was refocused on two initiatives. One group of scientists wanted to search for life elsewhere in the solar system, the leading candidates being the moons around the gas giants Jupiter and Saturn. Scientists have known for some time that Titan, a planet-sized moon of Saturn, has an atmosphere, a fact that has inspired some of the finest space art of the twentieth century. The *Voyager 1* and *Voyager 2* spacecraft that sped by Saturn in the early 1980s revealed that the atmosphere of Titan consisted mainly of nitrogen, as is the case on Earth, and methane, the material from which carbon-based organic molecules are formed. Stranger circumstances exist on Europa, one of Jupiter's many moons. Europa is covered with ice, which floats on top of an ocean of slush or water. In 1996 and 1997 NASA's *Galileo* space probe took pictures of Europa that looked strikingly like the ice floes that cover Earth's polar regions. The images, moreover, revealed a red-brown substance along breaks in the ice floes that suggested an organic chemistry. The tidal forces produced by the mass of Jupiter could be sufficient to produce the internal warmth that, with water, is necessary to harbor life. Scientists, for obvious reasons, wanted a closer look.[123]

Another group of scientists pressed for funds to investigate nearby solar systems. For some time, scientists have dreamed of constructing instruments so precise that they could discern planets around other stars.[124] If life thrives in those places, it could be detected in predictable ways. Life not only exists on Earth, it transforms the planet in predictable ways. The presence of large quantities of free oxygen in a planetary atmosphere, for example, is a signature of life.[125]

After a few false starts, astronomers in the mid-1900s began to produce a rush of evidence confirming the presence of planets around other stars. It seemed for a while that extrasolar planets were as common as dust. In response to these developments, NASA officials prepared plans for a twenty-five-year project using space and ground-based instruments to search for life outside the solar system. "In the not too distant future," NASA administrator Daniel Goldin promised, "we will have the technology needed to image any planets that orbit nearby stars." A network of small space telescopes could be deployed in such a way as to create a telescope with sufficient power to conduct spectroscopic analysis of planetary atmospheres or color studies of faraway oceans. How inspiring it would be to discover Earth-like planets. "That would change everything," Goldin said. "No human endeavor or human thought would be untouched by this discovery."[126]

The most promising possibilities in the search for extraterrestrial life seemed to lie in areas such as these. At least that seemed to be the case until the summer of 1996, when news of "life on Mars" exploded on the national scene. Public awareness of extraterrestrial life had been heightened by the release that summer of the alien invasion movie *Independence Day,* a modern-day Martian invasion story interlaced with references to a variety of popular beliefs. The extraterrestrials (not from Mars) arrive in an enormous spaceship with the single purpose of blasting humans off the face of the Earth. Government officials have known for fifty years that the aliens exist but have suppressed this evidence from the public at large and even the president of the United States. The movie is a cartoon, with impossible scenarios, but so entertaining that it set box office records that summer season.[127]

For two years prior to the summer of 1996, scientists at NASA's Johnson Space Center along with a university-industry team had been looking for evidence of extraterrestrial life in a rather unusual way. According to the best evidence, a large impact had blasted rocks away from the surface of Mars sixteen million years ago. Some of the rocks eventually intersected Earth's orbit and fell through the atmosphere to the ground. One of the rocks, found by an American scientist in the Allan Hills region of Antarctica, was dated back to the beginning of the solar system, four and a half billion years ago, when the Martian climate was probably more wet and warm than it is today.

Therein ensued a fascinating detective story, in which scientists on Earth probed one of the rocks for evidence of various organic compounds and fossil-like structures that might have been produced by ancient microorganisms. The scientists

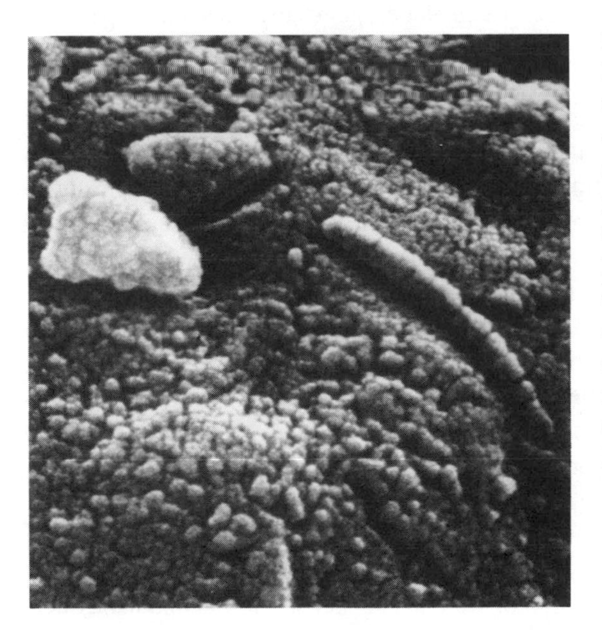

In the summer of 1996, NASA scientists announced that they had found fossil evidence of primitive life in a meteorite blasted from the planet Mars and sent on a trajectory that carried it to Earth. In the most famous (and least conclusive) bit of evidence, a high-resolution scanning electron microscope revealed a strange, tube-like structure on the rock from Mars. (NASA)

Space advocates hoped that renewed interest in the possibility of Martian life might resurrect support for a human expedition to that nearby sphere. In this illustration, the first visitors from Earth explore Noctis Labyrinthus, part of the vast Valles Marineris network of Martian canyons. (NASA)

planned to announce their findings in the 16 August issue of *Science* magazine, accompanied by a modest press conference at the Johnson Space Center. In its 5 August issue, however, the trade weekly *Space News* broke the discovery story with a small, three-paragraph report on the second page of the publication. Analysis of the meteorite, the weekly explained, "points to indications of past biological activity on Mars."[128]

The news hit government policy circles with the intensity with which moviegoers had greeted *Independence Day.* Had NASA discovered life on Mars? "If the results are verified," announced astronomer Carl Sagan, "it is a turning point in human history." President Bill Clinton called it "one of the most stunning insights into our universe that science has ever uncovered." He held a special press conference on the White House lawn to pledge that "the American space program will put its full intellectual power and technological prowess behind the search for further evidence of life on Mars" and announced that NASA's next Mars probe would land in less than one year on—what else—America's Independence Day. Space advocates cheered themselves with the hope that the findings would regenerate public support for the longstanding goal of a human expedition to Mars. As evidence that the story had gotten out of hand, NASA administrator Daniel Goldin issued a special news release: "I want everyone to understand that we are not talking about 'little green men.' . . . There is no evidence or suggestion that any higher life form ever existed on Mars."[129] All the plans for a low-key press conference at the Johnson Space Center were abandoned as Goldin and his associates rushed to prepare a nationally televised briefing with the principal investigators and other experts in Washington, D.C., on Wednesday, 7 August.

Additional interest was generated by the landing of NASA's *Mars Pathfinder* and its interplanetary roving vehicle *Sojourner* on 4 July 1997. The automated probe confirmed evidence of extensive flooding on a planet that now appeared in its geological history to be more like Earth than the Moon. Public interest in the mission was amplified by another summer season of extraterrestrial tales, including the movie version of Carl Sagan's novel *Contact.* The real probe demonstrated not only the creativity of spacecraft designers but also the power of a persistent image appropriately revived.

Within the scientific community, the search for extraterrestrial life on Mars did not enjoy a favored position. The scientific merits of other alternatives were equal if not superior to the chances of locating life on nearby Mars. In the public mind, however, Mars possessed a special status. The reports of flooding and biological activity, no matter how ancient or tentative, electrified people who had been raised on reports of Martian canals and seasonal vegetation. The prospect of Martian life, a possibility that had thrilled the general public for more than a century, continued to be front-page news. Even as serious scientists raised doubts about the likelihood of Martian life, space advocates continued to dream about flotillas that would one day take human explorers to that familiar realm.[130]

THE EXTRATERRESTRIAL FRONTIER

Space: the final frontier . . .

—Captain James Kirk

At Epcot, part of the large Walt Disney World complex near Orlando, Florida, a series of well-decorated buildings honor the accomplishments of various nations of the world. The exhibits celebrate, among other cultural contributions, Japanese art, Moroccan architecture, French cooking, and German beer. In the American pavilion a mural depicts the evolution of NASA's space shuttle. Next to it appear the words of historian and philosopher Ayn Rand: "Throughout the centuries there were men who took first steps down new roads armed with nothing but their own vision."[1]

Rand's statement commemorates the pioneering spirit, the notion that American accomplishments have been furthered by the presence of new frontiers into which people could carry fresh ideas, an experience thought to be one of the defining characteristics of American society. In the minds of its advocates, space exploration provides a means to continue this experience. Government support of space exploration would maintain the pioneering spirit by opening new frontiers, without which, space advocates argue, Americans would become indistinguishable from other nationalities and the American sense of purpose would disappear. Within the space frontier, Americans, as did their parents and grandparents, would become explorers and pioneers once more.

This is an ambitious vision. Most Americans experienced the early space program vicariously, as armchair explorers viewing television reports from expeditions substantially removed from their daily lives. Only a few hundred people actually ventured into space. As the space program matures, its advocates claim, humans in large numbers will join in the adventure. The later space program will reactivate the vast migrations that molded America in the past. Humans will move into space in multitudinous throngs, repeating the process of exploration, conquest, and colonization.

This vision of space as the final frontier arose at a most inconvenient time for those proclaiming the notion. For some time, historians have recognized that the popular image of the American frontier contains more myth than substance.[2] The romanticizing of frontier values is a twentieth-century phenomenon, perpetuated by vehicles such as the Hollywood western.[3] The real frontier, especially from the point of view of American Indians, Mexicans, African Americans, and Chinese, was materially different. The promoters of space as the final frontier offered their special vision just as revisionist historians mounted a full-scale campaign to debunk the myth, undercutting the frontier analogy, at least within intellectual circles, where ideas about public policy mature. Promoters not only had to overcome public skepticism about the technical feasibility of interplanetary migration but also had to deal with increasing skepticism about the frontier values being proclaimed.

Historical analogies have played an important role in the promotion of space exploration. One of the most popular analogies draws on the public's memory of sea captains, who in centuries past crossed vast bodies of water to reach distant lands. The only event comparable to the first landing on the Moon, editorial writers at both the *Washington Post* and *Washington Daily News* agreed, was "Columbus' discovery of the Western Hemisphere."[4] To commemorate the five hundredth anniversary of the first voyage of Christopher Columbus to the New World, a NASA-sponsored organization prepared a comic book for young children explaining the similarities between the challenges Columbus faced to those encountered by modern spacefarers. "Just as Christopher dreamed about opening a new trade route to the Far East, we can dream about a clean and beautiful Earth, about other space routes to Mars and colonization of our neighbor, the Moon."[5] The 1986 report of the National Commission on Space, whose members were charged by the president and Congress to set out civilian space goals for the twenty-first century, opened with the Columbus analogy:

> Five centuries after Columbus opened access to "The New World" we can initiate the settlement of worlds beyond our planet of birth. The promise of virgin lands and the opportunity to live in freedom brought our ancestors to the shores of North America. Now space technology has freed humankind to move outward from Earth as a species destined to expand to other worlds.[6]

In explaining his plans for gravity-assisted spacecraft that could cross the expanses of space between Earth and Mars, Buzz Aldrin drew on lessons from fifteenth-century maritime explorers. After his astronaut career, Aldrin, pilot for the *Apollo 11* lunar module that first touched down on the Moon, became a promoter of interplanetary exploration. Aldrin reminded members of his spacefaring society that European mariners of the fifteenth century had used the tropic winds that blow westward across the equator to reach the New World: "The new routes did not follow direct courses but instead looped along curving paths that sometimes appeared to carry the mariners away from their objective." The trade winds provided a pathway between the two continents, "making possible the great age of discovery."

Having established this analogy, Aldrin proposed that reusable spacecraft, called cyclers, be set into permanent orbits between Earth and Mars. The cyclers would use the force of gravity to whip by each planet, accelerating to the required velocity for the outward voyage. Smaller spacecraft would intercept the cyclers to move people and supplies home. Earlier routes of discovery provided the analogy that Aldrin needed to explain his plan: gravity provided a free and inexhaustible source of motion for interplanetary travel, just as trade winds had done for maritime explorers. "Like a ship sailing the trade winds," cyclers would follow a broad elliptical path rather than a more direct route between the two planets.[7]

Memories of the debate between Queen Isabella and her advisors over financing the Columbus voyage repeatedly appear in the consideration of modern space policy. "The many uses of space technology will make our investment in space as big a bargain as that voyage of Columbus," said then-governor of California Ronald Reagan. At the time, Reagan reminded his audience, support for the Columbus expeditions "was denounced as a foolish extravagance."[8] It is true that the Talavera commission set up by King Ferdinand and Queen Isabella of Spain had in 1490 issued a recommendation against financing the voyage. Although the monarchs did not follow the advice of the commission, the legend that the queen borrowed money for the expedition by pledging her crown jewels as collateral is untrue. That piece of folklore has nevertheless played a prominent role in modern space-policy debates. When space advocates first presented plans for a lunar voyage to President Dwight Eisenhower, they employed the Columbus analogy to justify the undertaking. Eisenhower reportedly countered that he was "not about to hock his jewels" to support a lunar expedition. Regardless of whether Eisenhower actually said any such thing, the story became an emblem of his modest space effort.[9] Queen Isabella came up again at a meeting of the Cabinet Council on Commerce and Trade in December 1983 as President Reagan and his advisers considered the proposal to start work on an Earth-orbiting space station. Budget director David Stockman, an opponent of the station, announced that the administration would never control the deficit if it continued to support such questionable projects. Attorney General William French Smith countered that Queen

Space advocates promised that extraterrestrial exploration would renew the frontier spirit they claimed had made America great. In 1986, as chair of a special national commission, former NASA administrator Thomas Paine used the frontier analogy to transmit the commission's vision of the next fifty years in space. "America was founded and rose to greatness through the courage of its pioneers," Paine claimed. Space would provide America with its next great frontier. (Courtesy artist Robert T. McCall)

Isabella must have heard the same story from her advisers. The simple reference to the Columbus myth was enough to disarm Stockman's objection and make everyone, including the president, laugh.[10]

In their attempts to justify the modern space program, advocates have worked hard to sustain the image of discovery fostered by previous explorers. Americans have a "continuing urge to chart new paths and to explore the unknown," NASA administrator James Beggs said in announcing NASA's effort to win political support for the new space station. "That instinct drove Lewis and Clark to press across the uncharted continent. It guided Admirals Peary and Byrd to the icy wastes of the poles. It drove Lindbergh alone non-stop across the Atlantic and sustained twelve Americans as they walked on the moon." The compulsion to probe unknown frontiers spurred the creativity of Americans. "If we ever lose this urge to know the unknown," Beggs argued, "we would no longer be a great nation."[11]

According to space advocates, frontiers foster innovation and encourage revolutionary discoveries. When the urge to explore is curtailed, advocates warn, civilization decays. A key step in the opening of the space frontier would be the establishment of bases on the Moon, described in works of imagination both fantastic and real. (Lockheed Martin Missiles and Space Photo Archive)

During the early part of the twentieth century, explorers like Robert Byrd, Roald Amundsen, and Robert Scott probed the icy regions of the South and North Poles. To many, it represented the last great era of discovery on the surface of the earth. "The lonely explorers like Ronald [*sic*] Amundsen, Robert Scott and others who endured the rigors of Antarctica in the early decades of this century are analogous to our space pioneers today," Beggs argued in a 1984 speech. Scott had died in his effort to beat Amundsen to the South Pole, an event that transfixed the attention of armchair explorers around the world. "This is as close as you and I are going to get to setting foot on another planet," explained a member of the U.S. team conducting research on the Antarctic ice.[12]

Defenders of the U.S. space program appeal to popular images of the American West. Speaking at the July 1982 landing of the space shuttle *Columbia,* President Ronald Reagan announced that the conclusion of the flight test program was "the historical equivalent to the driving of the golden spike which completed the first transcontinental railroad."[13] Defending his proposal to send a human expedition to Mars, President George Bush announced that "throughout our history, America has been a nation of discoverers." It would be hard to imagine Thomas Jefferson "sending a robot out alone to describe the wonders of the American Rockies and the Pacific coast."[14] During the easy-money years following the decision to go to the Moon, NASA actually funded a project that paid social scientists to determine the usefulness of using the railroad analogy to defend the U.S. space program.[15]

Where do these ideas come from? To the critics of such analogies, it is pure Turnerism, the most influential historical doctrine in the United States during the first half of the twentieth century.[16] Widely disseminated to the generation of Americans who managed the early space program, Turnerism remained popular with the educated public even as academic historians tried to debunk the doctrine. Frederick Jackson Turner was a young history professor at the University of Wisconsin when he delivered his 1893 paper on "The Significance of the Frontier in American History." He traced many of the distinctive characteristics of American society to the influence of free land across an open frontier. Inquisitiveness, inventiveness, and individualism were American traits forged on the frontier, Turner argued. Once created, these traits persisted even after the actual conditions of frontier life had disappeared. Turner traced the rise of American democracy and extended suffrage to social conditions on the frontier. Frontier life, according to Turner, bred a love of liberty that found its expression in the political doctrine of self-rule, and migration to the frontier provided a powerful engine for the cross-fertilization of ideas and cultures that promoted America's sense of national identity.[17]

As Turner's paper observed, the American frontier closed in 1890. Quoting a brief official statement from the superintendent of the census, Turner noted that the distinctive, ever-moving line of settlement that had characterized America since its founding ceased to exist as of that year. Gone with it was the source of "this perennial rebirth, this fluidity of American life, this expansion westward with its new opportunities."[18] Historians attacked Turner's thesis, insisting that American ingenuity and democracy could be traced to experiences other than the frontier.[19] But to advocates of space exploration, the Turner doctrine, however dimly understood, became the basis for a new adventure.

The more academic historians sought to discredit the myth of the frontier, the more space advocates exploited it.[20] Advocates of space exploration offered the extraterrestrial frontier as a place to energize the human spirit. All of the talk about technology spin-offs and economic benefits faded in comparison to this aim. To the proponents of space exploration, modern civilizations need frontiers in order to maintain human ingenuity. Anticipating the colonization of Mars, the editors of *Life* magazine predicted that "a frontier ethic that celebrates courage, independence, imagination and vitality will merge with a technological bias that from necessity mothers inventions."[21]

In the minds of space advocates, new challenges foster cultural revitalization. Space exploration offers such a challenge, and its advocates take comfort in the thought that similar animations of human spirit followed previous epochs of discovery. Even an institution as skeptical as the *Washington Post* has embraced this point of view. Commenting on John Glenn's first orbital flight in 1962, editorial writers at the *Post* likened the opening of the space frontier to the enlivenment occasioned by an earlier age of discovery: "There is something in the very air of this space age that is not unlike the climate of another great age of discovery which took place in the fifteenth century." Europe at the end of the fifteenth century was gripped in a period of depression and anxiety, the *Post* observed. Quoting from the historian Samuel Eliot Morrison, editorial writers announced that Western Europeans felt "exceedingly gloomy about the future." Their influence was shrinking, efforts to recover the Holy Sepulchre at Jerusalem had failed, Christianity was losing ground to Islam, the Ottoman Turks had overrun most of Greece, Albania, and Serbia. The *Post* continued:

> Then came an event that to Fifteenth Century Europe must have been quite as astonishing and breath-taking as the voyage of the Friendship VII. Into Lisbon harbor, came the Nina, sailing before a wintry gale to bring news of the discovery of the new world. That news changed the spirit of Europe. In Morrison's words: "New ideas flared up throughout Italy, France, Germany and the northern nations; faith in God revives and the human spirit is renewed."

Revolutions in science, philosophy, and religion followed, which the *Post*'s editorial writers optimistically ascribed to those early voyages of discovery.

Five hundred years later humans were gripped once more in a decade of doubt, suspicious of technology and fearful of a future that harbored the possibility of atomic war or environmental destruction. "So must these ventures into our space environment revive and renew the human spirit," the *Post* promised. The argument so impressed the *Post*'s editorial writers that they repeated it when the *Apollo 11* astronauts landed on the Moon.[22]

Assertions of spiritual rebirth through space exploration appear in the works of astronomer Carl Sagan. The most influential disseminator of popular science in the twentieth century, Sagan shot to national attention in 1980 with the release of his thirteen-part *Cosmos* television series. Possessed with an uncommon ability to share a sense of wonder about the universe, Sagan became the leading spokesperson for the effort to explore the immensity of space and search for extraterrestrial life. His philosophy of human rebirth through space exploration was most articulately presented in the 1994 book *Pale Blue Dot,* in which he traced the spiritual erosion of modern life to two great developments. The first is the closing of terrestrial frontiers. We are all wanderers, he maintained, from ice age humans who crossed the Bering Straits to Polynesian argonauts in outrigger canoes, from European sea captains to American pioneers: "This zest to explore and exploit . . . is not restricted to any one nation or ethnic group. It is an endowment that all members of the human species hold in common."[23] Humans, Sagan continued, first wandered as hunters and gatherers and continued to migrate as explorers and pioneers. Only recently, for a brief period in the lineage of the species, have humans confined themselves to established settlements. In spite of the material advantages to be found in villages and towns, humans remain restless. "For all its material advantages," Sagan noted, "the sedentary life has left us edgy, unfulfilled." Humans have not lost their urge to roam: "The open road still softly calls, like a forgotten song."[24]

Sagan suggested that the urge to explore is necessary to survival, an instinctive drive built into our behavior as a result of natural selection. Town and villages do not last forever, and the people who crave to move on protect their descendants against the catastrophes that inevitably befall those who remain in one place. The human experience has been diminished by this loss of mobility, not just among Americans lamenting the loss of the Western frontier but among humans everywhere.

Joining this restlessness of place, Sagan suggested, is a new desperation of spirit. For centuries humans took comfort in the knowledge that the earth sat at the center of the universe, that the Sun and Moon and stars rotated around the earth and that God had created humans in His own image for a special purpose. Science devastated those beliefs. Earth is a tiny blue dot rotating around an inconspicuous yellow star on the outer reaches of one of a hundred billion galaxies. There are certainly other planets, probably other intelligent beings on them, and possibly other universes that operate according to different laws of nature. Humans were not created by God and, according to Sagan, the best available evidence does not support the need for a Grand Designer.

"Human beings cannot live with such a revelation," Sagan quoted British journalist Bryan Appleyard as saying.[25] The great demotions, as Sagan called them, have created a more mature view of nature, but they have also devastated the human spirit. Maturity is painful; it is easier to think like a child. In the past, when humans believed themselves part of a greater purpose, they could accept moral codes passed down from people presenting themselves as the secular agents of the creator. Humans could follow the exhortations of religious and secular authorities. The apparent insignificance of the earth in the cosmos weakened those codes. It is difficult for humans to respect strict moral codes when those doctrines are based on patently false cosmologies. The new view of the universe undermined the leadership of religious and secular authorities who seemed to speak for no one but themselves and bred a sense of hopelessness.

Sagan believed that a new spirit of discovery could revitalize the human condition. The very science that created this sense of despair could create a new state of wonder: "Once we overcome our fear of being tiny, we find ourselves on the threshold of a vast and awesome Universe that utterly dwarfs—in time, in space, and in potential—the tidy anthropocentric proscenium of our ancestors."[26] Phenomenal discoveries will soon be made about our own solar system, and instruments possible with existing technologies will allow us to probe objects orbiting nearby stars. Humans may leave Earth and transform other planets. They may, as impious as it sounds, join in the process of creation by concocting livable planets out of lifeless spheres. All this in Sagan's view would create epochs of discovery far more enriching than those previously known.

In the decade following the 1969 landing on the Moon, many Americans turned inward and scaled back their expectations. The public lost interest in space exploration. Environmentalists talked about limits to growth and learning to live with fewer resources. Opinion leaders warned about the dangers of technology, and President Jimmy Carter delivered a nationally televised address on the national sense of malaise. This was exactly the sort of national despondency that those pushing space frontiers wanted to sweep away.

"History offers many trenchant examples of what happens when the urge to explore and the development of new technology are forcibly curtained," said NASA administrator James Beggs, commenting on the American withdrawal. Delivering a lecture before the Royal Aeronautical Society in 1984, Beggs once again reminded his audience of the experience of the fifteenth century, this time turning attention not to the accomplishments of Christopher Columbus but to the experience of the Chinese. Seventy-five years before the Columbus voyage, Ming emperor Yung-lo authorized a series of voyages to contact Western people and interfere in their affairs. Before the ships could reach Europe, conservative Chinese leaders prohibited private contacts with foreigners and forbid private voyages. Europeans moved out; Chinese turned in. By the technological and economic standards that came to dominate the world, the decision stunted Chinese civilization for centuries to come. By abandoning our exploration program, another NASA administrator added later, "we risk making the same mistake the Chinese emperors made more than 700 years ago."[27]

This is an admittedly ethnocentric view of the world. It gives little credit to the accomplishments of alternative cultures. In the minds of space advocates, however, it is how the world works. Frontiers imply conquest. History favors ethnocentrism. "The process of pushing back frontiers on earth begins with exploration and discovery, which are followed by permanent settlement and economic development," Beggs bluntly observed. Confrontations between technologically advanced civilizations and inward looking ones inevitably work to the detriment of the latter. When societies collide, the exploring culture invariably wins. That is why the histories of inward-looking peoples are so frequently written in the language of their conquerors.[28]

Advocates of space exploration embrace a frontier philosophy that to some seems sternly paternalistic. Dominate or perish, they say. For many, it is a matter of national survival. When John F. Kennedy accelerated the space race with his decision to go to the Moon, he did so because he wanted to preserve the American way of life. "Only if the United States occupies a position of pre-eminence can we help decide whether this new ocean will be a sea of peace or a new terrifying theater of war," Kennedy argued in defending his space policy at Rice University in 1962. "No nation which expects to be the leader of other nations can expect to stay behind in this race for space."[29] The exploration of space would go ahead, he assured his audience, whether or not the United States participated in it.

For others, moving into this new frontier is required to preserve the human race. A species cannot remain on a single planet for any extended period of time, Sagan observed. On a single planet, it will certainly perish—its demise assured by astronomical events such as asteroid strikes or home grown catastrophes: "Every surviving civilization is obliged to become spacefaring—not because of exploratory or romantic zeal, but for the most practical reason imaginable: staying alive." NASA administrator James Fletcher offered a similar argument in 1975. Any people who turn their back on the future will lose control of their destiny, he said. "Like Darwin, we have set sail upon an ocean: the cosmic sea of the Universe. There can be no turning back. To do so could well prove to be a guarantee of extinction." The great rocket pioneers Robert Goddard and Konstantin

Tsiolkovskii both agreed that the navigation of interplanetary space was essential for the continuation of the human race.[30]

To all of its advocates, space frontiers promise to keep the spirit of inquiry alive. "Frontiers summon the creativity, imagination, and inventiveness of the human mind," said Walter Hickel, governor of Alaska and a frequent space booster. "Civilization needs big projects, the kind that ignite the mind and inspire the soul." Imagination inspires new ideas in science and technology, James Beggs maintained. It nourishes art and literature and promotes the notions of freedom and self-fulfillment that people in democratic societies hold dear. "Small wonder," Beggs said, "that those nay-sayers and disbelievers who have ignored imagination and its potential to shape our destiny leave only a few, faint footprints on the sands of history."[31]

Promoters of space frontiers place a great deal of faith in the inspirational effects of exploration. "Looking back to the early navigators," NASA administrator Thomas Paine said in 1969, "the thing that impresses you is not the culture that they carried to continents like North and South America, Africa, Australia, and the Far East, but the effect of the culture that they brought back to Europe."[32] They brought back a global perspective that transformed the exploring nations and dominated civilization for the next five hundred years. They did not simply prove that the world was round (in any case, few educated people in Columbus's time subscribed to the flat Earth doctrine). Rather, explorers brought back a view of the world that encouraged the development of new technologies, such as sailing ships and navies that could master the seas. This New World view encouraged global commerce and global migration. It inspired scientific discoveries that would have been impaired in a more repressive, inward-looking society. In the geography of the seafaring world, Europe sat at the center of the map.

In a burst of Turnerism, Paine insisted that frontiers were responsible for the rise of democratic governments around the world. As Europeans settled new continents such as North America, they experimented with new forms of government needed to "conquer and organize a new continent." Historians agree that the absence of European institutions provided a fertile ground for the development of liberal democracies such as those that arose in Canada and the United States. Paine suggested that democratic ideas inevitably filtered back to Europe. The development of democratic governments in North America, he said, "set a new standard for governments around the world" and inspired the adoption of democratic reforms elsewhere.[33] The notion that frontier conquest promotes democratic government is not a new idea, but neither is it without controversy. Many factors encouraged the development of democracy in Europe, of which the experience in far-away America was only one.

"As we sail the new ocean of space," Paine insisted, "we are carrying out the same kind of exploration that the early navigators did when they set forth from Western Europe in their first ocean-going vessels." Fresh social transformations would surely follow as humans ventured away from the Earth.[34] Advocates of space exploration have had little difficulty imagining that these things will occur. However fuzzy or inaccurate their knowledge of past history, they have not hesitated to apply it to the space frontier.

As Americans prepared to land on the Moon, enthusiasts made plans to settle space. Oddly enough, the pioneering impulse intensified among the enthusiastic as interest in space exploration declined among the public at large. The more indifferent the general public became, the more imaginative the settlement schemes grew.

During the 1960s, when money for space exploration still flowed freely, NASA officials commissioned a number of studies in preparation for the establishment of bases on the Moon.[35] At the time, most people working on the space program believed that the Apollo missions would provide the first step in a long-range program of lunar exploration. Temporary lunar bases would follow, beginning in the early 1970s. NASA hoped to develop a twenty-five-thousand-pound module that could be launched using the Saturn 5 rocket and land softly on the Moon.[36] Initially, a single module would contain enough supplies for three astronauts to stay on the Moon for ninety days. Additional modules arriving from Earth would allow more astronauts to complete longer stays. One proposal contemplated the establishment of four temporary lunar bases, two at Grimaldi Crater, one on the far side of the Moon, and a fourth at the lunar south pole.[37]

By the late 1980s, explorers from Earth would be ready to establish a permanent lunar colony. One of the more imaginative proposals appeared in the 1968 movie *2001: A Space Odyssey,* in which screenwriter Arthur Clarke describes Clavius Base, located in the second largest crater on the visible side of the Moon. In an emergency, the colony could be entirely self-supporting. As did many scientists of that day, Clarke hoped that elements such as hydrogen, oxygen, and nitrogen could be produced from local rocks crushed, heated, and chemically treated on the Moon. Food was produced in an underground biosphere that also served to purify air. A variety of transportation vehicles, most moving on flex wheels or balloon tires, carried crews to various parts of the lunar surface. Eighteen hundred men and women lived and worked at the base, which, in the optimism of the day, was established by the U.S. Corps of Astronautical Engineers in 1994.[38]

No such bases were approved as the Apollo missions wound down, in spite of the sheer number of studies. Budget cuts in the late 1960s forced NASA to shut down the production line for the Saturn 5 rockets that would have launched the first modules toward the Moon. The failure to detect lunar water in the initial samples brought back from the Moon discouraged planners who hoped that local compounds could be used to support a lunar base. (Signs of water were detected by the *Clementine* satellite in 1996.)[39] NASA sent the last astronauts to the Moon in 1972, a consequence of falling public interest and constricting funds. The number of government-supported lunar studies dwindled.

Rather than discourage advocates of space frontiers, the cutbacks incited bolder proposals. In 1974 Gerard O'Neill excited the public with his vision of artificial colonies in space.[40] Shortly after the first landing on the Moon, O'Neill, a professor of physics at Princeton University, challenged his students to consider whether a planetary surface like the earth was the best place for an expanding technological civilization.[41] Over the next five years, he refined his colony concept. In 1974 he published his ideas in what author Michael Michaud called "one of the most photocopied science articles in history."[42]

Writing in *Physics Today,* O'Neill described how humans could move off the Earth into a multitude of artificially constructed colonies located at gravitationally stable points in the emptiness of space. The most efficient design, he argued, would be cylinders about four miles in diameter and sixteen miles in length. People residing on the inner edge of the rotating cylinder would live in an Earth-like environment, with lakes, mountains, trees, artificial gravity, and a blue sky spotted with clouds three thousand feet "above" the inner rim. Animals and plants endangered on Earth could thrive on these cosmic arks, but insect pests would be left behind, eliminating the need for pesticides. Light from the Sun would be directed into each cylinder from large moveable aluminum-foil mirrors, which would create night and day and seasons like those on Earth. Ample electricity would be provided by steam-turbine generators, powered by the Sun and providing a clean source of energy for transportation and personal use. "With an abundance of food and clean electrical energy, controlled climates and temperate weather, living conditions in the colonies should be much more pleasant than in most places on Earth," O'Neill prophesied.[43]

The first colony could be completed just after the turn of the century, O'Neill argued, in about twenty-eight years. With the manufacturing technology in place, the number of colonies could expand exponentially. A fully developed colony, he declared, could easily support a population of ten million people, plus desirable flora and fauna.[44] Continuing those calculations, O'Neill estimated that emigration to the colonies could reverse the population rise on Earth by 2050. In another thirty years, Earth's population could be reduced "to whatever stable value is desired"—perhaps 1.2 billion people.[45] Colonists would mine the Moon for materials to build the first colonies, then mine the asteroids. After exhausting the asteroid belt, they could tear up the moons of the outer planets. The raw materials available in the solar system, O'Neill offered in a fit of enthusiasm, could support a twenty-thousand-fold increase in the human race while reducing population pressures on the earth.[46] It seemed too good to be true, and probably was.

A number of obstacles stood in the way of development of the first space colony. To construct it, the sponsoring nation would have to move some five hundred thousand metric tons of metal, soil, rock, and water to the construction point in space. And someone had to move the first colonists from Earth to the new space frontier, a major challenge at a time when the cost of transporting just a trio of humans and their accompanying supplies from the earth to the Moon exceeded seventy-seven hundred dollars per pound in 1969 dollars.[47]

O'Neill solved these problems in an most ingenious way. First, he proposed that the bulk of the materials needed for the first space colonies be taken from the Moon. To launch them toward the construction site, he proposed a type of recirculating conveyor belt called a mass-driver. Magnetic impulses produced by electric energy would accelerate a twenty-pound bucket of lunar material to the velocity necessary to hurl its contents toward the appropriate spot in space.[48] As for the problem of transporting people and a few essential materials from Earth, he accepted the widely held notion that reusable launch vehicles would reduce transportation costs by a factor of ten. O'Neill's supporters argued that the first space colony, a rather spartan version, could be constructed for about $33 billion in 1972 dollars—roughly equivalent to the amount spent to send American astronauts to the Moon. Internal NASA studies set the price closer to $200 billion. NASA's inability to reduce the cost of space transportation by the much-touted factor of ten would have driven the price even higher had anyone attempted to fulfill O'Neill's dream.[49]

O'Neill's proposals were not unique, but the degree to which they captured the public imagination was. The idea of large artificial colonies in space had been advanced previously by an assortment of writers, from the famous to the obscure. At the beginning of the century, Russian space pioneer Konstantin Tsiolkovskii had envisioned dwellings in space that could house millions of people. British scientist J. D. Bernal advanced a similar concept in 1929, and Arthur C. Clarke helped popularize the idea in his 1954 children's novel *Islands in the Sky.* Dandridge M. Cole presented plans for space colonies formed out of hollowed-out asteroids in 1964, and Krafft Ehricke, a member of the von Braun rocket team, issued his call for an "extraterrestrial imperative" in 1971.[50]

Unlike earlier proposals, which attracted a narrow audience, O'Neill's images splashed upon the public scene. They attracted interest from the mainstream of American politics to the cultural fringes of radical thought. His concept was embraced by visionaries who wanted to pioneer space, environmentalists concerned about overpopulation and dwindling resources, futurists who saw technology as the solution to human problems in industrial civilization, and a variety of

Eventually humans would move into space in large numbers. According to the grand vision of space exploration, humans will construct artificial colonies suspended in gravitationally stable points of empty space and convert hostile planets into habitable spheres. In 1974 physics professor Gerard O'Neill put forth a proposal for a multitude of space colonies that would allow the human race to continue its expansion while relieving population pressures on Earth. (NASA)

space-age groupies in occasional need of psychiatric help. The California-based counterculture group responsible for producing the environmentally correct *Whole Earth Catalog* promoted the idea. Congress held hearings and NASA supported a variety of studies.[51]

Excited by the prospect of pioneering the high frontier, Americans throughout the 1970s in ever-increasing numbers filed into spacefaring clubs. O'Neill's vision spawned the L-5 Society, named after one of the gravitational stable regions near the orbit of the Moon at which objects such as a space colony could remain indefinitely. The main purpose of the L-5 Society, formed in 1975, was "to arouse public enthusiasm for space colonization."[52] The society attracted adherents whose exuberance about space colonization irritated people laboring on practical U.S. space activities. As a response, industry and government leaders in 1975 formed the more conservative National Space Institute, at its head the aging space warrior Wernher von Braun, which sought to mobilize grass-roots support for NASA's more conventional exploration plans. In 1987 the two organizations forgot their differences and merged into the National Space Society.[53]

Strange and wondrous groups continued to form. Distraught at the cancellation of their *Star Trek* television series, science fiction fans organized local clubs and federations as a means to keep their enthusiasm for galactic fantasies alive. Through a massive letter-writing campaign, "Trekies" convinced the government to name the first space shuttle test model after the starship *Enterprise.*[54] Following the broadcast of the *Cosmos* television series, 120,000 individuals joined Carl Sagan and Bruce Murray in forming The Planetary Society. The society collected signatures of notable and ordinary Americans for its Mars Declaration, a statement advocating the exploration of Mars as an important step "toward the long-term objective of establishing humanity as a multi-planet species."[55]

As of May 1980, by one estimate, there were nearly forty major interest groups promoting space exploration in one form or another. Local chapters, astronomical societies, and science fiction fan clubs pushed the total number of organizations close to five hundred.[56] They included groups promoting capitalism in space, groups advancing the role of women in space, groups set up to privately fund space activities, and groups prepared to train space pilots and pioneers. There was even a political action committee for space.[57]

O'Neill's vision attracted otherwise dissimilar individuals to the promise of space colonization. The more he spoke, however, the more his vision to move billions of Earthlings to environmentally correct space colonies looked like a dreamy hangover from the radicalism of the 1960s. Political interest in his proposals faded in the late 1970s as the supporting policy issues—overpopulation and the energy crises—passed from the public screen. By 1980, the frontier agenda in space had returned to more conventional objectives: establishing outposts on the Moon and Mars that might in turn lay the groundwork for future colonization.

Many books and reports appeared during the 1980s advancing various scenarios for the accomplishment of the latter goals. The most lavishly illustrated, if not widely read, was the 1986 report of the National Commission on Space, chaired by former NASA administrator Thomas Paine. Members of the commission recommended that the government establish an outpost on the Moon by 2006 and a human outpost on Mars by 2015. By the third decade of the twenty-first century, both the lunar outpost and the Mars base would be permanently occupied. "While that seems far away now," the report noted, "many of the people who will live and work at that Mars Base have already been born." The footholds established there would provide experience and technology "for human exploration still farther into our Solar System."[58]

Members of the commission proposed an elaborate infrastructure in space to make these expeditions possible. There would be an Earth-orbiting space station, a lunar-orbiting space station, and a station around Mars. A special spaceport at one of the gravitationally stable points near the Moon would prepare humans for the journey to Mars. There would be transfer vehicles designed to take humans between stations in space. There would be lunar landers and Mars landers and cycling spacecraft and special spaceships that could with a burst of speed catch the cyclers as they flew by. Paine's committee argued that much of the material needed to construct the lunar and Martian outposts could be found on those bodies. Analysis of lunar samples brought back by Project Apollo suggested an abundance of oxygen in the lunar soil (in the form of metal silicates), and Paine hoped that future expeditions might locate water ice and other volatile compounds in permanently shadowed craters near the lunar poles. On Mars, the commission members observed, all the necessary oxygen, hydrogen, nitrogen, fertilizer, and methane needed to start a permanent settlement could be extracted from that planet's puny atmosphere.

Paine's report contained wondrous illustrations of this new frontier: men and women tending fruit trees and vegetables in a lunar biosphere, a Martian lander arriving at a twenty-first-century Martian settlement, astronauts in space suits servicing a transfer vehicle at a gravitationally stable spaceport, and the same transfer vehicles using Earth's upper atmosphere in aerobraking maneuvers to slow their velocity after disgorging passengers onto one of the fast-moving cyclers. In another illustration, a specially designed robot worked to mine propellants from Phobos, a moon of Mars. Preliminary studies indicated the presence of water, carbon, and nitrogen on this tiny moon. "If so," the commission suggested, "there is an orbiting fuel depot just 6,000 miles above the red planet to top off the hydrogen and oxygen tanks of visiting spacecraft."[59]

Robots and mining occupied a central role in most settlement plans. Lunar and Martian colonists would spend much their time extracting and manufacturing the resources necessary to live off the land. One imaginative scheme envisioned the creation of self-replicating factories. Once set free on cosmic bodies, they would proceed to reproduce themselves indefinitely. In one of the last papers he wrote before his death in 1992, Paine predicted that the settlement of space would stimulate a new generation of industrial robotics. "Initial labor costs on Mars," he wrote, "which will be a hundred times those on Earth, [will stimulate] the development of reliable robotic production systems with a hundred times the productivity of terrestrial factories."[60]

The benefits of these developments would reach far beyond the earth, members of the commission wrote. In truth, very few of the tangible goods manufactured on the Moon or Mars would return to Earth. Nearly all the surface activities would produce materials necessary to sustain pioneers. This detail helped to reveal much of the underlying motivation for the venture. Extraterrestrial colonies would allow the human race to leave "the precious and fragile planet where it was born," and extend life "to the far reaches of the inner Solar System," commission members maintained. The ultimate purpose of space pioneering, it seemed, was to pioneer space, a case of circular reasoning if ever there was one.[61]

Throughout the 1980s engineers and politicians debated the practical details of the pioneering dream. Should humans work to establish a lunar colony or proceed directly to Mars? "Moon firsters" argued that the lunar colony would provide much-needed experience with frontier technologies necessary to move farther from home.[62] Examining the lunar option for NASA administrator James Fletcher, astronaut Sally Ride suggested that humans could return to the Moon by the year 2000 and set up a lunar outpost by 2005.[63] Within ten years of the first landing, an industry task force observed, a community of one hundred pioneers could be living and working on the Moon.[64] The settlers would explore their new home and set up scientific instruments such as a radio astronomy observatory on the back side of the Moon, shielded from interference from Earth. Like pioneers before them, they would look for ways to make their expeditions pay. Experts were especially intrigued by the possibility of mining the moon. Solar flares, experts suggested, had deposited on the lunar surface quantities of helium 3, which could provide a rich source of fuel for fusion reactors should that technology ever take hold. One lunar booster argued that just sixty thousand pounds of helium 3 per year returned to Earth would satisfy the energy needs for the whole planet.[65]

To others, development of a lunar settlement seemed like a waste of time. The United States had been to the Moon. If the objective of a lunar base was to prepare for Mars, why not get on with the larger goal? In the spring of 1981 a collection of space-interest groups organized a conference at the University of Colorado in Boulder to examine whether "a manned Mars mission was a viable option for our space program."[66] Out of the meeting emerged the so-called Mars Underground, a congregation of students, space boosters, and aerospace professionals devoted to making (as the title of the book emerging from the conference revealed) *The Case for Mars.* Two more books and conferences followed.[67] In 1986 NASA joined the discussion with its own Mars conference and the following year established the Office of Exploration for the purpose of coordinating agency activities and convincing Congress and the president to approve the endeavor.[68]

Much of the practical work contained in the various studies concerned the best way to get to Mars. In her 1987 report, Sally Ride proposed a series of three short sprints, with ten- to twenty-day stays on the planet's surface and an overall journey of no longer than a year. This would lay the groundwork for an outpost sometime around the year 2010. NASA engineers developed plans for more elaborate missions, some more than three years in length. Debate over the most efficient way to get to Mars led to government infighting. White House officials grumbled about conservative NASA bureaucrats and commissioned outside experts to develop more imaginative proposals. A deftly illustrated report by astronaut Thomas Stafford drew on suggestions from outside officials that nuclear propulsion could cut the one-way transit time to Mars from 224 to 160 days.[69]

Amidst the clamor and debate, the long-term goal remained steady. Humans would depart Earth, settle other planets in the solar system, and eventually move to the stars. A report from the Mars Underground predicted that the first human child would be born on Mars in the year 2020, when the population of the outpost reached one hundred persons. By 2081, authors of the chronology speculated, two hundred thousand colonists would live on Mars.[70]

For all of its appeal, however, Mars remained a very inhospitable place. The 1976 *Viking* lander revealed a cold, dry desert with little atmospheric protection from sterilizing ultraviolet rays. Humans had not yet settled the Antarctic on Earth, which was absolutely balmy by comparison. How did Mars enthusiasts plan to handle the hostile environment on Mars? No problem, they replied. They would simply transform the planet into an Earth-like environment by altering conditions there. Of all of the recommendations for pioneering space, few were as imaginative as the proposals for terraforming Mars.

Terraforming was once the preserve of science fiction writers. In the April 1937 issue of *Astounding Stories,* Ross Rocklynne described a successful attempt to move fifty-two million cubic miles of frozen water from the asteroid belt to Mars.[71] A series of stories published under the pseudonym Will Stewart in the 1940s explained the use of "paragravity generators" to attach atmospheres to previous lifeless bodies. Stewart, whose real name was Jack Williamson, gave the process the name that has remained with it since.[72] In 1950, Robert Heinlein described the terraforming of Ganymede, one of the giant moons of Jupiter, by several thousand colonists from planet Earth.[73]

In 1961 the young Carl Sagan published an article in *Science* magazine containing a plan for making Venus habitable, one of the first serious proposals for altering the environment of planets. Although Sagan's plan was flawed (he overlooked the problems posed by the density of the atmosphere), the article gave scientific respectability to the concept.[74] By 1975, NASA was ready to give its official blessing, sponsoring a study that examined the possibility of altering the environment of Mars to make it more habitable.[75] The Mars Underground, one of the many space advocacy groups to emerge during the 1970s, was born out of the interest of people anxious to explore and transform that planet.[76]

Terraforming was too futuristic to receive serious attention in government councils as White House officials inched closer to approving an actual Mars mission in 1989. Nonetheless, the concept continued to draw public interest.[77] In May 1991 *Life* magazine ran a cover story on the subject of terraforming Mars. Relying upon scientific opinion, the editors presented an ambitious 150-year scheme for transformation of the cold, dry planet. Orbiting solar reflectors would melt the polar ice caps, and Martian factories would produce greenhouse gases and ozone substitutes. As the planet warmed, nitrogen and water would seep out of the Martian soil and the atmosphere would thicken. This in turn would cause further warming. Clouds would appear, and the color of the sky would shift from pink to blue. Oxygen for the newly forming atmosphere could be extracted by Martian factories from carbon dioxide, carbonate rocks, and deposits of iron oxide. Pioneers would plant tundra plants and hearty evergreens as the mean planetary temperature approached the freezing point of water. Rain would fall and agriculture would thrive, but the maturing atmosphere would need more oxygen to allow humans and animals to live outside. With enough oxygen-producing factories and vegetation, the editors predicted, the planet could be made totally suitable for human habitation by the year 2170. Streams, lakes, and oceans would cover the surface of a new moist green globe.[78]

The cost would be high, but not beyond the reach of industrialized nations. Large subsidies would be required before the Martian colony became self-sustaining. *Life*'s editors optimistically predicted that investment costs would peak at $45 billion per year during the early buildup stage—an impressive sum but a fraction of what the world spends annually on national defense or government-assisted health care. The technical problems would be formidable, but not insurmountable. Engineers would need to develop inexpensive rocket ships and cheap sources of energy (fusion reactors would help considerably). Although it would require much planetary fine-tuning and the possibility of undesirable side effects would always exist, terraforming is not beyond the realm of possibility. As Carl Sagan observed, "We need look no further than our own world to see that humans are now able to alter planetary environments in a profound way."[79]

On 20 July 1989, to commemorate the twentieth anniversary of the first landing on the Moon, President George Bush challenged the United States to commit itself to a sustained program of exploration that would lead to the permanent settlement of space: "From the voyages of Columbus—to the Oregon Trail—to the journey to the Moon itself—history proves that we have never lost by pressing the limits of our frontiers." To begin the process, he called upon Congress to join him in supporting an outpost on the Moon and the first human expedition to Mars. In a later speech he set a goal of 2019 for the first Mars landing.[80] Congress, however, refused to fund these efforts. None of these things happened. As Carl Sagan asked, was this "a failure of nerve or a sign of maturity?"[81]

The conventional excuse for the failure of the Space Exploration Initiative blamed the lack of money, a consequence of federal budget deficits and competing priorities. With NASA committed to the space station and space shuttle, said Barbara Mikulski, chair of the Senate committee handling the agency's budget, there was simply not enough money to conduct lunar and Mars expeditions. This, of course, could be viewed as a loss of nerve. The United States certainly could afford to go to Mars, provided that politicians had the courage to raise the funds and commission the technology necessary to complete an undertaking whose benefits lay far in the future. President Bush called the refusal an act of short-sightedness by people who lacked vision.[82]

Alternatively, the failure to allocate funds can be viewed as a sign of maturity, evidence that at least some Americans grasp the reality of frontiers. People who want to send humans to Mars do so for one fundamental reason—to get off of planet Earth. They believe in the manifest destiny of humans as a spacefaring race, an argument that has special appeal in the United States, a frontier nation just one hundred years ago. If scientific discovery was all the Mars advocates wanted, they could get it much faster with robots. Robots, however, would not open a new human frontier.

The image of the frontier is America's creation myth. For many (but not all) Americans, it explains where they came from and why they are special among the peoples of the world. According to this myth, America was essentially an unoccupied land of boundless opportunity. Hardy, independent pioneers settled the wilderness through their own ingenuity and resources and created a new civilization. Unencumbered by old traditions, they formed simple democratic communities with governments that became a model for the entire world. The work was hard but satisfying. Generation after generation repeated this experience until the frontier was gone. The independence and ingenuity created by the fading pioneering experience remained in the American culture, however, waiting to be reapplied in new frontiers.[83]

The image, alas, is factually wrong. It is based upon a romanticized interpretation of history as far removed from reality as the Buffalo Bill Wild West Show was from the real events it sought to portray. The American West was not an empty land waiting to be settled from the perspective of Indians and Hispanics who already lived there when white settlers arrived. Pioneers depended extensively upon eastern subsidies and capital, especially government support for railroads. In business enterprises such as the gold and silver rushes, failure was as common as prosperity. The laws of economics that favored large companies employing wage workers were not suspended in the West. Territorial governments were no less corrupt or more democratic than those in the East, and as many principles of democracy emerged from the pens of intellectuals residing in the eastern United States and Europe as from the frontier experience.

The frontier metaphor conjures unpleasant as well as inspirational images. What were the president's speech writers thinking, asked historian Patricia Limerick, when they allowed Ronald Reagan to compare the fourth landing of the space shuttle to the driving of the golden spike? To meet the 1869 deadline, railroad workers laid the track so quickly that much of it had to ripped up and replaced. Leland Stanford, representing the Central Pacific, proved so unfamiliar with the elementary details of railroad construction that he could not drive the golden spike into the ground. The Union Pacific Railroad, half-sponsor to the transcontinental enterprise, went bankrupt twenty-five years later, and the Central Pacific tried to avoid payments on its government loans. Says Limerick, "A reference to the Golden Spike, to anyone who is serious about history, is also a reference to enterprises done with too much haste and grandstanding, and with too little care for detail."[84]

If the analogy is correct, what does the real frontier tell us about space? As people who have labored in the enterprise already know, space pioneering is likely to be hard and unglamorous. Space travel, like aviation before it, consists of long periods of boredom and repetitive work. In that sense, it may not be much different from life on the western frontier. "Nobody wants to be a cowboy," lamented one western employment specialist. "It's hard work, it's dirty work, it's round-the-clock work." It is something most Americans want to watch from a distance.[85]

Frontiers are rarely utopian in spite of efforts of their advocates to portray them as such. Commenting on the challenges of founding a lunar colony, Thomas Paine assured his supporters that it would "sweep aside old world dogmas, prejudices, outworn traditions, and oppressive ideologies." Konstantin Tsiolkovskii predicted that in space colonies "human society and its individual members [would] become perfect." Gerard O'Neill predicted that life in space would permit "most of the human population to escape from poverty" and that the environment of space colonies could "be optimized for good health."[86] By their apparent openness and lack of rules, frontiers attract utopian thinkers. History suggests, however, that new settlers bring society and all its imperfections with them.[87]

Frontiers are also a metaphor for ungrateful dependence. American colonists depended upon British troops for protection from French and Indian wars and were notoriously reluctant to pay their share. When the British imposed the Stamp Act as a means to recover their investment, American colonists organized the boycotts and demonstrations that eventually led to independence. In a similar fashion, space colonists will depend upon earthly taxpayers for the resources necessary to maintain their settlements in a hostile environment. Space colonies will require huge subsidies in their formative years. How will Earthlings react when space pioneers grow restless with their colonial status? Will Earthlings glorify the independence of pioneers or, like eighteenth-century British politicians looking toward America, treat them like ungrateful children?[88]

Among environmentalists, the myth of the unlimited frontier is a thinly disguised attempt to exonerate "rape-and-ruin" policies of resource extraction. The history of the American West is also the history of mining spoils and clear-cuts, of destroyed fisheries and atomic testing. Proposals to extract oxygen and helium 3 from the lunar surface would require in essence the strip mining of the Moon. Who knows how much environmental damage humans could do with a whole new planet to spoil?

Space advocates use the frontier image as a happy metaphor, but to many Americans the notion of frontier opportunity is insincere. African Americans did not view the New World as a land of opportunity and invention, to say nothing of the frontier views of conquered Indians demoralized by generations of alcoholism and unemployment on American reservations. Frontier women did not enjoy the same freedoms as frontier men. Perhaps this helps explain the persistent support for space pioneering among white males, for whom the frontier myth remains a sweet song.

In spite of the relentless attacks of classroom historians, the romantic image of the frontier endures among the public at large. Many people continue to believe in it. Space advocates call upon the popular image of the frontier to garner support for their visions, even as historians attack them. The space frontier is an appealing analogy to many people in the United States, given their pioneering history. The American creation myth provides a level of vindication for space exploration that compensates for less-glamorous expectations. It is doubtful that Americans would pay hundreds of billions of dollars to go to Mars simply to gain some technology spin-off or to establish the interplanetary equivalent of an Antarctic research station. The frontier analogy, with all of its flaws, allows people to believe that space exploration will reopen one of the longest and most formative chapters in American history. Never mind that the reality of space colonization will differ considerably from the popular image of it. Space flight is a dream, and dreams do not have to be entirely real in order to motivate behavior.

STATIONS IN SPACE

> I think if you ask the public at large, and quite possibly most of the people within NASA, what a space station was, they would think in terms of the movie that came out fifteen or twenty years ago.
>
> —John Hodge, 1983

Having created an imaginative vision of humans leaving Earth and settling space, devotees faced the practical difficulties of actually doing so. An essential step in practically every settlement or exploration scheme was the creation of a permanent facility above the surface of the Earth at which humans could live and work. Such a station, as it was called, would provide important knowledge about the practical details of maintaining human life away from the shelter of the planetary home, to say nothing of its value as a transfer point for voyages deeper into space. The knowledge gained from developing a space station would also foster the technology necessary to build self-supporting bases in more remote places, such as the Moon and Mars.

In the centuries preceding the space age, nations seeking to extend lines of exploration and settlement built the terrestrial equivalent of space stations across the face of the earth. Frontier forts, way stations, trading posts, and base camps provided convenient means for advancing human presence into unconventional territory. Colonists and pioneers, mountain climbers and polar explorers commonly employed structures such as these for the purpose of marshalling material and fashioning sanctuaries in hostile realms. By the start of the space age, the image of the frontier fort or base camp as a jumping off point for more daunting adventures was well established in the public mind. Quite naturally, advocates of the high frontier assumed that conquest of the cosmic void would require similar structures above the surface of the earth.

Elaborate plans for stations in space appeared. Some proposed stations rotated so as to provide occupants with the comforts of artificial gravity; others moved quietly through the ether as their occupants studied the unique effects of life in a place with practically no gravity at all. To excite public interest, promoters portrayed space stations as very large structures, some housing hundreds of scientists and supporting personnel. In January 1984 President Ronald Reagan directed the National Aeronautics and Space Administration to build a permanent way station in space. "Develop a permanently manned space station and do it within a decade," he told officials at NASA.[1] Ten years later, not a single piece of U.S. station hardware floated in space. Constructing an Earth-orbiting space station turned out to be easier to envision than to accomplish. Once again, reality overpowered imagination.

Nearly all the pioneers of space flight had supported orbital stations, and for a practical reason. The laws of physics, given the limited thrust of available propellants, did not permit workable rocket ships with humans on board to be propelled directly from the surface of the earth to the Moon and back, and certainly not to the planets. One early pioneer calculated that a ship bound directly from the earth to the Moon would have to burn 105 tons of liquid propellant during the first second of flight just to heave itself a few feet off the ground. "Landing on the moon is beyond the borderline of what chemical fuels can do," observed Willy Ley. "The direct trip to the neighboring planets is even further beyond." Some sort of refueling station was obviously required. The space station, as Ley observed, would provide the much needed "cosmic stepping stone for spaceships which were too weak to reach another planet directly."[2]

Fortunately for the advocates of space exploration, the conceptual foundation for the orbital station was already well established in the public mind. The experience was especially well set in North America, where frontier forts and way stations had speckled the face of the newly settled country. From the garrison houses of New England to the station settlements of Kentucky and Tennessee, from the brick and earthen fortresses of the Atlantic coast to the simple stockaded trading posts of the West, Americans had built forts and stations as an initial step toward settling unfamiliar territory. Many present-day cities bear the names of the forts around which colonial settlements grew—the Spanish fortification at St. Augustine in Florida, the French forts on Mobile Bay along the Gulf of Mexico, and the English Fort Pitt at the confluence of the Monongahela and Ohio Rivers. Frontier forts offered safe havens from the dangers of the wilderness, attracted commerce and trade, and provided jumping-off points for ventures into the wilderness. The impulse to settle new lands, said one commentator, transformed America into "the most internally fortified territory in the world."[3] By the end of the nineteenth century, fort building was firmly associated with conquest and settlement in the American mind. Space pioneers did not need to explain the underlying association between frontier stations and space exploration, and rarely did, because the association was so deeply rooted in the national experience.

The necessity for establishing way stations in space was reenforced by the obsessive attempts during the first half of the twentieth century to reach the last unexplored regions on Earth. Efforts to reach the South Pole and ascend Mount Everest attracted as much attention from armchair explorers as later voyages to the Moon. In 1911 the Norwegian explorer Roald Amundsen and four companions set out for the polar plateau. They were followed along a slightly different route twelve days later by a party led by Englishman Robert Scott. Amundsen located the South Pole on 14 December; Scott and his four compatriots reached the pole thirty-five days later but perished on the return journey when their supplies ran low.

The success of Amundsen's plan, as with nearly all polar expeditions to follow, owned much to his skill in establishing an adequate base camp on the outer edge of unfamiliar territory. Amundsen's base camp was prefabricated and shipped to the Ross Ice Shelf some 750 miles short of the pole. During the year that preceded his successful journey, Amundsen used the camp as a base from which to test equipment and set up caches of food and fuel along the proposed route. The base camp also served as an observatory to study the polar environment, a workshop to refashion equipment tested in the field, and a shelter to provide protection from the wind and cold.[4] Eighteen years later, the explorer Richard E. Byrd established Little America on the site of Amundsen's abandoned base, from which Byrd conducted the first airplane flight over the pole.

The public impression of base camps as adjuncts to exploration was further amplified by efforts to climb the highest mountains in the world. Because of the great distances and harsh conditions involved, mountaineers constructed elaborate pyramids of people and supplies designed to put a few skilled individuals on the Himalayan summits. The first successful ascent of Mount Everest took place in 1953, by Edmund P. Hillary and his Sherpa guide Tenzing Norkay. A supporting team of climbers and porters organized a large base camp at sixteen thousand feet, from which a series of nine additional camps were established at ever-increasing altitudes. Seven and a half tons of supplies were carried from Kathmandu to the base camp at Thyangboche. Those supplies supported a final cache of 650 pounds used in the final assault. It was not uncommon for Himalayan expeditions to employ five to six hundred porters to move the basic loads.[5]

Space pioneers envisioned similar schemes for the exploration of the Moon and planets. Hermann Oberth, the German rocket pioneer, stated that a space station would be necessary as a "springboard" for flights to the Moon. He drew directly from Antarctic experience in framing plans for his *Weltraumbahnhof*.[6] Willy Ley, one of Oberth's disciples, argued that a "terminal in space" would be necessary to marshall the equipment needed to explore the Moon and nearby planets. In the March 1952 issue of *Collier's* magazine, Wernher von Braun explained how giant rockets would lift thirty-six tons of material into space with each flight, allowing the construction of a large space station and the assembly of three large spacecraft bound on the first journey to the Moon.[7] More flights would add to the pyramid of equipment and supplies and allow for the eventual establishment of a lunar base. Such pyramid schemes appeared frequently in the plans of space pioneers.[8]

The creators of these early images were keenly aware of the advantages to be gained by using extraterrestrial stations as the new fortresses of the space age. Forts, after all, are military facilities, and throughout history, fortifications have been used to defend settlements and protect territorial claims. In North America, the need for fortifications intensified because of the completing claims of European nations bent on controlling common territory, to say nothing of the anger expressed by the native population that already lived there. Indian wars, colonial wars, and revolutionary wars all spawned the need for fortification.

In contemplating the nature of space, many pioneers believed it would be subject to the same sort of territorial claims. Nations might lay claim to the Moon and planets, planting their flags on extraterrestrial territories just as they had done on terrestrial lands.[9] In this event, the space station would serve a valuable purpose, standing as the twenty-first century equivalent of a fortress guarding the entrance to a harbor or waterway. Sitting at the gateway to heaven, a space station would fortify the claims of constructing nations to operate freely in the cosmos much as European nations seeking to settle the American frontier had used defensive works along colonial waterways to do the same. Like frontier forts, space stations would limit the claims of competing nations. "Whoever is first to build a station in space can prevent any other nation from doing likewise," the editors of *Collier's* magazine bluntly warned in 1952.[10] Circling the earth, the space station would cover far more territory than any fortress fixed on the earth. Rocket pioneers devised a variety of schemes to take advantage of the new military high ground. Scientists working for the Nazi government formulated plans for a giant solar mirror, a "sun gun" that could incinerate enemy forces on the ground. Officials in the U.S. Department of Defense pursued the possibility of using manned orbiting laboratories (MOL) as spy stations in the sky. Oberth predicted that "whenever work on a space station is started, it will undoubtedly be for military reasons."[11]

During the years leading up to the first orbital flights, visions of stations in space attracted as much attention as plans for flights to the Moon and Mars. In the 1944 edition of *Rockets* by Willy Ley, plans for a space station figured prominently. "The realization of the station in space is the realization of space travel in general," Ley announced. "Trips to the moon, around the moon, and even to the other planets are no longer difficult if they are made from that station."[12] In 1952 Wernher von Braun wrote "Crossing the Last Frontier," unquestionably the most influential English-language article on the coming space age, which promoted his vision of a large, Earth-orbiting space station.[13] Through their writing and lectures, Ley and von Braun exposed the ideas of space enthusiasts to a wide audience. The fact that they were able to do so in terms already familiar to an exploration-attentive public considerably strengthened the attractiveness of their ideas. The imagery of space-age fortresses and base camps fit so well into the thinking of that time it hardly needed to be explained.

The U.S. military never built an Earth-orbiting space station during the formative years of space flight. No space base served as the jumping-off point for the first expeditions to the Moon. Between the visions of space flight and its actual execution, a considerable gap emerged. Space travel proceeded down a course much different from that conceived by its pioneers.

The concept of an orbital station as a space-age fortress received its first practical test during the 1960s when the U.S. Department of Defense received permission to move ahead with plans for its Manned Orbital Laboratory. To MOL partisans, space was the ultimate high ground from which military campaigns of the future would be won or lost. To test the feasibility of the concept, military officers drew up plans for a small, fifty-six-hundred-pound research laboratory in which two soldiers could conduct a variety of surveillance activities and experiments. The U.S. Air Force even went so far as to recruit a corps of military astronauts.[14]

As the decade progressed, the applicability of the fortress analogy waned. Shortly after entering office in 1969, members of President Richard Nixon's administration canceled the MOL project. A number of factors doomed the idea. Technological advances allowed the development of remotely controlled satellites that could complete surveillance activities at a fraction of the cost of stations with humans on board, rendering bulky fortifications far less effective than satellites in space. During the 1960s the main combatants in the Cold War agreed to forgo territorial claims in outer space and accept unrestricted overflight of their territories by orbiting objects, decisions prompted largely by the knowledge that everyone's spacecraft were equally vulnerable to attack. This removed much of the motivation behind the development of large stations and maneuverable spacecraft that could control access to the new high ground.[15]

Although these events undercut the fortification analogy, the image of a space station as a base camp for distant exploration remained strong. But as with the military plans, practical developments undermined this analogy as well. In 1962 NASA engineers decided to forgo any sort of orbital assembly or refueling point around Earth as a prelude for the trip to the Moon. Instead, they devised a radical scheme that substantially reduced the weight necessary to take humans to the lunar surface and back. Lunar orbit rendezvous combined with the development of engines that burned liquid hydrogen allowed the United States to reach the Moon from Earth with a three-stage Saturn launch vehicle weighing only thirty-two hundred tons at launch, fully fueled. This sort of rocket technology was easily within the grasp of NASA engineers. Space-station advocates agreed to the plan reluctantly. Although they understood that it offered the only hope of reaching the Moon by the president's end-of-the-decade deadline, they also knew that the plan diminished the rationale for a space station as a gateway to the Moon. As NASA deputy administrator Hans Mark later observed, the price paid for Apollo was the lack of any space-based infrastructure once flights ended. "Apollo was essentially a dead-end from a technical viewpoint," he said.[16]

As if to further undermine the supporting analogy, mountain climbers experimented with Alpine-style dashes to Himalayan summits that minimized the need for elaborate pyramids of camps supplied by small armies of porters. As in space, lightweight equipment and new techniques transformed earthly exploration. In 1978 Reinhold Messner and Peter Habeler scaled Hidden Peak, a 26,470-foot companion to nearby K2, with only a single depot above their minimal base camp. As if to emphasize their point, Messner and a companion pulled a sled across Antarctica to prove that the new approach would work there too.[17]

Having launched the nation on the journey into space, station advocates did not abandon their hope for an orbital facility just because the supporting analogies dimmed. Advocates continued to maintain the need for stations as stepping stones to cosmic flight. Stations would be needed not just around the earth, they said, but also around the Moon and throughout the solar system. By identifying additional functions for space stations to perform and preparing other images to excite public interest, advocates hoped to overcome the continuing reluctance of political leaders to finance a station that met the vision of the pioneers.

The most durable aspect of the space-station image as presented to the public was its size, an aspect maintained not only by practical engineers but also by creators of fantastic stories. Both promoted the notion that any space station worthy of public support would be a massive structure—so large that people on Earth would be able to see it fly by.[18] In the minds of its advocates, creation of an Earth-orbiting space station would rank among the great engineering accomplishments of all time. Public interest was nurtured with images of big stations in space.[19] This created a discomforting dilemma for station advocates in search of government support for their plans. The station designs that generated the most public interest were large, multipurpose facilities that typically cost more than politicians were willing to spend, the stations the government could afford to build were often too small to arouse public interest. It was a situation guaranteed to produce anxiety. Politicians choked on impressive space station plans and sneezed at modest ones.

In his "Crossing the Last Frontier," von Braun proposed the construction of one of the most impressive space stations of all time, 250 feet wide and orbiting 1,075 miles above the earth's surface. Shaped like a wheel, the station would make a complete turn on its axis every twenty-two seconds so as to give the sensation of gravity, one-third of the force felt by humans on Earth, to occupants in its outer rim. Chesley Bonestell painted a large picture of the station to accompany the article. The illustration, one of the most reproduced works of twentieth-century space art, revealed a hublike wheel with graceful lines passing over the earth near the Panama Canal far below. Astronauts in space suits worked on a nearby observatory and on the space wheel; others unloaded supplies from a winged space shuttle recently arrived from Earth. Space taxis shaped like overgrown watermelons carried astronauts between the winged spacecraft and space station and satellites orbiting nearby.

To simplify construction of the space wheel, von Braun suggested that it be fabricated out of inflatable nylon and plastic. The material could be collapsed for its trip into space, then inflated once in orbit. Twenty sections, each an independent unit, would be assembled into the complete wheel. Constructing and supplying the orbital facility would require frequent shuttle flights. Said von Braun, "There will nearly always be one or two rocket ships unloading supplies."[20]

Accompanying a separate article by Willy Ley in the same issue of *Collier's* magazine, a painting by artist Fred Freeman showed a cutaway of one section of the wheel. The large space station crew worked on three decks inside the outer rim, much like sailors on a large submarine. Elevators carried workers to the station hub; regulators and pipes kept the station cool. Von Braun estimated that the space station would require five hundred kilowatts of electricity, which he proposed to generate through a condensing mirror and generator. A highly polished metal trough ran the circumference of the wheel, concentrating the rays of the Sun onto a steel pipe containing liquid mercury. Heat from the sun transformed the mercury into hot vapor, which in turn drove a turbogenerator.[21]

Stanley Kubrick improved substantially upon the image of the large wheel with the release of *2001: A Space Odyssey.* The phase of the movie devoted to modern times opened with the sight of a space shuttle preparing to dock at a very large space station. Kubrick's depiction of orbital activities set new standards for special effects. His station, with work on one of its twin hubs still underway, spun in an orbit two hundred miles above the surface of the earth and measured an astounding nine hundred feet across, as large as a modern aircraft carrier. It rotated more slowly than von Braun's wheel, one revolution per minute, producing a sense of gravity equal to that on the Moon. The facility housed an international crew of scientists and bureaucrats, welcomed passers-through on their way to the Moon, and offered amenities as comforting as those in a modern hotel.[22] In a further effort to equate size with significance, the creators of *Star Trek: The Motion Picture* designed an orbital facility that moved across the screen like a huge city. The facility served as a dry dock for the USS *Enterprise,* to which then-admiral James Kirk was returning. A smaller orbital transfer station hovered nearby.[23]

Engineers working to design space stations for the U.S. government did not avoid the proclivity for bulk. In 1969 NASA officials asked aerospace contractors to prepare preliminary designs for a permanent orbital facility to be deployed around the year 1977, hoping that politicians would be enthusiastic enough to fund one. The initial facility consisted of a 33-foot-wide cylinder with a crew of twelve. Even with its attached hardware, this configuration hardly approached the size of the designs to which the general public had been exposed. Confident that the initial facility would grow, NASA officials asked their contractors to assemble a group of

Just as terrestrial explorers used frontier forts and base camps to open up remote areas on Earth, space pioneers envisioned the construction of orbital stations as stepping stones to the conquest of space. Stanley Kubrick created a space station nine hundred feet wide for the 1968 motion picture *2001: A Space Odyssey.* This movie poster by Robert McCall depicts a winged space shuttle departing the docking port in the large orbital facility. (Painting by Robert McCall; ©1968 Turner Entertainment Co.; all rights reserved)

33-foot-wide cylinders into a space base that could house between fifty and one hundred occupants.[24] This produced some grand designs. North American Rockwell submitted plans for a pinwheel-shaped space base that grouped four of the cylinders around a common core. The base was organized along a central axis 359 feet long, a requirement made necessary by the placement of two nuclear-powered generators at the far end of the central shaft. Attached by booms to the central core, the cylinders stood a total of 234 feet apart from end to end. The base could be rotated so as to produce artificial gravity or dampened out for experiments requiring none at all. Though not a wheel, the resulting configuration was impressively large.[25]

In 1984 NASA finally received the political approval necessary to begin designing a real space station. The baseline configuration announced two years later was impressively large—more generous in all its dimensions than von Braun's original 250-foot-wide wheel. NASA engineers rejected wheel-shaped space stations and pinwheels in favor of a configuration that minimized the force of gravity. Contrary to general opinion, a stationary space station would not be entirely gravity free. Like any other heavenly body, a large structure would generate its own gravity and even slight changes in position would produce a gravitylike force. In an effort to minimize gravity, engineers placed the experimental labs and crew quarters as close to the center of mass as possible. One very long transverse boom supported the solar power generating system, and two vertical keels provided space for various experiment packages and the servicing of spacecraft and satellites. To illustrate the size of the impressive facility, NASA officials prepared a computer-generated illustration that superimposed the dual-keel design across a view of the U.S. Capitol. The station stretched from the House to the Senate, joining the two legislative chambers in what space enthusiasts hoped would be the political consensus necessary to complete the job.[26]

Space enthusiasts not only imagined that the station would be large but also assumed that it would last a long time. NASA engineers wanted to design a structure would last for at least twenty to thirty years, perhaps longer.[27] Once up, the structure would be modified and eventually replaced with something more advanced. As a long-lived facility, it would create a permanent human presence in space. From the time of its occupancy forward, humans would always live and work away from the Earth.

Station advocates carefully maintained this image. During the late 1960s, NASA officials won approval for an orbital facility called *Skylab* that was constructed out of equipment developed for the Apollo moon program. During 1973 and 1974 *Skylab* housed three crews of three astronauts each who spent a total of 171 days in orbit.

Clearly not a permanent facility, *Skylab* plunged into the atmosphere in 1979. NASA executives were careful to call it an orbital workshop, not a space station. "The ideal station would be permanent and large," confessed one NASA reference work. *Skylab* was "a worthy precursor to a larger, more elaborate station."[28] It was not the real thing.

When Congress in 1984 attempted to steer NASA back toward an orbital edifice that would house astronauts only part of the time (a "man-tended" facility), NASA officials objected once again. President Ronald Reagan had committed the United States to develop a "permanently manned space station," they pointed out. "The Space Station will, of course, be a highly automated system, and it will require many advances in automation techniques and robotics," NASA administrator James Beggs argued. "It is, however, the presence of man which makes it a unique national resource."[29]

To justify a large and permanent orbital outpost, advocates had to propose a facility that performed many functions. As they knew, a big space station would not be cost effective if used for only one task. Station advocates had to find many jobs for the crew to perform. The space station, according to popular conceptions of it, would be used at least as an observation platform, a scientific laboratory, a manufacturing facility, and an operational base from which the crew could service satellites and prepare spacecraft for missions farther from Earth.

Many functions helped justify the cost of the station, but they also complicated the station's design. In few areas did the reality of space engineering clash so dramatically with the image necessary to support it. Big stations had to perform multiple missions, but those missions were often better performed on platforms specifically designed for those purposes. Station planners, nonetheless, herded them onto the main facility.

Advocates insisted that a large orbital outpost could serve as an observation platform, the primary function of von Braun's 1952 wheel. "Technicians in this space station," von Braun wrote, "will keep under constant inspection every ocean, continent, country and city. Even small towns will be clearly visible." The desire to observe all portions of the earth drove von Braun to select a polar orbit for his space facility. With the earth turning below them, members of the crew would use specially designed telescopes attached to large optical screens and radarscopes to observe activity below with pinpoint detail.[30] Observations could be used to monitor natural events like cloud patterns, improving weather predictions. Most of the observations von Braun discussed, however, concerned national security. Well into the 1980s, space-station advocates believed the U.S. military would be an important customer for any orbital facility, given its potential to act as a spy-in-the-sky.[31]

Observations could also be made of astronomical phenomena. For years, scientists have recognized the advantages of astronomical observatories in space. Orbiting telescopes achieve unprecedented clarity from their vantage points above the murky atmosphere of Earth. Photographs in visible light are substantially impaired by atmospheric fluctuations, and only a small fraction of the infrared energy emitted by other heavenly bodies reaches the ground. Scientists devised plans for space telescopes that could collect visible and ultraviolet light, radio waves, X-rays, gamma rays, and cosmic rays. Astronomers agitated for space telescopes of all sorts.[32]

Station advocates viewed scientists as a powerful group whose cooperation broadened considerably the advocates' base of strength.[33] Unfortunately for the advocates, the space-based telescopes desired by astronomers required such long exposure times and exact pointing requirements that the telescopes could not be housed on the multipurpose facility. Other station activities, from the docking of spacecraft to the simple movements of people on board, would disturb the alignment of the telescopes as they probed the void. Von Braun recognized this and so proposed that his space observatory be placed on a remotely controlled platform some distance away.

This created something of a conceptual puzzle. If the space observatory was not physically attached to the space station, how could the two facilities be treated as one? Von Braun solved this problem by placing the space observatory in the same orbit at the station, hovering nearby. His observatory could be serviced from the station, affiliating the two in an indirect way. In the days before remote sensing, when sharp images could not be transmitted by machines, von Braun imagined that station technicians in space suits would have to manually load canisters of film and other special plates into the observatory. Technology rendered this unnecessary.[34]

NASA's Space Station Task Force, set up in 1982 to develop conceptual plans for the first permanently occupied facility, took this idea one step further. In order to win support from scientific groups, leaders of the task force included within the space station program two free-flying platforms. One would travel in the same orbit as the main facility and could be serviced by station technicians much in the same way as von Braun had envisioned. The other platform, however, would travel along an entirely different path, moving around Earth's poles. To reach it, station technicians would have to return to Earth and catch a space shuttle headed in an entirely different direction. Members of the task force attached the polar platform to the multipurpose space station in a most ingenious way. They included the appropriation for the platform in the budget for the main facility. NASA's whole space station program, as a result, contained both "manned and unmanned elements." Unless NASA build a multifunctional faculty, said the leader of its task force, "it really is not going to be justifiable."[35]

The space station would serve as a national scientific laboratory, its advocates said, uniquely situated but not conceptually different from government-funded research laboratories on Earth. The most fascinating research would take advantage of the unique properties of weightlessness, a condition impossible to reproduce on Earth for any significant period of time. Critics scoffed at the image of laboratory studies on weightless as "another two decades of original research on why astronauts vomit" in space.[36] To be sure, considerable research would be done on the human response to weightlessness, but the research proposed for various space stations went considerably beyond this.

Writing well before the first humans flew into space, Hermann Oberth predicted that materials with unusual properties, such as new metal alloys, could be created in the weightless environment of space. A continuously operating laboratory in space would accelerate the research necessary to develop such materials. Experiments could be conducted on plant growth under weightless conditions. Oberth wondered, for example, if plants unconstrained by the force of gravity would transform themselves into unusual shapes and sizes. "All of our scientific and technical knowledge is based on the existence of the force of gravity," he wrote. "It is most interesting to imagine a scientific and technical world without gravitation and to attempt to picture its consequences." An orbital laboratory would also permit extended research using nearly perfect vacuums or temperatures close to absolute zero.[37]

One of the more creative uses of the space station arose from the notion that it could be used as a factory in space. During the early 1980s, a number of space enthusiasts promoted the new frontier as the future realm of entrepreneurial capitalism. Preliminary experimentation had revealed a number of products that could be manufactured in and around an orbital facility. Drugs with a standard of purity not commercially attainable on Earth could be produced using an exotic technique called electrophoresis, possible only in the microgravity of space. Integrated circuits made from new materials such as gallium arsenide promised to revolutionize the computer industry. A huge new industry based on the possibility of growing protein crystals in space promised advances in the battle against diseases such as cancer and in the improvement of chemical processes such as gasoline production. Materials nearly free of contaminants could be produced using the extremely high vacuum of space.[38]

Swept along by excessive enthusiasm, station planners offered their orbital facility as the industrial park of the twenty-first century. Billions of dollars stood to be made in space, they said, and the United States could capture this market by taking the lead in establishing an orbital laboratory to complete much-needed research and set up the processing facilities needed to begin making products. In 1983 President Reagan met with a group of industrial executives familiar with the potential for entrepreneurial activities in space. What could the government do to speed up the commercialization of space? Reagan asked. More than anything else, they replied, they wanted a space station. The promise of space commercialization figured heavily in Reagan's decision to start work on the orbital facility.[39]

Actually using a multifunctional space station to perform these activities posed a number of practical problems. Once again, the gravity problem emerged. Scientists measure the absence of gravity in micro g, where one micro is equal to one-millionth part (10^{-6}) of the force felt by a person standing on the surface of the earth. Space-based facilities for microgravity work require levels below ten micros, approaching one micro where possible. The dual-keel design for NASA's space station was chosen because it promised to place the laboratory modules in the one-micro zone.[40] A space facility where everything is perfectly still can produce gravity reductions in this range. Machinery that vibrates will compromise the standard, as will clumsy people as they move about. As a simple matter of physics, more activities increase the number of micros likely to be felt. Under perfect conditions, a standard of one micro might be attained. Under actual circumstances, it could slip into the range of ten to one hundred micros.[41]

In part to solve this problem, spacecraft designer Max Faget proposed the construction of what he called an industrial space facility (ISF). After his retirement from NASA, Faget founded Space Industries, Inc., from which he sought support for an automated platform that could manufacture materials in space. The platform would be visited by astronauts, who would service the facility, change equipment, and set up new activities. The rest of the time, the platform would orbit quietly without people on board. Removing people from the processing facility enhanced the possibility of maintaining very low levels of gravity. It also reduced the cost of construction.[42]

In 1987 the president's Office of Management and Budget inserted $25 million in the NASA budget and instructed the space agency to investigate the possibility of supporting this commercial venture by leasing it from Faget. This caused great consternation among station advocates. At the time, the space station was fighting for its political life. Advocates angrily viewed Faget's proposal as a diversionary tactic that would allow politicians to kill the big, multipurpose station while shifting its most promising activities to a human-tended orbital facility. NASA officials raised elaborate objections to the ISF. With support from competing industries and the National Academy of Sciences, NASA executives beat back attempts to shift commercial activities off the multifunctional space station.[43]

Complicating all these proposals was the desire to use the space station as an operational base, a vision that had captivated exploration advocates for decades as they imagined a series of interplanetary service stations spreading out toward the Moon, the planets, and stars.[44] In 1980 planners at NASA's Johnson Space Center put forward one of the most imaginative versions of a station inaugurating this emphasis, calling it a space operations center. At that time, NASA did not have any politically approved plans to return to the Moon or go on to Mars. All human space-flight activities were confined to the space next to the earth. Those activities could be extended to geosynchronous orbit with a space operations center, Johnson Space Center engineers proposed. Astronauts at the operations center could assemble and check payloads bound for geosynchronous orbit. Working in space, astronauts could construct enormous communications platforms and high-resolution parabolic antennas too large to be launched from Earth. A reusable orbit transfer vehicle or space tug would be developed to push those platforms out to appropriate locations.[45]

The configuration proposed for the operations center, with its agglomeration of modules and trusses, looked like many designs of the day—with one exception. In the most widely distributed drawing, a large hanger protruded from the supporting framework. In the hanger, astronauts would service and repair satellites, a major purpose of the center. The designing engineers predicted that most science and industrial applications would be performed on free-flying platforms, such as orbital factories or space telescopes. "Some operations actually proceed best without the disturbance of human presence," the engineers admitted. Those facilities, however, would require periodic servicing and adjustment.[46] This could be done by astronauts in a space tug, servicing the facilities in their original orbits, or by bringing the satellites into the hanger for repairs. Science was not a major activity within the space operations center. Operations was.

The proposal for a space operations center was thrown out as soon as NASA officials began to seek political support for a large space station in 1982. NASA officials did not want scientists standing outside of the station looking in. Officials pushed the station concept back toward its original premise. Any station large enough to be worth building would have to accommodate many users.[47]

In 1984, having received permission to construct a large space station, NASA officials faced the practical difficulties of designing one that met popular expectations. This proved extraordinarily difficult to do. The image of a space station was well set in the public mind: it had to be big, it had to be multipurpose, and it had to provide a permanent outpost for future exploration. Given the budgetary realities of the 1980s, it also had to be affordable. The expectations created by these conflicting requirements doomed the program. Throughout the 1980s, the proposed space station kept shrinking. By 1994, the deadline set for completion of the facility, much of the enthusiasm for what Wernher von Braun once called "the greatest force for peace ever devised" had drained away.[48]

NASA officials tried hard to design a space station that met popular expectations. Explaining the vision to a Senate subcommittee, NASA administrator James Beggs recited the capabilities a multifunctional facility could provide:

> Properly conceived, a station could function as: A laboratory in space for the conduct of science and the development of new technologies; a permanent observatory to look down upon the Earth and out at the universe; a transportation node where payloads and vehicles are stationed, processed, and propelled to their destinations; a servicing facility where these payloads and vehicles are maintained and, if necessary, repaired; an assembly facility where, due to ample time on orbit and the presence of appropriate equipment, large structures are put together and checked out; a manufacturing facility where human intelligence and the servicing capability of the station combine to enhance commercial opportunities in space; and a storage depot where payloads and parts are kept in orbit for subsequent deployment.[49]

The actual design to emerge from this vision was large and complex. The so-called dual-keel configuration featured a rectangular metal truss as large as a football field (361 feet by 146 feet) on which NASA planned to hang a number of experiments and payloads. Stellar and solar observatories on the upper boom would point toward the heavens; Earth observatories on the lower boom would point toward home. Along the sides of the rectangle engineers planned to mount canisters that would house scientific experiments and small factories. Two rectangular service bays would be used to store equipment and service payloads.

On and around the core space station, NASA engineers planned to deploy a variety of exotic machines. A long robotic arm would creep around the truss like a spider, moving payloads and performing assembly work. Manipulators inside the service bays would assist astronauts repairing satellites. A busy little orbital maneuvering vehicle (OMV) would fly around the space station like an intelligent bug, repairing faulty equipment and retrieving satellites. An orbital transfer vehicle, mated with its own OMV, would carry payloads to and from geosynchronous orbit twenty-two thousand miles away. A self-powered, automated platform would conduct science experiments a short distance from the main station, while another funded out of the same budget would circle the poles.

A horizontal boom 503 feet long held the solar collectors and photovoltaic arrays that would generate the station's electric power, as well as radiators for dissipating the heat that power generation would create. Station designers planned to start with seventy-five kilowatts of power and expand to more than twice that level of power as the station grew.[50] Attached to the midpoint of the horizontal boom, in the center of the rectangular truss, sat the modules in which the crew would live and conduct laboratory experiments. With a useful life of thirty years, the station would evolve as its capabilities grew. It could eventually provide "a

To meet the expectations created by exploration advocates, large space stations had to perform many functions. The space station approved by President Ronald Reagan in 1984 and designed by NASA in 1986 would have served as a research laboratory, an observation platform, a microgravity manufacturing facility, and a service station for repairing satellites and launching missions deeper into space. Although it did not rotate so as to produce artificial gravity, it was impressively large—more than 500 feet long and 360 feet tall. (NASA)

staging point for spacecraft of unprecedented size," which could take humans back to the Moon or on to Mars. Engineers built "scars and hooks" onto the framework of the station on which the ambitions of the future could be placed.[51]

NASA officials planned to begin assembly of this station in 1993, place humans permanently on board in 1994, and complete the "full baseline configuration" by 1996. They were prepared to involve the European Space Agency, Canada, and Japan so as to produce a truly international facility. And they had agreed to complete all of this work for about $8 billion (in 1984 dollars).[52]

As ambitious as this design appeared, it still fell short of expectations from the past. The station, for example, would not house a large crew. The living quarters contained room for no more than eight astronauts, cramped in a fourteen-foot-wide habitat module.[53] John Hodge, the leader of NASA's Space Station Task

Force and later director of the development program, admitted at a congressional hearing that the public at large and "quite possibly most of the people within NASA" thought of a space station "as a very large rotating wheel with 100 people on it and artificial gravity." Writing for the popular *Science 83* magazine, Mitchell Waldrop observed that the actual station "will look more like something a child would build with an Erector set." It would not resemble a wheel, it would not rotate, and it would not soar very high. "2001 it's not," the magazine complained.[54]

Incredibly, as development work continued, the station shrunk. It grew smaller and smaller, moving even further from the expectations created by a century of wondering about outposts in space. Within a year of announcing their intention to construct the large, multipurpose, dual-keel space station, NASA officials scaled it down. In April 1987, confronting the fact that the proposed facility would exceed budget guidelines, NASA abandoned the vertical keels. That eliminated the capability of the station to operate as a service station in space. As the servicing capability disappeared, so did the co-orbiting platform. NASA officials hoped to resurrect these elements during a second construction phase, but the deferral became permanent as design problems swelled.[55]

By eliminating the vertical truss structure, NASA officials sharply reduced the functions the station could perform. Instead of a large, multipurpose facility, NASA was left with an orbital research lab whose residents would spend most of their time conducting experiments and making observations of the earth and heavens. In submitting her report on the future of the U.S. space program to NASA administrator James Fletcher, Sally Ride correctly identified the revised configuration as a "laboratory in space." Without much discussion, the original capabilities that had motivated space pioneers to dream about orbital stations disappeared. Said Ride, "Other capabilities, such as an assembly station or a fueling depot, will not be included in the initial phase, but could be accommodated later if a need for those functions is clearly identified."[56]

Any hope that those capabilities could be patched on to the initial facility evaporated when the station contracted again. In 1991 NASA officials issued plans for what they called the "restructured space station," also known as *Space Station Freedom.* The remaining horizontal truss was snipped back to 353 feet and the habitation and laboratory modules were reduced in size. The station was reconfigured, in the words of the advisory committee recommending the changes, "with only two missions in mind." The four-person crew would carry out life-science experiments and conduct microgravity research. Gone were the hopes for a large, multipurpose outpost in space.[57]

In 1993 the United States and Russia agreed to join their two space stations into one orbital facility. This was preceded by an episode in which the U.S. space station contracted once more. Upon entering office, the Clinton administration had requested another space station redesign. Not unexpectedly, this had produced a new configuration with less volume, less electric power, and a smaller crew.[58] The new *Space Station Alpha,* as it was dubbed, disappeared quickly as the Russian government agreed to join. Like a dieter rebounding after a long fast, the new *International Space Station* gained volume. The crew quarters where the Russian modules appeared grew a few feet larger and the central truss grew longer in order to accommodate a more extensive electric power generating system.[59] Russian participation halted nearly a decade of configuration decline, but it was not sufficient to restore hopes for the large, multipurpose facility from which the dream began.

The design to emerge from these efforts met few of the criteria established by people who had tried to imagine what a space station would be. It was not round; it did not rotate. It was not a fortress in space; it could not serve as a jumping-off point for voyages beyond. It would not perform many functions, being primarily a research laboratory. Its small crew would remain in orbit for only as long as funding prevailed, a precarious situation given the lack of excitement the scaled-back facility generated.

Why did the space station prove so difficult to build? NASA spent more than $10 billion designing it, an amount roughly equal to the 1984 cost estimate for the entire facility.[60] It took as long to design the station as it did to go to the Moon. The continuing agonies of the space-station program could not be blamed solely on engineering complications or cost overruns, factors that were in many regards mere manifestations of deep-seated conceptual contradictions.

Station advocates sought to build an orbital facility that was both cheap and grand. As NASA officials learned, it was hard to design a station that simultaneously met public expectations and budgetary constraints. The pioneers of space flight envisioned a station that would rival the other great engineering feats of the twentieth century, both in scope and cost. Scientists and engineers called together to advise *Collier's* estimated that their 250-foot-wide orbiting wheel would cost $4 billion, a huge sum in the currency of the day. This was more, they said, than the amount of money spent on the Manhattan Project.[61] Four billion dollars then, when adjusted for inflation, equals $21 billion in 1984 dollars. Yet when pressed for a cost appraisal during the White House review process, NASA officials backed away from an engineering project of that scale. They issued a preliminary estimate of $8 billion in 1984 dollars for the development of the orbital facility, a number based largely on the political reaction to various cost estimates. "I reached the scream level at about $9 billion," said John Hodge, leader of NASA's Space Station Task Force.[62]

The $8-billion cost estimate was based upon a number of assumptions that were technically feasible but difficult to fulfill. To construct a space station that fit

The more the government worked on the space station, the smaller it grew. The redesigned *International Space Station* provided for a much smaller facility that served as little more than an orbiting research laboratory. Once again, reality overpowered expectations. (NASA)

within the budgetary estimate, NASA executives in Washington, D.C., had to constrain the tendency of field officials to drive up the price with Earth-based infrastructure, overhead, and cost reserves. These extras contributed little to the capability of the orbiting facility but let the NASA field centers thrive. As the centerpiece of the human space-flight program for the 1990s, the space station provided a number of center directors with a lucrative opportunity for strengthening their institutions. Managers in the field pushed cost estimates higher; headquarters executives fought back. Field officials eventually won.[63]

One important factor that allowed NASA to issue a small estimate for such a large project backfired. NASA officials believed that a space station could be constructed in such a way as to grow incrementally. Using a modular design, NASA could deploy a small space station in the beginning and add more parts as funding allowed. "Space stations are the kind of development that you can buy by the yard," James Beggs pointed out in his effort to justify the low cost estimate.[64]

A space station that can grow "by the yard" can also shrink in the same way, as station planners learned. Every cost review produced an incrementally shrinking design. Modules, air locks, satellite servicing bays, solar collectors, experiment mounts, space tugs, coorbiting platforms, and scientific equipment fell off as cost reviews sped by. The 1985 letter by station overseer Philip Culbertson that initiated the first redesign came to be known as "scrub mother," the parent of all reductions to come.[65]

NASA executives premised the development of the space station on the availability of cheap and easy transportation to space. In Chesley Bonestell's famous painting of the rotating space station prepared for *Collier's,* a winged space shuttle hovers nearby. In *2001: A Space Odyssey,* Heywood Floyd, chairman of the National Council of Astronautics, arrives at *Space Station One* on a winged passenger shuttle.[66] President Ronald Reagan took the first step toward approving the space station at the 4 July 1982, landing of the space shuttle *Columbia.* James Beggs planned to fund much of the cost of the space station through funds freed as the space shuttle began to pay its own way. The orbital facility, he frequently said, was the "next logical step" after completion of the shuttle development program. As late as 1985, NASA officials were planning to launch as many as twenty-four shuttle flights per year. Construction of the space station depended upon the availability of cheap, easy access to space.[67]

All of that changed with the explosion of the Space Shuttle *Challenger* on 28 January 1986. Concerns about the cost and reliability of the space shuttle forced NASA engineers to make a number of important changes in station design. Engineers cut back on the number of shuttle flights needed to construct and maintain the facility, both to reduce costs and enhance safety. The cost of space transportation was not included in the original estimate of $8 billion in part because the cost of space transportation was forecast to be so low. At $400 million per flight, transportation costs drove up total program costs in ways never anticipated by early designers.[68]

Worries about the reliability of the shuttle raised concerns about the station crew. With shuttles always ready to fly, the crew had access to a rescue vehicle if some catastrophe occurred. In a serious emergency, the crew could retreat to what NASA engineers called a "safe haven" within the station until the shuttle arrived. Stripping away easy access required engineers to design an entirely new spacecraft, a "lifeboat" permanently attached to the space station that could drop the crew back to Earth following a major malfunction. Like other changes not anticipated, this drove up the cost of the space station once more.[69]

People imagining the space station had little difficulty envisioning the technologies needed to operate it. Engineers knew how to develop livable environments in pressurized spacecraft; they knew how to generate electric power in space. Experts knew how to stabilize large structures in orbit and write computer programs that could regulate the automatic functions of such craft. Even the 1952 editors of *Collier's* assumed that the technology for the space station was well understood. "Our engineers can spell out right now," they wrote, "the technical specifications for the rocket ship and space station in cut-and-dried figures."[70]

Technological specifications might have been understood, but NASA engineers did not want to build a space station with old technologies. They did not want to reconstruct *Skylab.* They wanted to build a complex, multipurpose facility with new technologies, taking a quantum leap into the twenty-first century and thereby producing knowledge that could be applied to bolder projects such as interplanetary spacecraft and bases on Mars.[71] The technology requirements for the space station considerably complicated its design.

Those who would actually have to operate the space station grumbled that the engineers were creating a facility that was technically infeasible and potentially unsafe. The assembly phase was too complicated, they said. The complex remote manipulator system would break down. Assembly and upkeep of the station would require too much extravehicular activity. Planned maintenance would require frequent spacewalks, and that did not include unanticipated repairs.[72]

Planners worried that a large, multifunctional facility would not work well. As Sally Ride wrote in her 1987 report on the future of space exploration, "a laboratory in space, featuring long-term access to the microgravity environment, might not be compatible with an operational assembly and checkout facility." In an initial facility, with a small crew and a limited number of experiments, competing uses could be compressed within a single structure. But as the size of the station grew, "branching" would surely occur, meaning that the government eventually would have to build separate space stations for each function. Station advocates faced enough opposition with just one orbital facility and chose not to broadcast the fact that future evolution might require more.[73]

From its beginnings, the space station was a concept at war with itself, built on terrestrial analogies with limited applicability in space. Even on Earth, the significance of base camps and forts diminished as the century progressed. Space-station advocates fought to combine on a single facility a large number of potentially incompatible missions. As advances in automation drove down the cost of separate facilities, the attractiveness of the multipurpose, centralized station faded. Even the appeal of the space station as a depot for voyages beyond declined. In 1990, when members of the so-called Synthesis Group released their report on human exploration of the solar system, they scarcely mentioned the station. Later that year, members of the Advisory Committee on the Future of the U.S. Space Program reached a similar conclusion. "We do not find compelling the case that a space station is needed as a transportation node for planetary exploration," they wrote. "First, many promising flight profiles do not appear to require such a node and, second, if they did, the need in our judgement is sufficiently far in the future that we would hardly know today what to ask of such a terminal."[74]

As the space program matured, the station was relegated to a role that amounted to little more than a research facility in orbit, disappointing those who still dreamed of huge space bases from which humans could launch expeditions to the planets and beyond. The vision of space flight suffered again at the hands of reality.

SPACECRAFT

It has proven to be much more complicated than I thought.

—Hermann Oberth, 1985

By the latter part of the twentieth century, advocates of space exploration had fostered an elaborate vision of interplanetary flight and extraterrestrial pioneers. Under the terms of this enticing vision, large numbers of people in extraordinary spacecraft would travel across the interplanetary frontier. Space flight, once the domain of test pilots and mission specialists, would receive ordinary citizens no better prepared than automobile drivers or airline passengers back on Earth. Everyone would fly into space. The experience would transform humanity.

This vision drew its inspiration from the history of aviation in the first half of the twentieth century. The advent of aviation spawned an outpouring of prophesy and public enthusiasm similar to that accorded space flight. At a time when only the brave and foolhardy flew, prophets of aviation predicted that air travel would become so commonplace that ordinary people in large aircraft unaffected by weather would travel from city to city across the face of the earth—a notion unimaginable to a populace stuck on muddy roads and horse-drawn carriages. The public went wild about the future of aviation, encouraging even bolder claims that became the basis for what air-minded people called "the winged gospel."[1] The fact that so many of the prophesies came true encouraged advocates of space travel to believe that a similar future awaited them.

For both aviation and space travel, the organs of popular culture promoted the popular view. Hollywood, which had churned out scores of aviation films, released movies and television programs depicting the effortless quality of space travel. Invariably, filmmakers and screen writers portrayed space flight as less difficult than it actually was.[2] Spacecraft of the popular mind covered vast distances at remarkable speeds and were affordable, reliable, and not much more complicated than a modern jet aircraft. With a little preparation, anyone could fly into space with virtually no personal risk. Science joined fancy in fostering the chorus of anticipation. Otherwise pragmatic rocket engineers issued extraordinary claims about the future of powered spacecraft, especially the NASA space shuttle, which was approved for development in 1972. Promoters predicted that the shuttle would cut the cost of space flight by a factor of ten. It would allow ordinary citizens to fly into space. It would be so easy to launch that a fleet of five could make fifty round trips into space every year.[3]

In 1970 a special panel of the president's Science Advisory Committee issued a sober forecast. Future developments on the high frontier, they predicted, would be determined by the outcome of efforts to diminish the cost and difficulty of space travel.[4] If rocket scientists could develop low-cost, reusable spacecraft, the dreams of space pioneers would come true. People in significant numbers would move out to explore the inner solar system. However, this was only one of several possibilities, as members of the committee cautioned. The most dramatic technological improvements in the final decades of the twentieth century might occur elsewhere—in the development of computers and remotely controlled spacecraft, for example. In that case, the future of human space flight would give way to robotic satellites and probes. Space travel under those conditions would not resemble the model supplied by aviation, which rose from the sands of Kitty Hawk into a huge personal transportation industry in hardly sixty years. Space travel would resemble nothing more than itself.

Since its beginnings, a central tenet of the spacefaring creed has been the belief that anyone should be able to travel into space. Space advocates were not content to imagine a few astronauts landing on outlying orbs or a hand-picked crew occupying a small research facility in some distant place. In order for the full vision of spacefaring activities to come true, ordinary people had to move into the cosmos, traveling in spaceships accessible to the common folk. The broad vision of humans moving into space and colonizing other worlds was not possible without cheap and easy transportation.

This vision clashed with the imagery used to promote the early space program. The first flights into space were conducted under conditions so stressful that only test pilots were allowed to fly. Although the courage of the first astronauts boosted public interest in the space program, rocket pioneers were anxious to move away from the image of space travel as the personal confrontation of high danger. They wanted to transform extraterrestrial flight into something more routine—not to distract attention from glory-hogging astronauts but so that ordinary people could imagine themselves flying into space.[5]

In clinging to this vision, advocates found comfort in the aviation experience, a powerful model for imagining that ordinary people could fly into space. Many of those who promoted the U.S. space program were earlier involved in aviation, in both industry and government research labs, fostering the milestones of atmospheric flight.[6] They knew that air transport, a familiar experience to millions of Americans by the second half of the twentieth century, had overcome the same sort of skepticism that affected the early space program. Like promoters of space exploration after them, the pioneers of aviation had struggled to convince a skeptical public that transport above the surface of Earth was real. Newspapers and magazines had greeted early attempts at powered flight with extreme skepticism. Turn-of-the-century stories about airplanes were lumped into the same category as reports of perpetual-motion machines and messages from Mars. Humans would fly, reported one magazine, just as soon as they could get the laws of gravitation repealed. The historic flight of Orville and Wilbur Wright near Kitty Hawk, North Carolina, on 17 December 1903 went virtually unreported in the mainstream press. Said Orville, "I think this was mainly due to the fact that human flight was generally looked upon as an impossibility, and that scarcely anyone believed in it until he actually saw it with his own eyes."[7]

To spread the gospel of flight, aviation pioneers had taken their evidence to the public at large. Wilbur awed Europeans with his flights over the French countryside in 1908, while brother Orville attracted attention at the War Department trials in Fort Myer, Virginia. The following year one million people watched Wilbur pilot one of their airplanes up the Hudson River from Governors Island to Grant's Tomb and back.[8] In 1910 the brothers formed the Wright Exhibition Company, whose newly trained pilots toured the country performing at county fairs and specially organized shows. Glenn Curtiss, a rival of the Wrights, formed a similar company that same year.[9] After the war, barnstormers took up the cause. Flying surplus military aircraft, barnstormers often worked alone, dropping into farmer's fields in search of passengers and flying stunts wherever a crowd could be found. "By decade's end," wrote historian Joseph Corn, "flying missionaries had exposed nearly every hamlet and crossroads in the land to the airplane."[10]

The message they carried was relatively simple: airplanes would become as commonplace as trains or steamships and would transform the lives of ordinary people. Said Orville Wright, "I firmly believe in the future of the aeroplane for commerce, to carry mail, [and] to carry passengers." Alexander Graham Bell, another air enthusiast, predicted that the next generation would "see the day when men will pick up a thousand pounds of brick and fly off in the air with it."[11]

These predictions, so skeptically viewed, eventually came true. The rapid development of aviation became one of the most impressive technological achievements of the twentieth century.[12] An airplane manufacturing industry rose up in response to the call of the Joint Army and Navy Technical Board in 1917 for the production of twenty thousand military aircraft. Airline entrepreneurs prospered after federal officials in 1925 established contract mail routes, providing a secure financial base for the business of carrying freight and passengers through the air. In 1936 only thirty-three years after the Wright brothers' 1903 flight, American Airlines put the first DC-3s into regular service. This workhorse of air transport allowed investors to make a profit simply by hauling passengers from town to town. Three years later Pan American Airways began transatlantic passenger flights with the Boeing Clipper. In 1954, with scarcely fifty years of aviation history behind them, workers rolled out the first prototype of the Boeing 707, inaugurating the era of jet transport. Fourteen years later Boeing introduced the 747 jumbo jet, each capable of carrying 490 passengers across the world.[13]

How many could have thought, standing at the dawn of the air age, that scarcely seventy-five years later powered aircraft owned by American firms would carry in a single year more passengers than the entire population of the United States? To commemorate the seventy-fifth anniversary of the Wright brothers' first flight, the Air Transport Association reported that in 1978 American carriers conveyed 275 million passengers through the air. Flying had become as commonplace as the pioneers of aviation had predicted. Indeed, anyone could fly.[14]

Not all of the prophesies prevailed. High expectations encouraged wilder claims, not all of which could be achieved. Among the more extreme was the belief that personal aircraft would become as common as the family automobile. Not only would everyone fly, but nearly everyone would own their own instrument for conducting it. Executives would conduct business from the air and commuters would take to the sky, leaving soot and traffic behind. "An Airplane in Every Garage" became the rallying cry for efforts to fully democraticize the experience of flight.[15]

This belief gave rise to experiments both amusing and sad. In 1926 Henry Ford introduced the "flying flivver," a small single-seat aircraft that attracted considerable interest until pilot Harry Brooks killed himself in one during a demonstration flight over Miami Beach. During the 1930s engineers at the government's aeronautical laboratory at Langley Field, Virginia, developed a foolproof two-seat aircraft that would not stall or spin. Unfortunately, the engineers gave up versatility in exchange for safety, a feature that earned the enmity of experienced pilots. Built for land lovers accustomed to driving automobiles across the two-dimensional space of the road, the aircraft could not perform the most rudimentary maneuvers necessary to move precisely through the three dimensions of the air. During the 1930s the U.S. Department of Commerce offered incentives to anyone who could develop an affordable personal aircraft that the average person could fly. The competition encouraged the submission of a number of hybrid designs, half-aircraft, half-automobile. One prototype arrived like a car at the front door of the Commerce Department on Pennsylvania Avenue. The pilot then motored to the National Mall and after a few adjustments lifted off for nearby Bolling Field. The prototype performed as poorly in rush hour traffic as it did in the air.[16]

The contribution of these inventions to the aircraft industry was less significant than the expectations they generated. Promoters of flivvers and other simple aircraft sought to break down the barriers of cost and complexity separating aircraft technology from ordinary citizens. Visionaries such as Henry Ford had successfully accomplished this with the automobile, creating an unprecedented degree of personal freedom. Anyone with a minimal amount of training could drive a car. The prophets of aviation sought to expand this freedom to include the air, and if it could be done in the atmosphere, imagine what would be possible in space. The drive to make aircraft and eventually space travel accessible would transform the lives of common citizens in ways unimaginable to their earthbound ancestors. The effort to let every middle-class American family own its own aircraft failed, but the expectations it generated lived on.

For space-flight advocates, realization of the dream required the development of spacecraft that could carry ordinary people. Unfortunately, early design choices stood in the way of this goal. In rushing to put the first Americans into space, the U.S. government shunned efforts underway during the 1950s to apply aircraft analogues to outer space. During that decade, the government began work on a number of hybrid aircraft capable of flight in both air and space. Flight engineers experimented with test aircraft that could rocket to the edge of space and studied airfoils called lifting bodies that could return from orbit like conventional airplanes. In 1958 the first X-15 experimental aircraft arrived at the Flight Research Center at Edwards Air Force Base in southern California. With the assistance of a B-52 drop plane, the X-15 rocketed to an altitude of fifty-nine miles, nearly half the distance attained by Alan Shepard's 1961 suborbital flight. Unfortunately for advocates of aircraft technology, the X-15 performed this feat more than one year after Shepard flew. By then, John Glenn had already orbited the earth. Anxious to push Americans farther into space, government officials rejected the aircraft model for space travel, instead adopting the capsule approach.[17]

If a group of people set out to design a spacecraft contrived to discourage public confidence in the ability of ordinary citizens to fly, that group could not have made a better choice than the space capsules designed for the early NASA flight program. The first astronauts traveled in conical objects that resembled warheads used to deliver nuclear bombs, from which their shape was actually derived. An image further removed from conventional winged aircraft would be hard to imagine. Like nose cones, the space capsules sat atop modified intercontinental ballistic missiles (ICBMs). Such missiles were never designed to take people anywhere and exploded with statistical regularity. Concerned with meeting their own deadlines, air force officers resisted changes in missile design that might have satisfied NASA safety standards.[18] In response, NASA engineers designed escape rockets that could separate a space capsule from a malfunctioning ICBM. Once in orbit, astronauts faced the difficult task of plowing back into the atmosphere at speeds in excess of seventeen thousand miles per hour. To prevent a human barbecue, NASA engineers installed an ablation shield that enveloped the capsule in a raging fireball as hot as the surface of the sun.[19] Back in the atmosphere, the astronauts could not fly home. Parachutes slowed their descent until the capsules dropped into the ocean, adding the possibility of drowning to the dangers of death by fire. This was not a craft designed to inspire confidence among ordinary ticket holders.

Tom Wolfe called the experience the human cannonball approach to space flight. Astronauts were launched into space by the brute force of military rocketry; they splashed into the ocean in a capsule they could hardly control. Lacking even the simple ability to guide their craft to a landing field, astronauts had to depend upon armadas of ships and aircraft to find them. In the space business, as Wolfe observed, recruitment was confined to those "whose profession consists of hanging their hides, quite willingly, out over the yawning red maw."[20]

In an attempt to put the Gemini astronauts on solid ground, NASA officials and their contractors tried to suspend the returning spacecraft from an inflatable paraglider. This would at least allow the bell-shaped capsule to skid back to Earth in what vaguely resembled an emergency landing. The paraglider looked and behaved like a paper airplane. Engineers planned to stow it in the *Gemini* capsule, where it would deploy during reentry, but tests produced so much wreckage that NASA abandoned the system in 1964.[21]

The methodology of early space flight was not contrived to make space flight a pleasant experience. Winged spacecraft capable of controlled landings and new propulsion technologies to ease the discomforts of launch were not available on the time line dictated by the Cold War. Once again, the history of aviation suggested that such obstacles could be overcome. Aviation had surmounted similar dangers as it matured. Danger was a close companion on the early exhibition circuit. Of the four pilots to sign two-year contracts with the Wright Exhibition Company, only one lived to fulfill it. Orville Wright injured himself seriously at the Fort Myer army trials after his propeller cracked, forcing a crash landing. His passenger, Thomas Selfridge, broke his skull in the impact and died. Dogfights and stunts did little to inspire public confidence in flying. "These accidents are dreadful," said aviation pioneer Glenn Curtiss. "Every time a flyer dies in a crash, we not only lose a precious life, but public confidence as well. How can we ever sell the idea that flight can be a safe means of travel?"[22]

Air flight improved as technology advanced. By the late 1920s, airlines were carrying passengers in transport aircraft like the Ford Trimotor. The fully enclosed cabins were a significant improvement over open-cockpit, two-seat biplanes. Even so, flying conditions remained primitive by modern standards. Like other unpressurized aircraft, the Ford Trimotor flew close to the ground, where turbulence inevitably occurred. Airlines kept a variety of implements on board to minister the needs of airsick passengers. In an emergency, customers could ultimately hang their heads out moveable windows. "Patrons immediately behind an airsick passenger learned to keep their windows closed," observed historian Roger Bilstein. "It was not unusual to hose out the entire interior of a plane after completing a turbulent flight."[23]

Bit by bit flying conditions improved. The Boeing 307 entered service in 1940, introducing passengers to the blessings of pressurized aircraft that could overfly troublesome storms. Tricycle-type landing gear improved takeoff efficiency and allowed passengers to find their seats without walking uphill. Piston-engine technology increased flight ranges eightfold while cutting costs in half. Gas turbine (jet) engines further increased speed and range.[24] Pioneers of space flight wanted to apply similar technologies to spacecraft. They wanted to put wings and landing gear on vehicles bound for the new frontier. The pioneers wanted ordinary people to fly and be changed by the experience. As with aviation, this would mark an end to the "barnstorming" era of space flight and the advent of technical maturity. It was a formidable challenge.

As if the desires of practical space-flight engineers were not enough, the promise of easy space flight was ground into the public consciousness by people producing works of fiction. Throughout the twentieth century, effortless space travel was no further away than the bookshelf or the television tube. By and large, producers of imaginative literature assumed that obstacles to space transportation would fall away in the face of modern technology. They certainly would be no more difficult to overcome than the challenges of aviation. Writers of fiction and popular science anticipated most of the important technological breakthroughs in space flight before those developments occurred. Famous spaceships traveled through the minds of ordinary people well before they ever soared in space.

To producers of imaginative literature, space transportation was frequently an afterthought to other, more important purposes. When H. G. Wells dispatched Martians to engage Earthlings in *War of the Worlds,* he gave little attention to the practical details of transporting them here. Flashes of hydrogen gas on the surface of Mars announce their departure; the first Martians arrive about ten days later.[25] Wells was more concerned with describing the clash between different cultures than explaining the mechanics of interplanetary transportation. The ultimate dismissal of transportation obstacles occurred when Edgar Rice Burroughs allowed John Carter to be transported to Mars as the result of a trance. Teleportation neatly avoided the technological obstacles to space travel and allowed Burroughs to get on with his story.

The first widely read writer of imaginative literature to deal seriously with the practical problems of space travel was Jules Verne. Verne's cannonball approach to space flight anticipated NASA's capsule system by nearly one hundred years. To dispatch his crew toward the Moon, he employed a large cannon capable of firing a nine-foot-wide aluminum shell. Verne understood that the force of the cannon blast would disturb the occupants of his bullet-shaped spacecraft and therefore installed a water bed in the spacecraft to absorb some of the shock. In fact, the crew's situation was much worse than Verne estimated. The force required to launch such a projectile toward the Moon would have subjected the occupants to g forces far in excess of those humans could endure. As Ron Miller, a chronicler of spacecraft history has observed, the g forces produced by the cannon blast would have reduced the spacecraft occupants to thin smears on the cabin floor.[26]

Having committed himself to a realistic treatment of space flight, Verne found himself in a literary quandary. Lacking similar artillery on the lunar surface, he could not imagine any realistic way to return his space travelers from the Moon. He therefore concocted circumstances that forced his lunar explorers to approach the Moon without actually landing on it. Verne then faced the difficult problem of a safe landing on Earth. He adopted a method whose credulity was bolstered only by the fact that NASA eventually adopted it: Verne's capsule splashed down in the Pacific Ocean to be recovered by the American navy.

Most writers who followed Verne treated space transportation as an afterthought to more pressing literary objectives. Where practical transportation barriers stood in their way, most writers simply dismissed them. This gave rise to inventions both practical and implausible. Writers of imaginative literature anticipated real developments such as the NASA space shuttle, but they also invented mysterious forces produced by devices such as antigravity machines and jumps into hyperspace.

The sum effect of these literary inventions helped convince an uncertain public that space transportation was a rather rudimentary affair that did not impose significant obstacles on the more exotic enterprise of interplanetary exploration. To persons growing up during the first half of the twentieth century, this was not an altogether unreasonable idea. Forty years after the Wright brothers' 1903 flight, pressurized four-engine aircraft shuttled passengers routinely between Europe and America. Was it so hard to imagine that similar changes might transpire by 2001, forty years after the first human flight into space? And what about the centuries beyond?

In the 1928 film *Frau im Mond* (*The Girl in the Moon*), producer Fritz Lang presented plans for a multistage rocket capable of flying humans to the Moon and back. The Lang rocket ship resembled in many details the multistage Saturn 5 rocket that delivered American astronauts to the surface of the Moon forty years later. It is hard to tell, given the history of the movie, whether fiction influenced science or science influenced the film.[27] In preparing the script, Lang drew on the technical advice of German rocket pioneer Hermann Oberth. Oberth in turn influenced Wernher von Braun, who directed the work that produced America's Saturn 5. Von Braun then used his charismatic expertise to promote the development of multistage rockets in popular outlets such as *Collier's* magazine, the movie *Conquest of Space* (1955), and the three-part Disney series released as the space age began.

The most impressive anticipation of near-term technology appeared in Arthur Clarke's and Stanley Kubrick's *2001: A Space Odyssey.* The space shuttle in that classic science fiction film flew just four years before President Nixon gave NASA the directive to start building a real one. Like many of the structures to fly from the minds of imaginative writers, Clarke's winged spaceship was impressively large. It measured two hundred feet from wing tip to wing tip. The space shuttle NASA eventually built had a wingspan of seventy-eight feet, still imposing compared to the thirteen-foot-wide Apollo command module. To boost the imaginary shuttle into orbit, Clarke employed a two-stage reusable system similar to the one NASA engineers hoped to build. After carrying the wing-shaped vehicle to the edge of space, the lower stage glided back into the atmosphere and landed at the Kennedy Space Center. The lower stage would be as easy to maintain as a modern jet liner, Clarke predicted: "In a few hours, serviced and refueled, it would be ready again to lift another companion."[28] NASA engineers set similar goals for their proposed space transportation system.

Features of the Clarke space plane made it nearly as comfortable as commercial airplane flight. Passengers suffered little sense of discomfort as the craft's steady source of propulsion generated only two gravities of force at its maximum point of acceleration, a vast improvement over Jules Verne's cannonball ride. On a second spacecraft, coasting toward the Moon, Clarke described an onboard toilet that spun like a centrifuge to provide relief for passengers during periods of microgravity flight. The spinning lavatory generated the equivalent of one-fourth g, enough to "ensure that everything moved in the right direction."[29]

Clarke predicted that a reusable space shuttle would reduce the cost of space flight substantially, a goal that proved easier to attain in works of fiction than in real life. Each flight of the Earth-to-orbit shuttle in *2001* cost a little more than $1 million, Clarke wrote. That worked out to $50,000 for each of the twenty seats, not exactly economy fare. Compared to the old Saturn 5 rocket, however, it was

Space enthusiasts predicted that space flight would become as easy as other twentieth-century transportation technologies, such as the airplane. Cartoonists and film producers helped promote this belief. Shortly after Charles Lindbergh crossed the Atlantic in his single-engine plane, Buck Rogers appeared in newspaper comic strips. The name "Buck Rogers" became synonymous with simple, inexpensive space flight.

a huge savings. In the uninflated dollars of its day, each launch-ready Saturn 5 rocket cost $185 million and never carried more than three people per flight.[30]

Although expert advice helped determine the shape of imaginary spacecraft, so did cultural trends. During the era of the underwater boat, many imaginary spacecraft resembled flying submarines. When automobiles developed fins, so did spaceships. Reflecting the fashion preferences of his day, Flash Gordon flew a spaceship designed in the art deco style.[31] That cultural trends created engineering problems hard to overcome is easily illustrated by the history of spacecraft design after World War II.

Contemplating rocket technology in the decade after the Second World War, filmmakers and story writers employed the familiar shape of the V-2 rockets that had crashed down upon European cities during that great conflict. The terror of military rocketry had seared the image of the V-2 into the public consciousness, intensified even more by the thought that nuclear warheads would be launched on vessels of similar design. When George Pal produced the classic *Destination Moon* in 1950, he dispatched his crew to the lunar surface in a rocket that looked like a German V-2. The most famous rocket shape of its time, the V-2 helped establish an image of aerodynamic grace among a generation of would-be rocketeers. The spacecraft was sleek and tall, shaped like a spindle with large fins at its tail, and rose from the Earth in a single stage so as not to compromise its graceful style.

A single-stage V-2–shaped rocket carried Earthlings from their doomed planet in the motion picture *When Worlds Collide.* In that film, scientists construct a space ark to carry forty humans and a menagerie of livestock to a planet where they can begin life anew. The technological challenges of such a planet hop are severe. The rocket must develop sufficient thrust to escape the gravity of Earth, decelerate through the atmosphere of the new planet Zyra, and deposit its heavy load on alien soil. Engineers decide to boost the spacecraft by propelling it along a mile-long slide using a rocket-driven undercarriage. Although the project directors worry about fuel, the rocket ship hops successfully between the two orbs.

Both spacecraft designs, although aesthetically pleasing, were technically infeasible. A fully fueled, single-stage V-2 had a maximum vertical range of about one hundred miles,[32] thousands of miles short of any extraterrestrial body, to say nothing of the propellant needed for landings or a return voyage. Screenwriters for *Destination Moon* solved this problem in an ingenious way. They developed an atomic power plant for the single-stage rocket, drawing upon the most promising energy source of their day. NASA actually tested a nuclear rocket engine nineteen years later, beginning in 1969. The NERVA (Nuclear Engine for Rocket Applications) project used a very hot nuclear core through which liquid hydrogen was pumped. The rapidly expanding gas provided the necessary thrust to propel the accompanying spacecraft. NASA canceled the project in 1972, citing budgetary cutbacks, but continued to study nuclear power as a source of interplanetary and translunar propulsion.[33]

Nuclear power seemed to hold the greatest promise for advancing propulsion technology. In 1969, NASA proposed the use of a nuclear-powered spacecraft to shuttle humans between a proposed Earth-orbiting space station and a proposed transit station circling the Moon. When White House officials sought new ideas on the best way to speed humans to the planet Mars, they called on a group of outside experts who in 1990 again laid out the virtues of nuclear power.[34] Popular culture helped reenforce the notion that more efficient propulsion technologies lay just over the horizon if only the government would take the risk and invest in them.

The vision of effortless space travel was expeditiously advanced with the 1977 release of George Lucas's *Star Wars,* one of the most popular science fiction films of all time. Of the numerous spacecraft depicted in this motion picture, none did more to reenforce the vision of accessible flight than the *Millennium Falcon.* Hoping to escape his frontier outpost on Tatooine and travel to the Alderaan system, Luke Skywalker is introduced to Han Solo in a rear booth of the Mos Eisley

Cantina. With the assistance the Wookie Chewbacca, Solo works as an independent freighter pilot and owner of this wondrous spaceship. The character of Han Solo, played adroitly by Harrison Ford, combines the bravado of a mercenary pilot with the technical skill of a garage mechanic, the latter recalling the "hot rodders" of the 1950s who worked to improve the performance of their own machines, stock automobiles whose mechanical alteration could vastly improve the capabilities of engine and drive train. The alterations permitted unlicensed drag racing and legendary flights from law enforcement authorities.

As portrayed in *Star Wars,* personal spacecraft technology is so simple that Solo can maintain and improve the *Millennium Falcon* himself. "She may not look like much," Solo admits, "but she's got it where it counts, kid. I've made a lot of special modifications myself." Given its patchwork appearance, Luke Skywalker doubts that the custom-modified spacecraft can even fly. "What a piece of junk," he replies. Solo is indignant. "It's the ship that made the Kessel run in less than twelve par seconds!" he responds. "I've outrun Imperial starships and Corellian cruisers."[35]

The success of the *Star Wars* tales was due in large part to Lucas's ability to take familiar images from American popular culture, such as stock-car bravado, and lay it in an intergalactic setting. Audiences instinctively respond to such imagery. The film helped solidify the notion that the space frontier would provide a tableau for the completion of old fantasies. In this sense, the *Millennium Falcon* represents the ultimate fantasy of aerospace pioneers—a personally accessible spacecraft that is relatively easy to operate and maintain.

In one major respect, the *Millennium Falcon*'s capabilities exceeded the hopes of contemporary rocket engineers. Solo's ship sped through space by jumping to hyperdrive. As Solo points out, "Travelling through hyperspace isn't like dusting crops."[36] To cross the great distances imposed by galactic travel, writers of imaginative literature have adopted devices that circumvent the laws of physics as understood by twentieth-century scientists. According to Einstein's physics, velocity and mass are interrelated. The faster one travels, the heavier a spaceship becomes. This effect is not noticeable at modest speeds but begins to intrude upon galactic travel as one approaches the speed of light. At such velocities, spacecraft mass grows rapidly. As it becomes heavier, the spaceship requires more energy to push it. An infinite amount of energy is eventually required to accelerate a spacecraft to the speed of light—about three hundred thousand kilometers per second. Simply put, there is not enough energy in the universe to propel a spacecraft that fast. Nature appears to have established a cosmological speed limit for spacecraft traveling between the stars.[37]

In works of imagination, however, this constraining limit disappears. Visionaries have developed a variety of techniques for approaching and even exceeding the speed of light. To many, the speed of light is no more insurmountable than prior obstructions to human flight. E. E. Smith presented his classic *Skylark of Space* only one year after Charles Lindbergh's 1927 crossing of the Atlantic. Chuck Yeager would not break the sound barrier for another nineteen years, but the hero of the *Skylark* tale crosses 237 light-years in a mere twenty-four hours. *Skylark* began its serialization in a 1928 issue of *Amazing Stories* that also featured the first *Buck Rogers* tale.[38] Both served to launch the modern "space opera" with its vast interplanetary settings, formidable weaponry, and awesome spacecraft. Such spacecraft crossed innergalactic distances as easily as airplanes spanned the oceans.

Images of interstellar travel in fictional literature helped generate scientific interest in advanced propulsion systems. A figure no less imposing than NASA's associate administrator of Manned Space Flight argued that humans would eventually develop the means to span the galaxy. "The future of Mankind," George Mueller argued in 1984 after his retirement from NASA, "lies in populating first,

No fictional spacecraft did more to heighten expectations about interstellar travel than the starship *Enterprise.* First introduced in 1966 as part of the *Star Trek* television series, the various configurations of the USS *Enterprise* used antimatter engines to drive the spacecraft at "warp" speeds. ("Star Trek" ©1966 by Paramount Pictures)

the Solar System, from there developing the technology to visit the stars and to begin populating the Universe as we now know it." Mueller called it the human destiny and maintained that it could be done. He insisted that history would vindicate his view. In the 1930s, most people treated serious proposals for flights to the Moon as science fiction. Thirty years later humans had accomplished that feat. Now experts were contemplating the technical requirements for interstellar flight. "Once again," Mueller observed, "the idea will be widely regarded as science fiction. But I am convinced that the more we learn, the more the question becomes not *whether* we can go to the stars, but *when.*"[39]

Mueller believed that interstellar travel would be expedited by propulsion technologies that could be developed soon. Such technologies crossed the boundary between science fact and science fiction. Some of the proposals were primitive, such as the notion that an interplanetary spacecraft could be propelled by exploding small atom bombs behind it. Following the landings on the Moon, the British Interplanetary Society proposed the use of nuclear fusion. The promoters of Project Daedalus, named after the mythical Greek craftsman who built wings for himself and his son Icarus to escape imprisonment in the Labyrinth, envisioned a large reaction chamber into which small pellets of deuterium and helium 3 would be thrust. Beams of electrons would cause the pellets to fuse, producing the power that drove the spacecraft. Although only a small percentage of the pellet mass would fuse, the power produced would be sufficient to drive a large spacecraft to just over 12 percent of the speed of light. With such a propulsion system, a robotic probe could conduct a flyby of Barnard's Star in forty-five years flight time.[40]

One of the most promising methods of propulsion technology involves antimatter drives. While this sounds as fictional as antigravity devices, the concepts are physically real. Antimatter possesses physical properties the opposite of normal matter. The protons in the nucleus of an ordinary atom carry a positive electrical charge, whereas antiprotons carry a negative one. When introduced into the same chamber, protons and antiprotons annihilate each other. All the antimatter (and accompanying matter) is converted to energy, producing a spectacular burst of power.[41]

Substantial but not insurmountable obstacles exist to the development of actual antimatter engines. Small quantities of antimatter are routinely created in large research laboratories. Once produced, antimatter is notoriously difficult to store because it reacts violently with ordinary substances. Storage is not impossible, however. Research scientists have suspended antimatter in magnetic fields where it does not touch ordinary matter. As a rocket fuel, antimatter could take the form of small balls of frozen antihydrogen channeled into a magnetic nozzle. According to one calculation, a one-ton payload could be propelled using an antimatter drive into an orbit in the Alpha Centauri system (our nearest stellar neighbor) in just twenty-five years.[42]

If this sounds like science fiction, it is. Although hundreds of scientific papers have been written on the technology of interstellar travel, most people learn about exotic propulsion systems through works of imagination. The most famous spaceship to fly through twentieth-century fiction uses an antimatter drive. Publications directed at the infinite curiosity of *Star Trek* fans describe the operation of various *Enterprise* spacecraft in delicious detail. The technical manuals for the various *Galaxy*-class spacecraft may be prophetic in some particulars. According to the writers, antimatter is generated with relative efficiency at major Starfleet fueling facilities using a combination of solar and fusion power. It is produced as antihydrogen in slush form. Magnetic fields continue to be the preferred method of storage. These details are indistinguishable from known scientific principles. The

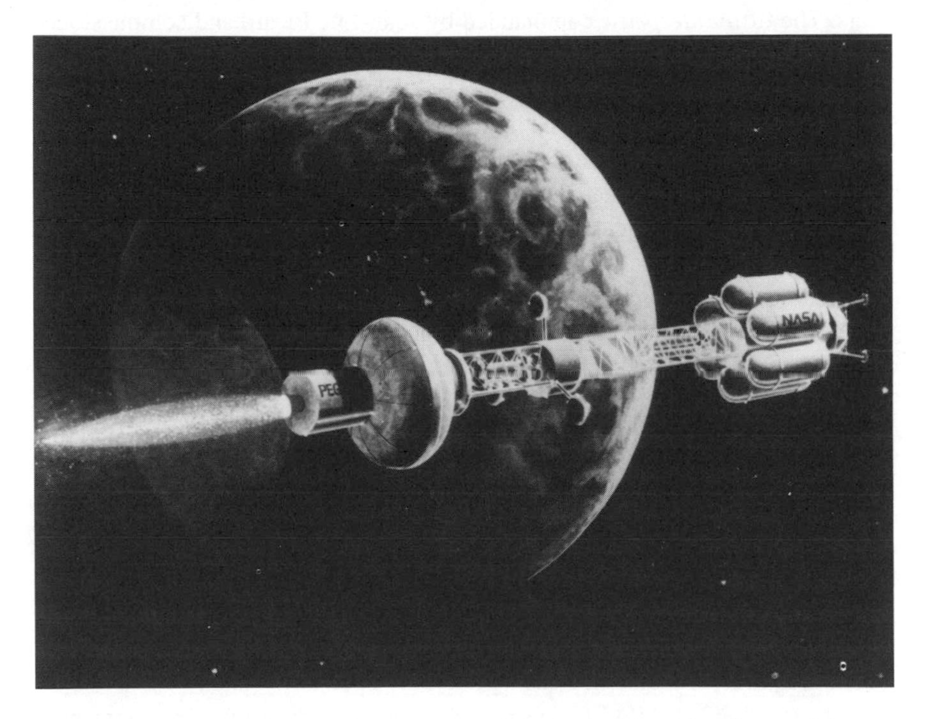

Rocket scientists experimented with fantastic forms of propulsion. Electric propulsion vehicles using magnetoplasmadynamic thrusters such as the one depicted here were studied during the late 1980s as a means for moving cargos to Mars. Rocket engines powered by small fusion-bomb explosions or liquid hydrogen pumped through a very hot nuclear reactor core gave added hope to enthusiasts who dreamed of speeding through the solar system and beyond. (NASA)

combustion chambers on the *Enterprise* spacecraft, however, are not. According to its creators, the combustion nozzles use a fictitious material known as "dilithium crystal," which antimatter passes safely through, eliminating the need for magnetic injectors.[43]

From these brushes with reality, *Star Trek* writers venture into a familiar fantasy. As in earlier galactic sagas, writers plot paths through the speed-of-light barrier that scientists cannot find. In the *Star Trek* series, this takes the form of "warp" speed. Theoretical work underlying superluminal flight was completed by the middle of the twenty-first century, according to the writers. The first prototype capable of exceeding the speed of light for a brief instant was produced in 2061. The key to avoiding the limits of physics, the writers speculated, lay in the process of nesting layers of acceleration against each other with exponential effects. The first practical engines to use this method helped establish the Alpha Centauri colonies. The spacecraft crossed the span of 4.3 light-years in just four years. The fifth *Enterprise,* commanded by Jean-Luc Picard and commissioned in 2363, can maintain a cruising speed of warp 6, or 392 times the speed of light. For brief periods of time it can accelerate to velocities four times that fast. As the Milky Way is 100,000 light-years across, a complete transit at cruising speed would require some 250 years. Limits on the velocity of the *Enterprise* arose more from story lines than technology, the writers explained. Making the ship go too fast "would make the galaxy too small a place for the *Star Trek* format."[44]

Who knows how much of this will come true? At the outer reaches of scientific investigation, inquiry merges with imagination. Faster-than-light spacecraft, antigravity machines, and transporters ("beam me up, Scotty") violate the known laws of physics. Antimatter engines, ion drive accelerators, and hyperspace do not. The casual reader might be forgiven for confusing the real with the fictional.

Before their lift-off in an imaginary rocket to Mars, visitors to Disneyland are informed that the speed at which the spacecraft travels was considered science fiction only a few decades earlier.[45] Imaginative voyages help create challenging expectations. As John Mauldin observes in his serious review of interstellar travel, "When a theme as common as travel to the stars is taken for granted in a large but fictional literature, the public tends to assume that the concept will become certainty, even that particular methods will be developed fairly soon."[46]

In the late 1960s NASA officials took the first step toward the development of spacecraft in which ordinary people could fly. NASA proposed and won approval to build a fleet of four space shuttles that could travel to and from an Earth-orbiting space station "in an airline-type mode." The shuttles were envisioned as part of the larger Space Transportation System that would expedite travel deep into space. For use beyond the range of the space shuttle, NASA hoped to deploy a chemically fueled space tug and a nuclear-powered transit vehicle.[47] Only the space shuttle was approved.

Expectations for the new space shuttle were high. "Toward the end of the Seventies you will no longer have to go through grueling years of astronaut training if you want to go into orbit," predicted Wernher von Braun. "A reusable space shuttle will take you up there in the comfort of an airliner." The shuttle was designed to make space flight accessible to the common folk. "I'm convinced that by 1990 people will be going on the shuttle routinely—as on an airplane," said NASA planner Robert Freitag.[48] Although the shuttle would take off like a rocket, it would land like an airplane. This was an important milestone in making space travel seem familiar. "When I was a kid reading *Buck Rogers*, the spacecraft all looked like bullets or saucers, with sweeping fins and fancy tail skids," said astronaut Michael Collins. "We are beginning to see Buck's dream emerge in the squat but elegant space shuttle."[49]

Significantly, NASA engineers planned to give the shuttle wings. They sought to take NASA out of the cannonball business and put it back into the realm of conventional flight. For those who began their careers in the NACA, this was particularly important. As the story goes, Johnson Space Center director Bob Gilruth approached spacecraft designer Max Faget and told him to "get off this blunt-body, parachute stuff. It's time we thought of landing on wheels."[50] Wings would allow the spacecraft to land in the manner familiar to all pilots. Faget wanted to put real wings on the shuttle, the type that stick straight out from an aircraft fuselage. At lower altitudes, straight wings would give the spacecraft an enormous amount of lift. Such a spacecraft would land practically like a glider, at the relatively low speed of 150 miles per hour. A shuttle with straight wings could land at a conventional airport, like New York's Kennedy or Washington's Dulles. As a safeguard against trouble, it could glide around and land slowly.

Airplane-type landings would help make the space shuttle accessible to ordinary people. Astronauts falling back from space in a blunt-bodied capsule faced deceleration forces in excess of 7 g's, to say nothing of the indignity of landing in a large body of water. Passengers on the shuttle would experience forces less than 3 g's and would depart the spacecraft on a concrete runway. John Young, who flew the first shuttle into space, announced, "If you're alive and breathing, you can fly on the shuttle."[51]

Most important, the shuttle would be reusable. The need to drop rocket ships into the ocean after their first and only use frustrated efforts to make space flight more accessible. "There's no way that you can make a railroad cost-effective," observed one NASA executive, "if you throw away the locomotive every time."[52] The space shuttle, as designed, could be used again and again, cutting the cost of space transportation significantly. Cost expectations for reusable transportation systems approached hyperbole. NASA administrator Thomas Paine predicted that "by 1984 a round trip, economy-class rocket-plane flight to a comfortably appointed orbiting space station can be brought down to a cost of several thousand dollars." The cost of a trip to the Moon would fall "to the $10,000 range."[53] Officially, NASA committed itself to reduce the cost of Earth-to-orbit transportation "by a factor of ten." At that time, remotely controlled rockets cost about $1,000 per pound to deliver their payloads to low Earth orbit. The nonreusable Titan III C, for example, delivered its twenty-three-thousand-pound payload to low Earth orbit at a 1972 cost of $24 million, or about $1,000 per pound.[54] NASA officials estimated that a reusable space shuttle could do the same job at one-tenth the cost. That required NASA to develop a shuttle with a fifty-five-thousand-pound payload that could be launched for no more than $5.5 million per flight in 1972 dollars. Officially, NASA committed itself to an estimate of $10 million per launch—still impressive compared to the cost of launching vehicles such as the Saturn 1B.[55]

Given its low cost, NASA expected that the space shuttle would become the common carrier to space. Various publications called it a "cargo plane" and a "space truck," and an aerospace industry publication predicted that the shuttle "will eliminate the need for expendable launch vehicles."[56] Both NASA and the U.S. Department of Defense agreed to use the shuttle as their primary vehicle for placing payloads in orbit. NASA officials tried to convince European nations to forgo development of the *Ariane* rocket, a competing technology, on the grounds that the shuttle would make nonreusable rockets obsolete.[57]

The shuttle would prove so popular, NASA officials predicted, that it would practically pay its own way. To make the shuttles cost-effective, NASA had to launch them at least twenty-five times per year. One study set the figure as high as fifty-two flights per year, or one launch every week. Most of those launches would be for paying customers—other government agencies, such as the Department of Defense, commercial firms, and other nations wanting to put payloads into space.[58] The money NASA received from these groups would help defray the cost of operating the shuttle. Spreading the fixed costs of shuttle operations over a large number of flights helped reduce the per launch price, which in turn would attract more customers. On paper, the plan looked convincing.

Critics of the shuttle program failed to construct a persuasive rebuttal, and President Nixon approved development of the system in 1972.[59] NASA's expectations seemed reasonable, given popular beliefs at the time. Popular science and works of imagination had laid the groundwork for the belief in cheap, reusable spacecraft. In addition the airplane industry had made enormous advances by 1972, and it was hard to believe that similar changes did not wait in space. NASA's own record of performance in sending humans to the Moon, a task believed infeasible only a few decades earlier, encouraged confidence in the new goals. NASA officials played on those popular expectations by setting extraordinary goals for the shuttle program.

NASA's expectations melted away in the face of reality. By the mid-1980s, a few NASA watchers had concluded that the shuttle would not meet its original goals. It took the explosion of the space shuttle *Challenger* to communicate that message to the nation at large.[60] Flight procedures were anything but routine. The shuttle orbiter, although reusable, was much too complicated to permit a launch every week or two. The largest number of launches NASA achieved in any one year was nine in 1985, far short of the expected level of twenty-five to fifty-two. Flight costs remained exceptionally high. The $10 million per launch estimate escalated to $57 million and then to $225 as NASA gained flight experience. In 1992, NASA set the average recurring cost of each flight at $412 million. Even adjusting the $10 million figure to account for inflation, which produces a per launch cost of $40 million in 1992 dollars, experience sat well above original cost goals. In reality, NASA found itself operating a spacecraft that was just as expensive to fly as the old expendable rockets the shuttle had replaced. Pound for pound, the shuttle cost about the same per flight as the Saturn 1B that had lifted American astronauts to the *Skylab* orbital workshop during the 1970s. After the *Challenger* accident, NASA abandoned the goal of using the shuttle as the common carrier for all space missions large and small.[61]

In one respect, the shuttle did meet expectations. It flew like a spacecraft and landed like an airplane, an impressive accomplishment. Its airplanelike configuration allowed ordinary people to fly in space. Unfortunately, the first civilian to accept this challenge, Christa McAuliffe, boarded the space shuttle *Challenger.* NASA abandoned the civilian-in-space program shortly thereafter.

From the start, the effort to build an easy-to-fly spacecraft was in trouble. NASA engineers wanted to build a fully reusable shuttle, mating the shuttle orbiter to a winged booster as large as a Boeing 747 jumbo jet. One crew would fly the booster while a second would pilot the orbiter. After lifting the orbiter off the surface of the earth, the first crew would return to a landing strip near the launch site, where the booster would be refurbished for the next flight. The orbiter would propel itself into space and eventually return in the same manner, landing like an ordinary airplane.[62]

The fully reusable design was very expensive. One internal NASA memo set the initial development costs at $10 to $13 billion. This did not fit under the appropriations ceiling that White House aides imposed on NASA during the early 1970s. In order to get the shuttle program approved, NASA executives proposed a shuttle design with startup costs estimated at only $5.5 billion, which required NASA to substitute two liquid-fueled boosters for the reusable first stage. The boosters would carry the orbiter to an altitude of about forty miles and fall away, after which parachutes would break their descent to Earth. NASA planned to reuse the boosters after retrieving them from the ocean. Flight engineers were understandably nervous about reusing rocket engines that had been dunked in salt water and were discouraged that they had lost their airplanelike first stage.[63]

Worse news followed. Under increasing pressure to cut shuttle costs further, NASA executives abandoned the use of boosters powered by liquid propellants and substituted solid fuel. Whatever their shortcomings, liquid-fueled boosters had one major advantage: they could be shut down if problems arose. Solid-fueled boosters, on the other hand, would burn continuously for the first two minutes of flight. Once the solid rocket boosters were lit, the shuttle was going to go somewhere for two minutes.[64]

Even the design for the shuttle wings disappeared. Engineers at the Johnson Space Center preferred straight wings, which would provide more lift during the crucial landing phase. Defense Department officials, however, favored delta-shaped wings. Landing a spacecraft with delta wings is tricky. The spacecraft comes in much faster, dives in order to pick up speed, then flares up seconds before landing. A delta-winged orbiter needs longer runways and cannot land at conventional airports; its higher landing speeds put extra pressure on failure prone brakes. Critics complain that it has the gliding characteristics of a pair of pliers.

Practical experience showed how hard the development of advanced spacecraft could be. NASA wanted to develop a reusable Earth-to-orbit shuttle that could be flown frequently at one-tenth the cost of conventional launch vehicles. Although the space shuttle is a technological marvel, being the first spacecraft to reenter the atmosphere and land like an airplane, it proved neither cheap nor easy to fly. (NASA)

Nonetheless, the Defense Department insisted on delta wings and NASA needed their support in order to get the project approved.[65]

During the 1960s, NASA flight engineers understood that going to the Moon would be difficult to do. With sufficient funds, they were able to manage the risks and complete the mission on time. The space shuttle created a different challenge. Funds were short and space advocates announced that a reusable spacecraft would be easy to fly. Making such an announcement was relatively easy to do, given the image of spaceflight in the popular mind. Executing it proved much harder. In reality, cheap and easy were on a collision course with each other. Development of safe, reliable flight technologies for orbital flight requires massive subsidies, experimentation, and the ability to replace mistakes with new designs. That costs money, which NASA did not have.

Rather than announce that an easy spacecraft would require large development outlays, space advocates perpetuated the myth that the era of cheap and easy space flight had arrived. Dissent was largely ignored. The public, as a result, was unprepared for a catastrophe. "When the shuttle prangs," one realist warned, "it will be the media event of the decade."[66]

In producing a low-cost spacecraft, NASA officials took a risk. The risk was not difficult to calculate. In the wake of the *Challenger* accident, NASA officials announced that the space shuttle contained approximately 750 objects that, if they failed, could endanger the life of the crew and set the overall probability of a catastrophic failure at between 1 and 2 percent.[67] Flying the shuttle is worth the risk, but it is not something the same person would want to do every day. It is clearly not as safe as flying a commercial airliner. "Anyone who has lived with large rocket engines," observed astronaut Michael Collins, "understands that . . . a thin and fragile barrier separates combustion from explosion."[68]

Throughout the early 1980s, NASA officials and their allies promoted the shuttle flight experience as something increasingly routine. Expressions of confidence increased as one rose up the government hierarchy. After hearing an estimate to the effect that the probability of a catastrophic failure was only one in one hundred thousand, physicist Richard Feynman asked incredulously, "What is the cause of management's fantastic faith in the machinery?" Feynman was a member of the commission that investigated the *Challenger* accident. As he pointed out, probabilities on that order would allow NASA to launch one shuttle each day for three hundred years expecting to lose only one.[69]

The space shuttle was sold as a means of attaining low-cost, routine access to space. This would not have been possible without imagination. No one successfully stepped forward to deflate this expectation because the image of routine flight was so strong. Reliability standards of incredible magnitude had been attained by the airline industry, the analogy on which the shuttle program drew. Works of imagination continuously exposed the American public to the dream. To achieve the goal of low-cost access to space, NASA would have had to change its way of doing business. It would have had to invest heavily in new technologies.

Imagination helps people maintain beliefs that in some cases are not objectively true. The choice between wretched reality and comfortable fantasy is difficult to make, especially when reality requires arduous organizational change. This is not to say that space advocates formed a conspiracy to mislead the public. To the contrary, the evidence suggests that people within the government believed their estimates—especially those who were removed from the day-to-day work of flying objects through space. Many of the misgivings that NASA executives possessed about their ability to produce a low-cost, easy-to-fly spacecraft disappeared when the campaign to develop the shuttle got underway. For a brief period, space advocates were allowed to hope that the dream of easy space flight had come true.

LIFE ON EARTH

> The most significant achievement of that lunar voyage was not that man set foot on the Moon, but that he set eye on the Earth.
>
> —Norman Cousins, 1976

As politicians fondly point out, not a single tax dollar appropriated to the National Aeronautics and Space Administration has ever been spent in space. Except for the footprints and machinery left behind, the impact of the space program has been wholly confined to Earth. The civil space program has had a far greater impact on terra firma than on the extraterrestrial realm.

The early space program was created by a generation raised under the specter of the Great Depression. To that generation of Americans, technology offered the best hope of escaping the poverty that had restrained their parent's dreams. By the early 1940s, Americans of that generation had come to equate technology with social mobility, economic progress, and new products. Advertising and product display had drummed that message into the American consciousness by the time the space race began. In American culture, the space program became the ultimate manifestation of the connection between progress and technology. Space exploration promised to transform life on Earth, creating social and economic benefits unknown to previous generations. Spin-offs from space seemed to do just that. From electronic computers to communication satellites, government-supported technologies helped create a cornucopia of consumer products that ushered in the age of personal consumption.

Promoters of the consumer culture promised that humans would master the industrial arts. Advertising campaigns of the 1950s invariably depicted people in command of consumer products, a salving message at a time when new technologies seemed to overpower efforts at comprehension or control. Advertisers knew they could not sell products that resisted human control, a theme advocates of new government programs appreciated as well. For reasons that had little to do with the scientific issues involved, humans had to play a central role in the exploration of space. The "man-machine" debate had to be resolved in favor of humans in space, if for no other reason than to make space technology appear benign.

Based on the cultural norms of the day, advertisers invariably depicted the person in charge as a man. The debate over the role of automation in astronautics was resolved firmly in favor of men, in stark contrast to the emancipating promise of aviation, in which women appeared to play a central role. In fostering an image of space technology under masculine control, exploration advocates created a message that helped alienate a substantial sector of the American population.

In another, more important way, NASA helped lay the foundation for its own transformation. To be sure, modern technology created a cornucopia of consumer products unimaginable during the Great Depression, but space technology also allowed humans to view the world from afar, as a small blue-and-white ornament suspended in the cosmic void. Standing next to a rocket ship at Cape Canaveral, the future looks limitless. No problem seems too large for technology to conquer. From a distance, however, Earth looks small and fragile, its resources exhaustible. The view of Earth from afar helped to foster a new environmental consciousness, a movement to restrain technology in order to preserve the only site on which humans comfortably live. This in turn served to redirect the space program away from the Moon and planets, its original goals, and back toward the needs of planet Earth.

Just two decades before the era of space travel began, to a generation of Americans struggling through the Great Depression, the economic future indeed looked grim. Thirteen million Americans were out of work. Net income for American manufacturers had fallen by more than 66 percent. The gross national product rested at a miserly $75 billion, and federal tax revenues had dipped below $2 billion annually. This was hardly a scenario capable of inspiring confidence in large technological ventures such as a voyage to the Moon.

Many people believed in socialist reform as the most hopeful solution to economic depression. Socialists promised to spur recovery by wresting economic control from private entrepreneurs and placing it in the hands of government planners. In the United States, Franklin Roosevelt offered an economic New Deal that featured social security, price stabilization, and public-works employment—not socialism per se but sufficient to frighten capitalists of the day.

To its critics, socialism and its various manifestations proffered a dreary future. Writers such as George Orwell and Aldous Huxley helped popularize the view that government-created utopias would retard individual initiative and prevent economic growth. Orwell's *1984* depicted a totalitarian government ruling an ever-increasing population struggling to divide an economic pie of diminishing dimensions. In Orwell's novel, socialist reform offered little more than equality of misery.[1]

To combat government interference in its various forms, business leaders fashioned an alternative to socialistic doctrines. They promoted a consumer society blessed with a cornucopia of personal goods in which economic growth would be propelled by corporate initiatives satisfying a constant desire for new products. Technology and redesign would create the goods. The private sector would manufacture them, generating the payrolls that paid the bills. During the 1930s business leaders set out to sell this alternative to a Depression-weary public. Acceptance of this vision created the culture of consumption that dominated the U.S. economy in the decades that followed, and elements of that culture nurtured support for the U.S. space program.

Advocates of this consumer society had to sell the American public on the idea that people should buy goods they probably could not afford and often did not need. Business leaders employed advertising principles and product-display techniques in deliberate campaigns to promote this idea. For more than half a century, new products such as electric lights, telephones, and industrial machines were displayed at a succession of expositions and fairs. Building on this tradition, corporate and civic leaders promoted the vision of the consumer society at the 1939–40 New York World's Fair: "The fair's streamlined buildings rose from the ground like a utopian dream come true. . . . Its dazzling array of technological marvels attracted millions of Americans, eager to replace their memories of poverty and breadlines with visions of prosperity and material comfort."[2] Fair organizers displayed the latest consumer goods in theatrical fashion: automobiles, refrigerators, dishwashers, cameras with color film, and the most marvelous innovation of all, the television set. Exhibits entertained fairgoers with the vision of a consumer society in which the average family could live like kings.[3]

To entice Americans to consume, corporations repackaged products in creative ways. Market specialists had learned during the 1920s that product appearance could be manipulated in such a manner as to affect sales. Corporations hired industrial designers, who employed artistic techniques and new materials to repackage consumer products. In an effort to entice Americans to shop their way out of the Depression, designers altered the shape and color of toasters, vacuum cleaners, radios, and other consumer goods. Invariably the items became sleek, with flowing curves and shiny surfaces, designs derived from the shapes of airplanes, boats, and trains. "The streamlined form," author Donald Bush notes, "came to symbolize progress and the promise of a better future."[4] It implied the ability to speed ahead, away from the adversities of the Depression, toward a more luxurious life-style based on efficiency and technology. Industrialists further learned that they could render a perfectly useful product obsolete simply by changing its outward form, without the introduction of new technology. Planned obsolescence, often based on nothing more than stylistic changes, became part of the engine that ran the consumer society.

As a social phenomenon, the U.S. space program reenforced many of the messages that drove the consumer society in mid-twentieth-century America. This was a key factor in its early popularity. The space program promised more than the exploration of space, it offered a better world through technology. Speaking of the organization he led, NASA administrator Daniel Goldin asserted that the agency was "the one organization in American society whose whole purpose is to make sure our future will be better than our past."[5]

The space program appeared at a time when optimists wanted to believe that technology would create a better world. In formulating plans for his first theme park, Walt Disney sought to reenforce the prevailing notion that the future would be better than the past and that progress generally worked to further the common good. Disneyland opened to the public in 1955 with as much public interest as any other social occurrence of that decade. The Main Street exhibit through which visitors passed after entering the park represented a sanitized version of America's past, the way that Disney wanted to remember his boyhood home in Marceline, Missouri, shortly after the turn of the century. Tomorrowland exhibits symbolized the future of America some thirty years hence. The two periods marked the time of passage between the last and the next visit of Halley's Comet, a major astronomical event. What would America look like when Halley's Comet returned? "Tomorrowland was conceived with the unbridled optimism of a thriving post-war industrial society," Disney's associates wrote. "In Walt's vision, the future of 1986 would be a time of automation, a time of leisure and a time of limitless opportunity. Anything you dreamed could be created." Humans would fly to the Moon in giant rocket ships, travel freeways in sleek new sports cars, and navigate waterways in plastic boats. A Monsanto "House of the Future" displayed the latest interior design techniques. In 1967 Disney added the "Carousel of Progress," in which the General Electric Company traced the effect of electric appliances on an imaginary American household.[6]

Disney had run out of money while constructing his magic kingdom. To finish Tomorrowland, he called on corporate sponsors to prepare exhibits depicting the future. The message that Disney and his corporate sponsors transmitted was unambiguous: progress through technology would better the lives of average Americans. "Progress is our most important product," one corporate sponsor proclaimed. No references to the darker side of technology appeared, such as the nuclear-arms race or industrial pollution. The tiny gasoline-powered Autopia sports cars underwritten by the Richfield Oil Company chugged along Disney's miniature freeway at eleven miles per hour, adding to the California smog. Little publicity was given to the problems that plagued the guidance system for the plastic boats, a fault that caused park managers to remove the ride after the 1955 Christmas season. Walt Disney did not want to scare park visitors with foreboding visions of technology. For thrills, he sent visitors to the 146-foot Matterhorn Mountain bobsled ride.

To entice consumers to purchase new products, manufacturers had to portray friendly technology. No snapping toasters nor exploding microwaves appeared to frighten prospective customers. Such metaphors were left to critics of the consumer society. Through advertising and product display, American industrial leaders revealed gadgetry that was designed to beguile rather than frighten consumers. Early advocates of space flight reenforced this promise of benevolent technology. Cutaways of space stations and rocket ships revealed an dazzling array of technical gadgets, many of which later became consumer goods. An October 1952 article in *Collier's* magazine on a prospective flight to the Moon revealed astronauts at work inside a four-story lunar spaceship. In the kitchen area, astronauts prepared meals by removing a precooked delicacy from a freezer and inserting it in a "short-wave food heater," a forerunner of the modern microwave oven.[7]

When the space age began, government officials were obliged to deliver the goods. Corporate leaders had promised that investment in new technologies would produce new consumer products, and space exploration provided the proving ground for that claim. Demonstrating the "spin-offs from space" became an important justification for the newborn U.S. space program. During World War II, military research had produced an assortment of new products, including nuclear-powered generating plants, jet aircraft, radar, and penicillin. Fighting world wars as a means of enlarging consumer choice was a hard sell, however. Space research, on the other hand, offered a substitute that could produce much good with apparently little harm. If the space program had not delivered, it would have undercut the claim that investment in technology advances consumer comfort.

Critics of the space program ridiculed the spin-off argument, holding up as examples the $23 billion frying pan and Tang, the drink that went to the Moon. Teflon, a substance designed to smooth the path of objects reentering the atmosphere, was developed as part of the military space effort and later used to coat kitchen utensils, a major improvement over the iron frying pan. Tang was a powdery substance that when mixed with water produced a thin orange drink. While useful for space voyages, the product was in fact developed prior to the start of the civil space program. Friends of the space program refuted the critics. "Every time someone operates a computer, makes a long-distance call, watches television or uses an automatic teller machine, the benefits of space technology are being felt," argued Administrator Goldin. Space spin-offs had led to a cornucopia of consumer goods, its supporters said. "Every time someone undergoes a CAT scan, has arthroscopic or laser surgery or enters intensive care at a hospital, he or she benefits from NASA work."[8]

President Lyndon Johnson, in an offhand remark that revealed state secrets, observed that the knowledge gained just from space photography alone was worth "ten times what the whole program has cost." Johnson was referring to the development of reconnaissance technology, which allowed the United States to ascertain the exact size of enemy forces and monitor arms agreements. Estimates of enemy forces using indirect means, such as covert operatives, produced greatly exaggerated estimates. Had satellite technology not corrected those figures, the U.S. government would have diverted vast sums of money to defend against threats that did not exist.[9] Indirectly, in that sense, space research enhanced the funds available for domestic investment.

Had the space program merely produced tangible benefits, it would not have been as popular as it became. Spin-offs from space certainly enhanced the rationale for government investment, but they were not the only force propelling the American space effort onto the center stage of public interest. Like other products at that time, the American space program had to be sold. It competed for public attention with a variety of initiatives, both public and private. In part the early space program attracted attention because it embodied so well the subliminal messages motivating the consumer society.[10]

Presentation of the space program followed a pattern already well established by American merchants by the time the space effort began. Following the postwar spending binge, American manufacturers had found themselves with huge inventories and lagging demand. Planned obsolescence helped maintain public interest in consumption. "Our whole economy is based on planned obsolescence," one industrial designer observed. "We make good products, we induce people to buy them, and then next year we deliberately introduce something that will make those products old fashioned, out of date, obsolete." Television and radio allowed manufacturers to communicate product changes with unprecedented speed. By the late 1950s, when the space race began, Americans had become accustomed to the frequent introduction of products that only appeared to be new.

Inadvertently, the early space program adopted this pattern, regularly replacing rockets and spacecraft with new models every few years. In 1961 Alan Shepard rode a Redstone rocket on his suborbital flight from Cape Canaveral, the following year John Glenn traveled into orbit on a modified Atlas launch vehicle. The

Titan 3 rocket made its human space-flight debut in the spring of 1965 with the first manned Gemini flight, followed in 1968 by the Saturn 1B and later that year by the launch of the first Saturn 5 with humans on board. The single-seat *Mercury* capsule was replaced by the two-person *Gemini* spacecraft, which in turn gave way to the more spacious Apollo spacecraft with room for three. Once on the Moon, astronauts introduced a variety of gadgets, among them the first four-wheeled lunar roving vehicle and the first live color television broadcasts from the lunar surface, including the remotely controlled transmission of a lunar lift-off.

Automated probes produced similar variations. Fleeting closeup photographs from probes dispatched on a crash course with the Moon were replaced by high-quality photographs from orbiters and the first *Surveyor* spacecraft to make soft landings. Brief flybys of Venus and Mars were succeeded by spacecraft that orbited nearby planets and eventually landed on them. Fuzzy photographs from the *Pioneer* spacecraft that flew by Jupiter and Saturn were followed by *Voyager* images of incredible detail.

During the early years of the space program, NASA was constantly doing something new, thus maintaining interest in the effort among the public at large, as well as preventing worker complacency within. Once the space program began to mature, the introduction of new models occurred less frequently. NASA was still flying the space shuttle fifteen years after the first orbital flight in 1981, a long lifespan by aircraft standards, to say nothing of the proclivity of Americans for changing their personal transportation vehicles every few years.

NASA's propensity for introducing new gadgets during its formative years was driven by both technology and culture. The primitive quality of rocket and spacecraft technology prompted model upgrades. In addition, NASA embraced an engineering culture that encouraged employees to tinker with new ideas. The preference of the NASA engineer, said one insider, was "invent and built . . . and then go up and invent, build and watch something else work."[11] For a variety of reasons, NASA developed an internal culture that harmonized with consumer expectations. Part of the bargain with technology was the promise of products that at least appeared to be new, and NASA engineers obliged.

One of the greatest benefits of the early space program was the least tangible. The flights to the Moon proved that it could be done. Such demonstrations are a peculiar feature of the consumer society. Merchants in twentieth-century America face the difficult task of convincing consumers to buy more products than needed for their daily chores. Planned obsolescence is only part of the solution; the other part is to convince consumers through marketing that personal attributes can be acquired through the acquisition of goods. Products, in short, can make statements about one's abilities and self-worth.

This attitude requires a particular sort of personality. In 1950 social scientist David Riesman published one of the more influential treatments of the changing character of American society. In *The Lonely Crowd,* Riesman argued that modern capitalism fosters the development of what he called "other-directed" personalities, self-absorbed people whose personal sense of worth is defined by what their contemporaries think of them, in contrast with the more traditional "inner-directed" persons whose values and needs are implanted early in childhood and remain unshakable through life. Consumer societies depend for their prosperity upon a growing supply of other-directed people whose insecurity can be assuaged by acts of personal acquisition of ever-increasing magnitude.[12]

Project Apollo was the most dramatic manifestation of a national goal based on the other-directed characteristics of American society that the twentieth century produced. Here was a status symbol the nation could barely afford (4 percent of all federal spending by 1964) that it had to attain regardless of cost. The goal to send a man to the Moon was established principally to impress people in uncommitted nations with the worthiness of the American system (a commentary itself on the power of the consumer culture to attract adherents worldwide). Shortly before President Kennedy established the goal, a national commission lamented the lack "of urgency throughout our nation about the mortal struggle in which we are engaged."[13] According to the commission report, Americans had lost their sense of purpose; the space race helped create the image that they had found it again. "The space race," concludes historian Michael Smith, "was consummately other-directed, revealing a curious mixture of unsurpassed power and deep insecurity among American leaders."[14]

Other-directed acts of consumption, by their very nature, provide only a temporary sense of relief. They must be followed by other, more grandiose acquisitions in order to maintain the feeling of self-confidence they impart. In that sense, as an expression of national worth, the U.S. space effort was doomed from the start. Americans would reach the Moon, but that would not produce permanent satisfaction. The lunar landing would have to be followed by more expansive adventures—probably a landing on Mars—in order to forestall backsliding into insecurity. That accomplishment would have to be succeeded by even more impressive acts ad infinitum.

In the nature of races, whether for armaments or space or personal effects, one can never have enough. Eventually a day of reckoning occurs, when unpaid bills multiply faster than possessions. President Richard Nixon performed the governmental equivalent of cutting up the credit cards when in 1970 he refused to approve plans for a lunar base and expedition to Mars. For the next twenty-five years, the U.S. space program remained in a state of social limbo. It did not disappear, but it lost its capacity to perform the marvelous feats that a consumer-based society had come to expect.

Critics of the consumer society had expressed distress at the obvious gullibility of people engaged in consumption frenzies motivated by advertising campaigns. The notion of transference, for example, is transparently silly. Consumers do not take on the attributes of persons displaying products simply by purchasing them. That fact, however, did not interfere with the success of a long series of advertising campaigns promoting products such as automobiles, cigarettes, and beer. Critics sought to expose the fallacies that supported the consumer society and understand why otherwise skeptical Americans could so easily be misled. Books such as *The Lonely Crowd, The Waste Makers,* and *The Organization Man* probed the phenomenon. The title of one 1961 book expressed the dismay at American gullibility: *A Nation of Sheep.*[15]

Intellectuals likewise sought to understand the transformation of the U.S. space program. Attached as it was to the tenet of progress through technology, the program received its share of criticism too, often in the form of arguments directed against the central role given humans in the new endeavor. As originally conceived during the Eisenhower administration, the American space effort did not take human flight as its primary goal. Scientific discovery was the primary goal, an emphasis that disappeared after the presentation of the first seven astronauts and President Kennedy's decision to go to the Moon.

The early resistance to human preeminence even appeared within the human flight program. As originally designed, the *Mercury* space capsule had no window from which astronauts could look out. Engineers who designed the capsule did not envision the need for a window because astronauts would not be called upon to pilot the craft. In the words of author Tom Wolfe, the astronaut was little more than a "test subject" with biosensors attached, an organism designed to measure the effects of space travel on the human frame.[16] Hugh Dryden, who as director of the National Advisory Committee for Aeronautics made speeches on behalf of space exploration, nonetheless likened the early flight program to the circus stunt of "shooting a woman out of a cannon." The comment led to his disqualification as NASA administrator.[17]

Mercury astronauts objected to the fact that the capsule trajectory would be regulated entirely by machines. As veteran test pilots, they wanted some control over the spacecraft's path. This horrified the engineers, who doubted that even experienced test pilots could react with sufficient rapidity to adjust the trajectory of what was essentially a ballistic nose cone. The window issue became symbolic of this disagreement. The astronauts prevailed. The press cheered when John Glenn took over the flight of his *Friendship 7* spacecraft (with a window) after an automated control jet refused to work properly. Glenn's intervention provided a "human triumph over impersonal technology," they crowed. Said Glenn: "Now we can get rid of some of that automatic equipment and let man take over."[18]

Wavering engineers were converted to the virtues of human involvement by the crash program to go to the Moon. From that point forward, human travel into the solar system became the official justification for the civil space effort. Engineering decisions on U.S. spacecraft design favored an increasing level of human involvement. That left the studious justification for automation in the hands of space scientists and fellow intellectuals. Their attacks on human space flight took on the quality of the revenge of the nerds. In effect, critics were forced to argue that working with machines was more profitable than working with people. Machines were less cantankerous than astronauts with large egos.

Like critics attacking social trends as a whole, scientists and their allies honed in on what they saw as logical fallacies in the promotion of human space flight. Machines, they claimed, could perform any task in space that humans sought to accomplish. "For virtually any specific mission that can be identified in space, an unmanned spacecraft can be built to conduct it more cheaply and reliably," repeated one critic. Was there anything that a human could do in a spacecraft that a machine could not? "Yes," answered another critic, "but why would anyone wish to do it at such a high altitude?"[19]

The notion that humans were needed to compensate for the shortcomings of machines, in the view of scientists, had "very limited validity." To the contrary, critics asserted, humans were a hinderance on most space flights. They used up precious space. Machines that could be installed to gather information had to be replaced with equipment designed solely to keep occupants alive. What would humans do on spacecraft bound for the outer planets, or a probe investigating the surface of Venus? They could not stand on the gas giants even if they got there, nor could they survive the nine-hundred-degree temperatures of Venus "underneath its veil of sulfur rain."[20]

Critics pointed out that astronauts commonly made judgment errors while in space. During his orbital flight, astronaut John Glenn observed specks of yellow-green light that hovered like fireflies around his space capsule. The observation excited public speculation about the possibility of an alien encounter. The NASA psychiatrist who interviewed Glenn upon his return gave a deadpan response: "What did they say, John?" Glenn failed to correctly identify the flakes as steam expelled from his life-support system condensing and freezing during his passage across the dark side of the earth, a incident that cast doubt on the cognitive powers of humans in space. Encased in a space suit far from home and preoccupied with the immediate business of surviving the voyage, an astronaut is in a poor position to make helpful decisions. Moreover, most of the work astronauts perform is highly rehearsed and leaves little room for improvisation.[21]

"The histories of the space programs of the United States, the Soviet Union and all other countries provide overwhelming evidence that space science is best served by unmanned, automated, commandable spacecraft—the obvious and only important exception being the study of human physiology and psychology under free-fall or low-g conditions," said James Van Allen, the scientist who designed the experiment package for American's first satellite in space. Echoing broader attacks on public gullibility, Van Allen assailed the assertion that human adventures were necessary to maintain public interest and political support. Such an assertion, he said, "is an insult to an informed and intelligent citizenry."[22]

"NASA's fixation on man in space has actually become the curse of the space program," another critic protested. Scientists complained that funds lavished on NASA's human flight programs displaced resources needed for important space science missions. Humans were "a costly nuisance in space, admittedly unexcelled at fixing orbiting toilets, but orbital plumbing wouldn't be there to need fixing if (the humans) weren't on board." Scientists claimed that funds saved by grounding humans would be diverted to space science, a dubious interpretation of federal budgetary politics. Savings from the human flight program disappeared into any number of holes of which space science was only one.[23] Critics also attacked the notion that human space flight made life better on Earth. The most important technology spin-offs came from automated flight, they said, a result of remotely controlled equipment and the sophisticated computers necessary to support it. These advances far exceeded the value, said one physicist, of "such researches as the development of means for rendering urine potable."[24]

Officially, NASA executives claimed that space flight was best maintained by a combination of humans and machines. Engineers would push automation as far as it would go, but at some point human intelligence and versatility would necessarily intervene. "The more complex the mission, and the farther from the Earth it must be carried out, the greater will be the need for that human versatility," NASA's director of Space Sciences argued when the space program began. NASA administrator James Fletcher explained that automated systems were appropriate where a detailed definition of the space mission could be specified in advance. "But when the objectives and opportunities cannot be fully defined in advance, as in the case of exploration, or when the required operations are exceedingly complex . . . the presence in space of man with his unique intelligence and versatile physical capabilities can be an essential advantage," he wrote in 1971. The director of NASA's Institute for Space Studies argued (given the state of computer technology in 1971) that a machine with the cognitive powers of a single human brain would cost $10 billion, weigh one hundred thousand tons, occupy eight million cubit feet, and consume one billion watts of electric power. "It seems a safe bet," he concluded, "that in space exploration man will be superior to the machine for all difficult investigations, from now to the end of the 20th century."[25]

During the infancy of space flight, experts could not agree on the relative advantages of humans versus machines. Scientists and their supporters offered convincing arguments, but so did the experts promoting human flight. The president's Science Advisory Committee observed in its 1970 report that "at the present time, there is insufficient information available upon which to judge clearly between the two opposing views."[26] When expert opinion is divided, culture and imagination play an enlarged role. This proved to be the case with the role of humans in space. Throughout the mid-twentieth century, advertising campaigns promoting the consumer society emphasized human mastery over technology, an understandable approach given the public reaction at that time. Most new technologies, such as the jet engine and atomic power, were difficult to understand. Others, such as early computer software, proved exasperating to operate. Some were powerfully frightening, such as the technology of nuclear weaponry and ballistic missiles. Many commentators worried that technology was out of control, forcing society into choices that reduced human discretion. To sell Americans on the notion of progress through science, industrialists had to market the concept of human mastery.

Advertisers accomplished this in a number of ways. One of the most effective was the use of the helmsman, a figure who uses technology to master a challenging environment and in so doing elicits efforts to emulate that behavior. (During the 1950s the helmsman was a distinctly masculine figure.) The helmsman was used to promote a variety of postwar products, none more successfully than Marlboro cigarettes. The Marlboro Man was portrayed as a master of technology—a pilot, a race-car driver, a captain of boats, and, most effectively, a cowboy, the autonomous individual in control of a raw frontier. By smoking Marlboro cigarettes, the ads suggested, consumers too could become masters of their environments.[27]

Technological jargon was created to convey a sense of technological mastery. Deliberately unfamiliar phrases were created by advertisers to foster an illusion of understanding. Technojargon filled automobile advertisements in the 1950s as advances in transportation technology exceeded owner comprehension. "Torsion-Aire" suspensions and "HC-HE engines" allowed owners to discuss vehicle features without the slightest technical understanding of how those features actually worked.[28]

In the process of selling the space program, advocates of space flight inevitably made use of these techniques. The astronaut provided the figure of the helmsman, the lone eagle of Lindbergh fame recast as the space cadet of tomorrow. Presenting the original astronauts as moral supermen, which the press willingly

The space program was part of a larger effort to promote a society powered by an endless stream of consumer goods provided by technology and design. Business leaders displayed goods in such a way as to give the impression that they would speed consumers away from economic depression toward a more luxurious life-style. To allay fears of technology, advertisers were careful to show humans in charge. This space station cutaway shows astronauts at work inside a materials processing laboratory. (NASA)

did, ensured that the public would grow attached to the flight projects they endorsed.

As with advertising campaigns, technojargon was employed to make flight hardware appear more manageable than it actually was. Ready launch systems were "A-OK," not just okay, suggesting a standard of reliability that exceeded correct working order. Astronauts in orbit "initiated a retro-sequence" with their "retrograde package," a delicate way of announcing that they were prepared to ignite rocket engines that would strand the astronauts in space if the engines failed to start. The language of space-flight engineering gave an impression of total control. NASA used "systems analysis" to assure "operational control." The dangers of "lift-off" were managed by "Launch Control," and the lives of the astronauts were protected during their journey by "Mission Control." In space, the rocket ship was kept from tumbling by the use of "10 reaction control system engines" that provided "attitude control about 3 axes." Technojargon transformed ordinary Americans into space experts and helped convince them that the very risky business of space flight could be subjected to human mastery.[29]

A space program organized around machines was inconceivable at the beginning of the effort given the need within American society to demonstrate human supremacy over technology. No single act better satisfied the public need to dominate technology during this time than human faces in spaceships on routes to the Moon. Machines alone could not have accomplished this. At the same time that American astronauts landed on the Moon, the Soviet Union completed a mission in which a robotic probe landed on the Moon and returned lunar samples to Earth. The Soviet feat was largely forgotten in the aftermath of the Apollo expeditions. Said the 1990 advisory committee headed by aerospace executive Norman Augustine, "There is a difference between Hillary reaching the top of Everest and merely using a rocket to loft an instrument package to the summit."[30]

Intangible objectives such as the need to demonstrate human mastery have been a constant force in the promotion of space exploration. In spite of the pleading of scientists and fellow intellectuals, the more tangible goals of scientific investigation have not dominated the ethereal aims of exploration and conquest. During the flights to the Moon, scientists complained that their experiments received a low priority in comparison to the technical problems of transporting human occupants. "They don't care about a damn thing but that their machines work right," said one scientist about the spacecraft engineers. To the suggestion that a scientist replace one of the pilots on the lunar lander, the astronaut in charge of replied, "It sure as hell wouldn't make any sense to put a scientist on the next flight just to say we've got a scientist on board and then blow the whole thing trying to make a landing. . . . A dead astronaut-scientist is not going to do anybody any good."[31] When asked to state the reason for taking humans to the Moon, Neil Armstrong replied that the flight itself provided its own justification. "The objective of this flight is precisely to take man to the moon, make a landing there, and return. . . . The primary objective is the ability to demonstrate that man, in fact, can do this kind of job."[32]

Scientists and their intellectual allies found such lines of reasoning exasperating. Instead of elevating space exploration as a form of scientific rationality, it reduced space flight to a mystique. Support for such adventures seemed as inexplicable as the public's faith in progress through technology. Neither could be defended rationally. Some scientists blamed this apparent loss of reason on decades of promotion by science fiction writers and their counterparts in popular science. "The simple taste for adventure and fantasy expressed in that sentiment," lamented James Van Allen, "has been elevated in some quarters to the quasi-religious belief that space is a natural habitat of human beings."[33]

Reason and logic, however, do not always drive government policy. Public desire for particular products, whether items of consumption or government service, are motivated by yearnings that go well beyond the practical requirements of life. Like so many other commodities in the consumer era, human space flight has been offered as a means for satisfying social and romantic yearnings. In the long run, such adventures are what make life worthwhile.

Having advanced the premise that humans everywhere would benefit from space technology, advocates of exploration wrestled with the problem of who would participate in its production. Equal benefit from space technology did not mean equal participation in producing it. Although virtually everyone would benefit from space spin-offs, not everyone could travel beyond Earth. This exposed a fundamental contradiction in popular impressions about space. Given the culture of the day, the humans appointed to lead the march into space were invariably depicted as males, both on the ground and in space. Well through the twentieth century, space experts referred to missions with humans on board as "manned" space flight, and flight controllers and rocket engineers were invariably men. This symbolic distinction served to separate a substantial portion of the U.S. population from direct involvement in a technology that promised opportunities and benefits to all.

The conflict between promise and reality was heightened by the experience of the aviation movement in America. To a considerable extent, space exploration drew its strength from aviation pioneers. Airplane model builders, aircraft pilots, and workers from aviation laboratories nurtured the march into space. According to American mythology, the aviation movement provided women with one of the first major opportunities to rise from domesticity and act as the equals of men. The role of women was exaggerated but, nonetheless, they played key roles in promoting aviation technology, especially as pilots in that new realm.

The first women pilots served, in the words of Joseph Corn, as "the most effective evangelists of aviation in the period." During the formative years of aviation, women pilots promoted the notion that flight was both safe and easy. "More than the men who barnstormed around the country or crossed the oceans by plane, women pilots domesticated the sky, purging it of associations with death and terror," Corn observes.[34] Although outnumbered by men, women pilots proved on a number of occasions that they were equally capable of mastering this complex technology. In the 1936 Bendix Trophy air race, women pilots captured three of the first five places, including the coveted first prize. The Bendix race was the Superbowl of its day, a contest to determine the fastest flight time across the United States. Until 1935 women pilots were not allowed to compete with the men. In the second year of their participation, women pilots placed first, second, and fifth. In the fledgling aviation industry, women served as test pilots, aerial photographers, and pilots for business enterprises. Many, like Louise Thaden, the winner with teammate Blanche Noyes of the 1936 Bendix race, demonstrated and sold aircraft for commercial firms. "Nothing impresses the safety of aviation on the public quite so much as to see a woman flying an airplane," she observed. If a woman could handle the controls, "it must be duck soup for men."[35]

Beryl Markham, a British citizen, worked as pilot for the infamous Kenyan safari guide Baron von Blixen, scouting game from the air and landing on a variety of unimaginable terrains. In 1936 Markham garnered enormous public interest when she nearly became the first person (male or female) to fly alone from England to New York, a prodigious feat given the prevailing North Atlantic winds.[36] Charles Lindbergh's wife, Anne Morrow, was an accomplished pilot who flew with her husband around the world. In 1930 United Airlines hired the first stewardesses, women trained in nursing who were prepared to take charge of the passenger cabin in an emergency. United could have hired male stewards, following the tradition of railroad porters, but chose women instead in order to help allay the public's fear of flying. During World War II, females served as Women Airforce Service Pilots, or WASPs. They tested aircraft, delivered airplanes, towed gunnery targets, and flew cargo.[37] Flying seemed to give women a degree of freedom and mobility unimaginable to previous generations.

No woman had a greater impact on the expectation of equal opportunity through aviation than Amelia Earhart. An accomplished pilot, she became a media celebrity whose fame was exceeded only by that of Charles Lindbergh. Earhart won public acclaim in 1932 by repeating Lindbergh's triumph on the date of its fifth anniversary, becoming the first woman to fly across the Atlantic alone. A carefully orchestrated public relations campaign with flights of ever-increasing difficulty kept her in the public eye throughout the 1930s. She used her status to promote the idea that flying was safe and that women were equal to men in their flying ability. Earhart disappeared in 1937 during a long-distance flight around the world, an event that immortalized her fame.[38] To many Americans of that period, female pilots seemed like aliens from another world. Competent and confident of their abilities, they presented a form of female independence that would not become widespread for another fifty years. As a new technology, aviation seemed to forecast social trends. At least that is how it seemed.

Beyond aviation waited astronautics. In spite of the close association between the two, the precedent apparently established by female pilots did not spill into the realm of space exploration. It occurred in neither the mythology of space nor in the first astronauts corps. One searches in vain through the early literature on space exploration for female advocates, either fictional or real, with stature equal to that of women in aviation. Wilma Deering served as Buck Rogers's girlfriend, and Dale Arden remained perpetually engaged to Flash Gordon. In both cases, the status of these fictional heroines was defined by their relationship to the male protagonist, often serving to confirm his masculinity.

Women wrote science fiction, not an unusual development given the large number of female authors and editors at that time. Many literary historians credit the origin of modern science fiction to Mary Shelley, who penned *Frankenstein* in 1818. During the 1930s and 1940s C. L. (Catherine) Moore wrote interplanetary adventure stories, and Leigh Brackett produced works of science fiction before turning to Hollywood screenplays.[39] The stories produced by women such as Moore and Brackett, however, generally repeated the formulas established by their more numerous male counterparts. During the 1920s and 1930s, science fiction was viewed within the literary marketplace as a type of adventure story that appealed almost exclusively to pubescent males. Men were thought to be the primary consumers, and even stories penned by women aimed to gratify boyhood fantasies. The women who appear in these tales are generally defined by their relationship to their male heros, as objects to be rescued or educated, protected or tamed. Even the strongest females characters, such as the lovely princess Dejah Thoris in Edgar Rice Burroughs series on Mars, eventually submit to their dominant male.

Practically nowhere in early science fiction can one find memorable female characters of strength and independence. Not until the feminist movement was well underway did science fiction writers produce female protagonists of independent means. In the screenplay for the 1977 film *Star Wars,* George Lucas allows young Luke Skywalker to attempt a rescue of the beautiful Princess Leia Organa, then surprises him with the discovery that the rich and powerful princess is perfectly capable of rescuing herself. Two years later Hollywood produced *Alien,* in which the only match for a shrewd carnivore set loose upon a merchant cargo ship is Ripley, the tough, pragmatic crewmember played by actress Sigourney Weaver. This was a total reversal from the traditional approach. When Gene Roddenberry produced the pilot for the highly influential *Star Trek* television series, a morality play with many liberation themes, he promoted the topic of sexual equality by placing a woman second in command of the starship and dressing the crew in unisex uniforms. This proved too radical for the moguls at NBC, so after *Star Trek* began its weekly serialization, the woman returned as a nurse and the other female characters were given miniskirts to wear. Not until 1995 did a woman receive command of the fictional *Enterprise.*[40]

Well through the dawn of the space age, works of imagination reenforced conventional sexual stereotypes. Rockets and spacecraft were presented as expensive toys for men and boys. Women were portrayed as technologically clumsy and incapable of comprehending engineering technology, to say nothing of their presumed inability to command men or pilot spacecraft. When George Pal produced *The Conquest of Space* in 1955, he portrayed the all-male crew of his Earth-orbiting space station as little more than rowdy, sex-starved sailors at sea, a formula drawn from naval wartime films. Based on the *Collier's* series and the works of Ley and von Braun, the movie was the most serious attempt at space realism during the decade. In its treatment of women, however, it was light-years behind the future it sought to portray. A similar formula greeted viewers of *Forbidden Planet,* released one year later. In that movie the all-male crew of an investigating spaceship discovers an obsessive scientist and his virginal daughter Altaira on a distant planet, the sole human survivors of a lost expedition. Altaira is saved.[41]

In spite of their apparent aviation role, the requests of women for participation in the new space program were greeted by a succession of bad jokes. Responding to a question about the possibility of female astronauts, Wernher von Braun replied in 1962 that the men in charge of the rocket program were "reserving 110 pounds of payload for recreational equipment." When women took their case to Congress, the chairman of the investigating subcommittee explained that "the whole purpose of space exploration is to some day colonize these other planets and I don't see how we can do that without women." When asked for his opinions on the issue, astronaut John Glenn announced tongue in cheek that he would welcome females into space "with open arms."[42]

A realistic opportunity for women's involvement occurred in 1960 when the Lovelace Foundation for Medical Research began a secret program to test female astronauts. The foundation, headed by W. Randolph Lovelace, had been commissioned by NASA to conduct the physical screening tests for the first group of male applicants to the astronaut corps, a process vividly portrayed in *The Right Stuff.* NASA officials distanced themselves from the female tests, insisting that Lovelace was acting on his own initiative. Nonetheless, the U.S. space community viewed the tests with considerable interest, given the news that the Soviets were training a female cosmonaut.[43]

Dr. Lovelace arranged for Geraldyn Cobb to undergo the same type of physical tests given the Mercury astronauts. The twenty-nine-year-old Cobb, one of the most accomplished aviators of her day, had logged more than seven thousand flight hours, set a number of flight records, including a world altitude mark, and been named pilot of the year by the National Pilot's Association. Elated at her selection, she dreamed of leading women into space. Quietly, she trained hard for the first battery of tests.

In August 1960 Lovelace publicly announced the results of Cobb's performance. She had passed the Mercury astronaut tests and would, in the opinion of the physicians that examined her, "qualify for special missions." Encouraged by the results, Lovelace asked Jacqueline Cochran, one of the leaders of the women's aviation movement, to select more female candidates. Thirteen women pilots, including Cobb, completed the first round of physical tests. Cobb, meanwhile,

Though all Americans would benefit from new technology, not all would share equally in producing it. In 1960 Geraldyn Cobb passed the physical screening tests given to the Mercury astronauts. NASA canceled the testing program after more women pilots applied and passed the exams. Not until 1995 did an American woman pilot an American spacecraft. In this photograph, Cobb is shown ncxt to thc *Mercury* spacc capsule in which she hoped to fly. (National Air and Space Museum)

completed more trials, including the difficult jet orientation and centrifuge tests at Pensacola, Florida. The women proved that they were as capable of space flight as men, and in fact enjoyed certain physiological advantages, such as lower oxygen intake and less susceptibility to radiation poisoning.[44]

NASA administrator James Webb announced the results of Cobb's tests and appointed her as a NASA consultant. Cobb was inundated with requests for personal appearances, and a special article in *Life* magazine appeared. All of the women prepared for further tests.[45] Within NASA, however, controversy over the program grew. What had begun as a program to test the capability of women as space travelers was becoming a female astronaut program. NASA engineers were preoccupied with the task of getting men into space and did not want to divert resources to the preparation of a new group. NASA, moreover, had a rule that required astronauts to possess experience as military jet test pilots. This automatically eliminated women, because the military, in turn, had a rule that excluded them from going to test pilot school. NASA officials wanted the issue to go away before it spun out of control. In July 1961, as the women prepared for a new round of tests at Pensacola, NASA officials canceled the program.

Cobb and her colleagues fought back. Cobb gave speeches to supportive audiences and lobbied Vice President Lyndon Johnson, asking him to use his influence as head of the National Aeronautics and Space Council to make NASA reopen the program.[46] Jane Hart, one of the thirteen women to pass the physiological tests (and wife of Michigan senator Philip Hart) pressured members of Congress. In July 1962 a special panel of the House Science and Astronautics Committee held hearings on the issue. Said Cobb in her opening remarks: "There were women on the *Mayflower* and on the first wagon trains west, working alongside men to forge new trails to new vistas. We ask that opportunity in the pioneering of space."[47]

NASA officials refused to set up an additional training program for women astronauts and allowed Cobb's position as a special NASA consultant to lapse. In her place, Webb appointed Jacqueline Cochran. Although Cochran had helped finance the Lovelace tests, her position on female astronauts was well known. "There is no present real national need for women in such a role," she wrote Cobb, not so long as NASA could find a sufficient number of qualified men. Pushing too hard, Cochran warned, would "retard rather than speed" the acceptance of women astronauts. NASA retained Cochran as its consultant on this issue through the decade.[48]

The first woman to fly in space, Valentina Vladimirovna Tereshkova, piloted the Soviet *Vostok 6* spacecraft on 15 June 1963. Not until 1995 did an American woman pilot an American spacecraft. (Sally Ride, the first American woman to fly in space in 1983, was a mission specialist, not a pilot.) To commemorate the event, shuttle pilot Eileen Collins invited the women who had passed the secret physical exams nearly a quarter-century earlier to watch the launch as her guests. Seven of the eleven surviving women arrived, including Geraldyn Cobb.[49]

In rejecting the pleas of women for a place in the early astronaut corps, NASA officials were in fact maintaining practices that had existed for some time. In spite of the publicity accorded women pilots, they were not given as a matter of practice an equal role in the early aviation industry. Nor did most claim it. Women pilots at that time rarely used their position to promote equal opportunity in general. Their words and actions scrupulously avoided what would have been viewed as radical sentiments such as these.[50]

In aviation men needed women to promote the notion that flying was safe, whereas in space exploration men sought to maintain the idea that the venture was as dangerous as it seemed. Astronautical pacesetters who had taken large steps in advancing technology remained conservative in challenging social trends. As a result, they lost an important opportunity to build support for the U.S. space program among a substantial portion of the voting public. The price they paid is well recorded in the annals of American public opinion. During the 1980s, pollsters revealed a large gender gap in public support for space exploration. Responding to a Media General/Associated Press poll in 1988, 56 percent of the men interviewed said that the U.S. spent "too little" on space exploration; only 25 percent of the women responding agreed. Similar gaps were recorded by the Gallup organization. True to its roots, space exploration remained a program that drew its support disproportionately from men.[51]

NASA's record with minorities was not much better. Although agency leaders appreciated the value of technological demonstration, they seemed insusceptible to the value of social display. Not until 1978 did NASA recruit the first class of astronauts to contain minorities (three African Americans and one Asian American), as well as the first to contain women (six). NASA officials had resisted efforts to broaden the criteria for astronaut selection through the mid-1970s and reneged only under the strongest pressure. Support for the U.S. space program among African Americans, not surprisingly, remained low.[52]

Space exploration, in spite of its reliance upon a message of benefits for all, remained throughout the twentieth century an endeavor that appealed primarily to white males. This maintained the social tradition that had grown up around the venture, both in works of imagination and among practicing rocketeers. Space advocates did little to change this tradition when they had the opportunity, in spite of the example of aviation.

In the same manner that practice contradicted promise in the realm of equal opportunity, so the products of exploration affected expectations shaping another social trend: the environmental movement. Who would have thought that a space program founded on the promise of technology would produce images extolling its limits. Many memorable images emerged from the early space program: portraits of humans standing on the Moon and closeup photographs of planetary neighbors. As impressive as they were, no image had a greater social impact than the picture of the whole Earth in space. The first Apollo astronauts to circle the Moon returned with images of a crescent Earth rising above lunar plains. The final trio of explorers, who returned home in the last month of 1972, captured a frame-filling photograph that became, in the words of astronomer Carl Sagan, an "icon of our age."[53] At the bottom of the photograph rests Antarctica, surrounded by delicate storms. Rising above it sits the continent of Africa, where human life began. The lands bordering the Mediterranean Sea, from North Africa through the Arabian Peninsula, are clearly visible across the top. The oceans are blue, the deserts ocher, the forests green. No national boundaries can be seen.

Every perspective of Earth serves as a metaphor for the beliefs of its day. When humans were stuck to the surface of the planet, plowing crops on lands away from which residents rarely ventured, the world seemed huge. People so situated could imagine themselves at the center of creation, with Gods in the heavens and

Technological optimism spurred interest in the early space program; the space program in turn produced images that undercut the originating beliefs. On the outbound leg of the last mission to the Moon in 1972, the crew of *Apollo 17* took this picture of the whole earth. The photograph became an icon for the age, encouraging people to view Earth as a small and fragile globe with limited resources suspended in a cosmic void. (NASA)

Dante's hell beneath their feet. As humans traversed the seas and visited strange civilizations, new perspectives arose. The earth still seemed large, but a patchwork of nations embracing a multitude of beliefs appeared. Not surprisingly, the age of exploration supported the age of nationalism, as communities unified into nation states in order to preserve and extend their cultures. The view from space created a new metaphor. It became harder for intelligent people to stand on hills and imagine a landscape divided by artificial boundaries. From space, Earth looks small and whole, like a spacecraft traveling through the cosmos. Space exploration fosters an image of an earth without boundaries, one on which ideas and products can shoot around the globe.

"To see the earth as it truly is, small and blue and beautiful in that eternal silence where it floats, is to see ourselves as riders on the earth together," poet Archibald MacLeish wrote as the first astronauts circled the Moon. Ten years later, at a ceremony honoring outstanding astronauts, President Jimmy Carter observed that "of all the things we have learned from our explorations of space, none has been more important than this perception of the essential unity of our world":

> We saw our own world as a single delicate globe of swirling blue and white, green, brown. From the perspective of space our planet has no national boundaries. It is very beautiful, but it is also very fragile. And it is the special responsibility of the human race to preserve it.[54]

The view from space accelerated acceptance of Earth as a single interdependent system. To champions of this perspective, the image expedited a new way of thinking, one captured in the title of publications such as the *Whole Earth Catalog* and the *Mother Earth News.*

True to the vision, the world began to operate as a single system. Space-based communication networks, created by space-age technology, created what Marshall McLuhan called a "global village." People who heretofore had considered themselves members of national communities received news and gossip from networks that respected no boundaries. Industrialists constructed globally integrated economies that bypassed national barriers. Futurist Alvin Toffler characterized this development as the "Third Wave" of civilization. (The discovery of agriculture launched the first wave nearly ten thousand years ago; the second wave began with the industrial revolution.) The third wave, Toffler maintained, would produce global networks governing human affairs but would not promote the development of a single world society. To the contrary, societies would splinter as communication networks linked people with common interests heretofore separated geographically. "As the Second Wave produced a mass society," Toffler explained, "the Third Wave de-massifies us, moving the entire social system to a much higher level of diversity."[55]

Just as information and money bypass national boundaries in the postindustrial society, so does the detritus of technology. Pollution migrates across the earth, a fact visibly confirmed by photographs from space. As images from space became part of public consciousness, so did a new sensitivity to environmental issues. Many factors encouraged the worldwide rise of environmental concerns during the 1970s, but one of the most important was the panorama from space. The environmental movement required its adherents to view the earth and its habitats as a single ecosystem; images of the earth from space confirmed that it was.

Engineers like to think in terms of systems, an approach that emphasizes feedback loops and inclusionary analysis. Just as a thermostat uses a simple feedback loop to regulate the temperature of a room, so large systems employ complex feedback mechanisms to adjust their parts. In 1970 a professor from the Massachusetts

Institution of Technology presented a model of the entire earth to a group of concerned individuals known as the Club of Rome. The club encouraged researchers at MIT to calculate the interaction of five basic factors: population, agricultural production, natural resources, industrial production, and pollution. By treating the earth as a single system, the team predicted the way in which each intricate part reacted to changes in the rest.

The findings, published in 1972 under the title *The Limits to Growth,* created a firestorm of controversy. Armed with the vision of Earth as a single system (and aided by computer simulation techniques), the group investigated the consequences of unchecked consumption and population growth. The results were disturbing. "If the present growth trends . . . continue unchanged," the team reported, "the limits to growth on this planet will be reached sometime within the next one hundred years." At that point, the earth would no longer be able to support a large, industrially dependent population. A catastrophic decline in both population and industrial productivity would be the most likely result.[56]

Critics of the report lambasted the team for ignoring the effects of technology. The very technologies that had permitted the vision of a single earth, the critics argued, could rescue it from overconsumption and growth. Giant space-based reflectors could beam solar energy down to Earth, advances in robotics could permit the development of pollution-free industries, and humans could migrate to other worlds. The Club of Rome team belittled these arguments. Advances in technology, team leaders observed, "would only delay rather than avoid crises" and offered computer runs of their model to prove it. Only by severely limiting population growth and economic expansion could the human race produce a stable equilibrium and avoid ultimate collapse.[57] Conclusions such as these allowed radical environmentalists to scare the public with convincing warnings of impending doom. Global warming, ozone holes, and exotic viruses provided what many believed to be the first signs of an earth on the slippery slope of disequilibrium.

The most extraordinary notion to receive its impetus from the belief in the globe as a single ecosystem was the Gaia hypothesis. The hypothesis garnered its name from the Greek goddess of the earth, mother of the sky and the sea. In its more respectable form, the Gaia hypothesis suggests that the earth is a self-regulating system wherein features such as the climate and atmosphere are continually adjusted by the organisms that inhabited it in order to maintain conditions suitable for their existence. The average mean temperature of the earth, for example, has responded for millennia to the presence of greenhouse gases such as methane and carbon dioxide, which are naturally produced by termites and other living things. In the past, fluctuations in naturally produced greenhouse gases have helped stabilize temperature shifts on Earth and counteract fluctuations in the energy produced by the Sun.[58]

The more radical proponents of the hypothesis make the startling claim that the earth is actually alive. "The Gaia hypothesis is the first comprehensive scientific expression of the profoundly ancient belief that the planet Earth is a living creature," one wrote. According to supporters of this vision, the planet regulates itself in the same manner that animal bodies reflexively adjust to changes in temperature, body chemistry, or the introduction of foreign substances. Given this view, the earth contains systems that act like vital organs in the animal realm: "Regions of intense biological activity, such as the tropical rain forests and coastal seas, are seen as vital not only to their geographic regions but to the entire global environment, much as the liver or spleen is necessary to the survival of the body as a whole."[59] The living earth may also have the capacity to resist invaders, which like viruses in animal bodies threaten to overwhelm the planet through uncontrolled multiplication. A living Earth may view the human population explosion as such a threat, and so threatened, Mother Earth may strike back.

It is hard to imagine something as wild as the Gaia hypothesis by just standing on the ground. From space, the earth, with its seasonal bands of vegetation and moving clouds, looks like a dynamic body capable of self-regulation. It appears fragile, susceptible to alteration from the combined effects of the species that dominates it. From a distance, the atmosphere looks as thin as the skin on an apple. Astronauts who have traveled more than once into space have watched rain forests disappear. Is it so hard to then imagine that with insistent human tinkering the earth might be transformed into a less-hospitable place?

The images of the whole earth reshaped not only public consciousness but also the NASA space program. Given those images, the NASA program inevitably turned back from the heavens and reexamined the earth. In the mid-1980s, advocates proposed that space technology be redirected toward a more complete understanding of the only home that humans had. In her 1987 report to the NASA administrator, Sally Ride suggested that the space agency undertake a wide-ranging "Mission to Planet Earth." Although the United States had used satellites and spacecraft to study the home planet since the space race began, the emphasis on Earth studies was a major departure from NASA's orientation toward the outward bound. Repeating the philosophy that guided the new environmental consciousness, Ride announced that "interactive physical, chemical, and biological processes connect the oceans, continents, atmosphere, and biosphere of Earth in a complex way." A network of orbiting platforms, she suggested, joined with ground-based observations, would allow scientists to predict changes in the global environment before it was too late. NASA subsequently won approval for a series of Earth Observing System satellites to monitor the planet from above.[60]

Throughout its formative years, the space program drew inspiration from the promise of better living through technology. Advocates who promoted this vision proffered a streamlined future of comfort and opportunity and a cornucopia of goods so extensive that consumers could dispose of useful items simply because of changes in style. Technology seemed to offer solutions to dangers imposed by dwindling resources: new sources of energy, space-age materials, and products of unimaginable potential. As late as 1970 futurist Alvin Toffler was assuring Americans of a future so full of goods that people would live in a "throw away society."[61]

Whereas many advocates of extraterrestrial exploration extolled the spin-offs from space, others saw in space technology a means to escape a planet growing too small. It is no accident that the fascination with Gerard O'Neill's proposal for space colonies appeared in conjunction with the first images of the whole Earth. O'Neill himself admitted that he sought to resolve the desire for growth with a planet of limited size. Perhaps Carl Sagan's observation is correct: every long-lived technological civilization in the universe is eventually forced to become spacefaring in order to survive.[62]

Given the growing concern with the dangers of technology, it was probably inevitable that environmentalists would come to protest the space program itself. What had heretofore been viewed as the most beneficial of technologies came under attack. Advocates of exploration, who had always viewed space flight as a benign technology, were shocked when environmentalists sought to halt the launch of the space shuttle *Atlantis* in 1989. The immediate cause of their complaint was the fact that the Atlantis would carry a space probe (the *Galileo* mission to Jupiter) powered by a radioisotope thermoelectric generator (RTG) with radioactive fuel. Remembering the *Challenger*'s demise, environmentalists warned that a catastrophic accident might release substantial quantities of plutonium 238 into the atmosphere. Five previous U.S. and Soviet space probes using RTGs had plowed back into the atmosphere after mission malfunctions, and one (perhaps three) had disintegrated. NASA's reassurance that its triple-containment system would prevent any future dispersal did little to assuage protesters' fears. They worried that NASA had transformed its technology into a broader class of "risky systems" that humans concerned about the future of the earth could better live without.[63] With the protest, environmentalists turned against the technology that had been born in so much optimism decades ago.

CONCLUSION

IMAGINATION AND THE POLICY AGENDA

I will always love the false image I had of you.

—Author unknown

The argument offered in this book can be summarized as follows. The rise of the U.S. space program was due in part to a concerted effort by writers of popular science and science fiction, along with other opinion leaders, to prepare the public for what they hoped would be the inevitable conquest of space. Conquest meant that humans would explore the Moon and planets, establish settlements, and eventually move out toward the stars. In constructing this romantic vision, advocates took fantastic ideas and laid them upon images already rooted in the American culture, such as the myth of the frontier. The resulting vision of space exploration had the power to excite and entertain, or, as in the case of the Cold War, to frighten. The vision prevailed over lesser alternatives and moved onto the public policy agenda not because of its technical superiority but because it aroused the imaginations of people who viewed it.

In the beginning practical experience with new ideas is typically limited. People are free to imagine wondrous consequences of new endeavors. Communication of those ideas often takes the form of gospel—literally, the "good news" that will change the world. Where practical experience is limited, acceptance of the gospel must be based on faith, intuition, or the attractiveness of the vision. Imaginations are free to soar.[1]

Practical experience with space flight was entirely lacking during the years in which the initial vision was revealed. The promise of space travel took the form of a revelation not yet experienced by people who would be affected by it. The "good news" of space travel led to great expectations, firmly supported by other aspects of American culture. By being first in space, proponents proclaimed, the United States would win the Cold War. Space exploration would promote trust in government. Scientists would discover life on other planets and resolve other great mysteries of the universe. Space travel would rekindle the frontier spirit. Space travel would be easy and cheap. Humans would leave Earth and colonize the cosmos. Space exploration would revolutionize life on Earth, creating a cornucopia of consumer goods based on high technology. As it came to pass, the reality of events confuted the expectations created by pioneers. Most of the prophesies contained in the gospel of space flight did not come true, certainly not on the timetable proposed. The prophets of space flight, having created a rich vision of their endeavor, had to grope with the fact that much of the actual space program failed to fulfill first expectations.

The vision of space travel drew its strength from its capacity to excite and entertain. Those same strengths contributed to its failure. No real program of exploration, forced to confront real laws of physics and the mysteries of the universe, could satisfy the expectations that created it. Is this merely a consequence of space exploration, or is it a metaphor for a variety of problems confronting democratic governments in the modern age?

The history of space travel might be dismissed as a weird dream, the result of an overactive imagination unrestrained by concrete events, except that the pattern reappears with eerie familiarity in other areas. Speaking of the history of aviation, Robert Wohl reports:

> Strange as it may seem to us today, weary veterans of crowded commercial airliners and depressing airports, the invention of the airplane was at first perceived by many as an *aesthetic* event with far-reaching implications for the new century's artistic and moral sensibility. Long dreamt about, enshrined in fable and myth, the miracle of flight, once achieved, opened vistas of further conquests over Nature that excited people's imagination and appeared to guarantee the coming of a New Age.[2]

Similar claims heralded the peaceful uses of atomic energy and the so-called War on Poverty undertaken by the U.S. government in the mid-1960s.[3]

For many years, students of public policy have treated culture and technology as a backdrop to the real action that takes place on the governmental stage. Like a painted screen hung at the back of a stage, culture and technology were thought to provide a frame of reference for the actual play, not to influence it directly. Culture did not receive as much attention as politics and personality, in part because the former seemed to fluctuate less often than the action on the stage.[4] This is changing. Not only are scholars paying more attention to the affects of imagination and culture on society at large, but those same people are also beginning to probe the way in which such forces directly affect public policy.[5]

We know, for example, that no policy can long exist in a political culture in which popular beliefs and myths question its feasibility. For a policy to succeed, people must be able to imagine the activity taking place and possess a view of the world into which this image comfortably fits. The social welfare state was not possible so long as people imagined poverty to be an act of God, as inevitable as the weather. That early point of view supported little more than individual acts of charity and Poor Laws designed to keep the destitute from becoming a nuisance.[6] Science and the enlightenment created a new world view in which people came to believe that circumstances such as poverty and hunger could be overcome. The malleability of poverty was promoted through works of imagination, such as Charles Dicken's *Oliver Twist,* as well as through academic tracts.[7]

Likewise, in the U.S. space program, pioneers of this new adventure had to convince a skeptical public that space travel could come true. Works of imagination helped transmit this message. Some were fictional, such as thc portrayal of a lunar expedition in the 1950 movie *Destination Moon* or the Rocket to the Moon attraction that opened at the Disneyland theme park in 1955. Others were nonfictional, such as the various *Collier's* articles on space exploration and atomic war. Together they altered the public's vision of space exploration from one of skepticism into something that could actually occur.

New directions in policy often require major shifts in cultural beliefs. Such shifts do not happen often, but when they do, they can profoundly alter public policy. The acceptance of new policy initiatives is often foreshadowed by a period of advent in which attitudes toward social and natural phenomenon change. The effect of culture shifts on American conservation policy has been demonstrated in a convincing way.[8] Few voices in early America extolled the beauty of wilderness or proclaimed the necessity of preserving it. Most people at that time viewed wild areas as savage, uncontrollable, and evil. So long as Americans retained an image of nature as something repugnant, little support could be mustered for its conservation. The conservation movement that swept America during the early twentieth century required a radically different view. Art and imagination made this possible.

Prior to the age of television and glossy art, many Americans experienced natural wonders through the display of landscape art. Landscape paintings were exhibited in great galleries much as movies would be shown in decades that followed. Beginning in the early nineteenth century, American artists painted and displayed large landscapes that romanticized the wilderness as a place of great natural beauty. By exaggerating geological features and emphasizing the luminosity of light, artists inspired a reverence for American natural wonders that was every bit as powerful as the pride of Europeans in monuments of antiquity such as the Parthenon. In 1823, Thomas Cole gave up his career as a portrait artist and began to paint landscapes of the Catskill Mountains in New York, giving rise to the Hudson River school of American art. Following his first journey into the American West in 1858, Albert Bierstadt became professionally successful by painting natural wonders such as the Rocky Mountains and Yosemite Valley. Thomas Moran accompanied the Hayden expedition to the Yellowstone Plateau in 1871, and his paintings of the plateau, the Tetons, and the Sierra Nevadas helped build support for the national park movement.[9] Congress appropriated twenty thousand dollars for Moran's paintings of the Grand Canyon of the Yellowstone and the Grand Canyon of the Colorado and placed them in the U.S. Capitol.[10] American natural wonders became a source of national pride.

The movement to exalt the American wilderness was further advanced by literary naturalists like Henry Thoreau and James Fenimore Cooper.[11] Through their art and fiction, Frederic Remington and Owen Wister generated an imaginative view of the frontier that glorified the Wild West.[12] With his 1902 novel *The Virginian,* Wister created the American Western, one of the dominant art forms of the early twentieth century. By the time the twentieth century began, a major cultural shift was underway in the American mind: people began to view wild areas as places to be preserved, not destroyed. Without this cultural shift, government support for the conservation movement could not have taken place.

Culture shifts typically prepare the public for a change in government policy that some precipitating event triggers. In their work on policy agendas, Frank Baumgartener and Bryan Jones depict long periods of apparent policy equilibrium broken by punctuating events that allow public officials to initiate policy change.[13] This is a familiar scenario, repeated in a number of policy areas. In the years that led up to the Civil War, an influential work of imagination helped shift public attitudes toward slavery. Published in book form in 1852, *Uncle Tom's Cabin* put a personal face on the policy debates of that time.[14] The novel by Harriet Beecher Stowe dealt with the political issues surrounding the Compromise of 1850, a pre-Civil War law requiring the return of fugitive slaves to their owners in the South. Stowe set out to write her novel as a way of making "this whole nation feel what an accursed thing slavery is."[15]

The novel describes the horror of slavery on a Kentucky plantation and a mother's attempt to save her child by escaping to freedom in Ohio. The indebted plantation owner proposes to sell the child Harry to a New Orleans slave dealer in order to reduce plantation debts. The mother Eliza plots escape across the ice-clogged Ohio River in order to prevent the impending separation. Every mother in nineteenth-century America had watched a child suffer through terrible childhood diseases, and many, including Stowe herself, had watched at least one die.

Stowe touched on this agony of motherhood by reminding her readers that every day the American slave trade stripped children from their mothers. She pleaded with readers to remove this "great and unredressed injustice," a source of pain with which millions of families could immediately identify.[16]

Uncle Tom's Cabin did much to define public perceptions of slavery and the purpose of the Civil War. A Southern reviewer, who forcefully objected to the images in the book, nonetheless agreed that the novel had succeeded in "filling the minds of all who know nothing of slavery with hatred for that institution."[17] Upon meeting Stowe, President Abraham Lincoln is said to have remarked, "So you're the little lady who started the war."[18] His testimony to the power of the novel was exaggerated, but not unwarranted. The novel helped create instabilities in attitudes toward slavery that punctuating events, such as the firing on Fort Sumter and Lincoln's own election, translated into a policy of war.

Historically, cultural shifts such as these take a very long time to occur. Attitudes toward social welfare took centuries to change; the rise of the conservation ethic in the public mind took place over a period of one hundred years; the slavery debate consumed the United States for much of half a century. The culture shift that gave rise to space exploration, during which space pioneers converted science fiction fantasies into images of something real, occurred in less than ten years. The access to American popular culture created by television, amusement parks, and popular magazines allowed this cultural transformation to take place far more rapidly than had previously occurred.

In the same manner that imagination helps create culture shifts before policies change, so it helps to resolve disputes during policy debates. In a society saturated with information, experts frequently appear to disagree. Where expert opinion is divided, amateurs must turn elsewhere for guidance. Imagination can affect the resolution of policy disputes by putting a personal face on the positions taken by divided parties. The history of the effort to deinstitutionalize the mentally ill shows how this can occur.

Prior to the 1960s, mentally ill persons were often housed in large state institutions, grim places resembling prisons more than hospitals. A group of reformers sought to close the facilities and replace them with community mental health centers, and a few reformers went so far as to suggest that the "crazy" behavior of inmates was a normal reaction to the grim conditions of incarceration in large state-run institutions. Being a temperamental lot, artists have often expressed reservations about confinement of the mentally ill.[19] In 1962 Ken Kesey published *One Flew Over the Cuckoo's Nest,* a novel depicting conditions at an Oregon state mental institution from the point of view of one of the inmates. The story pits the freewheeling Randle Patrick McMurphy, who possesses the power to cure the inmates from within, against the representative of institutional authority, Big Nurse Ratched.[20] The novel, which subsequently became a Broadway play and an award-winning Hollywood movie, helped undermine public confidence in state-run mental health institutions and make deinstitutionalization what psychiatrist Paul McHugh has called the "cultural fashion" of the day.[21] At that time, the empirical evidence on the virtues of deinstitutionalization was mixed. Expert opinion was divided. The artistic portrayal of the issue, however, captivated the public mind. It fit so well into the philosophy of the counterculture movement then sweeping the American scene that many people accepted it without question.

Culture and imagination alone did not prompt the push toward deinstitutionalization. Medical technology, which produced drugs for the treatment of mental disorders, as well as changes in the methods of financing mental health care, also favored the dismantling of large state-run institutions. Imagination, nonetheless, played a critical role. As in the case of space exploration, when imagination started pushing in the same direction as technology and economics, sweeping policy change was unavoidable. Tens of thousands of mentally ill persons were released from state-run institutions, often to roam city streets as the homeless poor.[22]

Imagination even affects the type of administrative methods the government employs. Once again, experts disagree. Many experts are convinced that public bureaucracies are destructive of democratic values; others believe that bureaucracies are essential to the operation of the modern state.[23] In the realm of imagination, however, the contest is not in dispute. Works of imagination overwhelmingly favor nonbureaucratic methods of administration.

Stupid government bureaucrats are a staple fare in modern fiction. The Ministry of Love (i.e., torture) exists solely to perpetuate the jobs and power of the people who run it in George Orwell's *1984.* Colonel Cathcart in *Catch-22* endangers the lives of the pilots under his command so that he can win a promotion. A government bureaucrat releases hundreds of demons onto New York streets after learning that a "Ghostbusters" team has captured and stored toxic spirits without a proper license. Portraying government agents as stupid or evil is a well-established formula neatly guaranteed to please audiences seeking to be entertained.[24]

The formula works because it appeals to a deeply-held ideal in American culture. As Alexis de Tocqueville noted during his nineteenth-century visits, frontier Americans depended more upon each other than upon public officials to resolve mutual affairs. Community barn raising became an icon of this attitude among Americans so inclined to believe.[25] Even as the circumstances of frontier life disappeared, the ideals that those circumstances fostered remained, and works of imagination helped to perpetuate them. The writings of Mark Twain, the preeminent American novelist, contain a prominent anti-institutional streak.[26] The American Western, and the frontier myth it helps to maintain, reenforce the idea that the highest moral virtues emerge from individuals working together in a pregovernmental "state of nature."[27] With so many works of imagination preaching this theme, it is hard to maintain the idea of the good bureaucracy.

The very act of imagining organizations influences their operation. People develop mental images that help them understand how things work. Some of the images are based upon metaphors; other are based upon ideas. At the turn of the century, industrialists visualized business organizations using the metaphor of the machine. They sought to engineer the structure of the firm. It should come as no surprise, then, that industrialists who viewed the factory as a complex machine treated their employees as cogs, buying and selling labor as they would a fan belt or an electric motor. Gareth Morgan calls this "imaginization."[28] Such ideas and metaphors have a self-fulfilling quality. People use them to organize reality. When the popular culture presents the idea of bureaucrats as self-serving idiots, then the public tends to treat their organizations that way. Among the public at large, "bureaucracy" is a pejorative word. It stands for impersonal, inattentive, and unresponsive service. It is synonymous with rigidity and bloat.[29] This image owes much of its persistence to works of imagination.

If works of imagination help shift public attitudes in preparation for policy changes prompted by punctuating events, can a work of imagination, by itself, provide the event? The most dramatic example of such an event occurred during the first decade of this century, when a single work of fiction motivated the Congress to create the federal system for regulating food and drugs. For some time, reformers had promoted the virtues of federal regulation of the nation's food supply. Their requests for legislation were easily checked by the resistance of industry lobbyists, who commanded congressional majorities. In 1906 Upton Sinclair published *The Jungle,* a novel portraying the lives of immigrants who labored in Chicago's stockyards and meat-packing plants.[30] Sinclair's work was supported by a socialist weekly, *The Appeal to Reason,* where the novel first appeared in serial form. Through a work of fiction, the publishers hoped to arouse public sympathy for the plight of workers and advance a program of socialist reform.

The characters in Sinclair's novel suffer unimaginable tragedies arising from disease and broken health and the hopeless search for employment. Readers were not moved by Sinclair's descriptions of human exploitation so much as they were terrified by the background information he presented on the meat-packing industry, especially the loose system of local inspection that allowed diseased cattle and hogs to slip into the nation's food supply. Sinclair related industry folklore, such as tales of workers who slipped from factory planks above boiling vats to emerge somewhat later as Durham's Pure Beef Lard. Scarcely a dozen pages in the book described such incidents, but they gripped the attention of the public and made the twenty-seven-year-old author an instant celebrity. "I aimed at the public's heart," Sinclair observed, "and by accident I hit it in the stomach."[31]

As interest in the book rose, sales of meat products fell. Industry efforts to reassure the public through forums such as the *Saturday Evening Post* failed. President Theodore Roosevelt invited Sinclair to the White House and dispatched a special government investigating team to Chicago to assess the charges. The Neill-Reynolds report confirmed the worst and intensified calls for reform. With no other means to restore public confidence in their products, industry officials accepted federal regulation. Less than six months after the release of the novel, Roosevelt signed legislation establishing the modern system of food and drug regulation. The novel had had a direct impact on public policy, enlisting a new audience in such as way as to reshape an old debate.

Works of imagination enlarge public interest in policy debates by putting a human face on otherwise torpid issues. Fiction personalizes complex issues, allowing the public to reduce complicated controversies to familiar values such as love and tragedy. By enlarging the participation in a policy debate, works of imagination can alter the balance of power between proponents and opponents of change. As E. E. Schattschneider noted in his classic work on political strife, losers typically seek to escape their minority status by asking the inattentive to join the debate. "The most important strategy of politics," Schattschneider observed, "is concerned with the scope of conflict."[32] People with the skill to involve an otherwise inattentive audience through the power of imagination possess the power to change public policy.

At some point in time, the policies that imagination and culture help create must confront reality. In some cases, given a creative presentation of that reality, actual events can satisfy expectations. The popularity of the 1991 Gulf War in the United States was due in large measure to such a presentation abetted by a friendly reality. The war was short-lived and fast-paced, and the opposing nation led by a character of villainous style. Smart bombs and stealth raids made combat look like a high-tech computer game. On the U.S. side at least, machinery rather than soldiers stood in harm's way. The news media joined in the display, using devices that gave the war the character of a television drama rather than of a news broadcast.[33] In stark contrast to the conflict in Vietnam, the Gulf War seemed to satisfy rather than irritate public expectations.

Space exploration followed a more common pattern. As space activities became familiar, faith in the immediate attainment of the originating vision declined. "The dream of space travel is glorious," sociologist William Sims Bainbridge observed, "the contemporary reality is dismal."[34] Completion of the initial flights to the Moon inaugurated a period in which knowledgeable people began to sense that the United States was not going to complete the grand vision of exploration and settlement, certainly not soon. "It's over," said one critic of big space endeavors in 1994. "Turn out the lights, have a nice life, you're out of here."[35] Author Paul Theroux offered a similar glimpse of the real future:

> Forget rocket-ships, super-technology, moving sidewalks and all the rubbishy hope in science fiction. No one will ever go to Mars and live. A religion has evolved from the belief that we have a future in outer space; but it is a half-baked religion—it is a little like Mormonism or the Cargo Cult. Our future is this mildly poisoned earth and its smoky air. . . . There will be no star wars or galactic empires and no more money to waste on the loony nationalism in space programs.[36]

Works of imagination have become so pervasive in American culture that the latitude of the government to satisfy them grows narrower by the day. Politicians are obliged by the nature of their jobs to satisfy public expectations, but the expectations that imagination creates grow more and more unattainable. Such a situation undermines government and generates public distrust with amazing speed.[37]

What happens to the bearers of the original vision when this occurs? In part, the bearers can retreat into fantasy once more. Images of space travel did not retreat from the American consciousness so much as they moved back into the realm of fantasy that early space pioneers had struggled so hard to overcome. The symbolic watershed in this development was the decision of the Walt Disney corporation in the early 1990s to close its Mission to Mars ride. The direct descendant of the original Rocket to the Moon attraction that opened during the first year of park operation, Mission to Mars had helped keep alive the belief that space travel was real. As public excitement about actual space travel waned, so did interest in the ride. At Walt Disney World, Disney executives replaced Mission to Mars with Alien Encounter, a total fantasy featuring an extraterrestrial being accidently beamed into a spaceship taking an imaginary voyage. "One way for an attraction to remain timeless is for it to be based in fantasy, rather than reality," a Disney representative observed.[38]

Because subjective expectations are rooted in imagination, they often are not objectively real. Even when they are real, expectations may change as society moves on. Either of these situations is sure to create a gap between expectations and reality, a nightmare for officials trying to carry out public policies.

Policy advocates can encourage excessive expectations by overselling the programs they promote. The head of the Atomic Energy Commission, Lewis L. Strauss, predicted in a 1954 speech to the National Association of Science Writers that atomic energy would create electricity so cheap that the power industry would not have to meter it.[39] A series of aviation advocates promised that airplanes would become so inexpensive for the average American that by the second half of the twentieth century a personal airplane would replace the automobile in the family garage.[40] In launching the development program for the nation's space shuttle, President Nixon repeated the oft-heard claim that the new transportation system would reduce the access cost to space "by a factor of ten."[41] Government programs launched with great expectations breed imagination gaps as the prophecies fail to come true.

From the beginning of the effort to persuade Americans that space flight is real, critics have labored to create a more realistic view. In the marketplace of ideas, however, competition between points of view is not decided wholly on the basis of facts. In the early years of space travel, the known facts were few. People accepted the grand vision because it was more attractive. It attracted adherents, as this book has sought to show, because it was more exciting, more entertaining, and more compatible with other beliefs already well established in the public mind.

This is a formula for disappointment—not just in outer space, but among many public policies as well. Gaps between expectations and reality invite discontent. And what are advocates to do? Advocates of old policies do not abandon their efforts just because expectations fail; they freshen those expectations with new dreams. In some ways, the reaction of policy advocates so disturbed resembles that of small groups of true believers forced to confront an opposing reality. Long studied by social scientists, such groups adopt a number of behaviors designed to maintain their beliefs as their prophecies fail.[42] Such groups may increase their level of proselytizing or search for new ideas that confirm old behavior.[43] They ignore unfulfilled predictions, issue rationalizations, and discourage members who want the group to modify its beliefs.

All public policies have a subjective dimension rooted in imagination that affects their development. For many, such as the U.S. space program, the subjective dimension is as important as forces such as money and politics. Such programs typically begin amidst a mood of great euphoria and expectation. The more heavily the program draws upon imagination, the more intense the period of euphoria is likely to be. As experience accumulates, euphoria fades. Eventually the expectations of early advocates confront reality. For policies heavily rooted in imagination, this is likely to be an especially turbulent period. Program advocates may experience a loss of public support. The willingness of the news media to glorify the program may wane. Politicians may withdraw funding and cancel projects. As the fears and yearnings that motivated the policy disappear, the new reality creates a stimulus for policy change.

The period of turbulence is likely to be a crucial one for the future development of the policy. If advocates can fashion new images that still attract public interest, government policy may enter a new level of maturity. Alternatively, it may wither away. America aviation, for example, emerged from its barnstorming era to become one of the dominant technologies of the twentieth century. It did not fulfill the euphoric expectations of aviation pioneers, but it transformed society in significant ways.[44]

The U.S. space program faces such a challenge. The reality of space travel has depleted much of the vision that originally inspired it. Space-flight engineers have not developed technologies capable of achieving the dream; advocates have not formulated alternative visions capable of maintaining it. At the same time, no alternative vision of sufficient force has appeared to supplant the original dream. Advocates still embrace the original vision of adventure, mystery, and exploration. They continue to dream of expeditions to nearby planets and the discovery of habitable worlds. The dreams continue, while the gap between expectations and reality remains unresolved.

NOTES

INTRODUCTION: THE VISION

Epigraphs: Prov. 29:18 King James Version; see also Lyndon B. Johnson, "Remarks Upon Viewing New Mariner 4 Pictures from Mars," 29 July 1965, *Public Papers of the Presidents of the United States* (Washington, D.C.: GPO, 1966), 805. Joel Achenbach, "Uncle Sam's Second Mortgage on America," *Washington Post,* 8 February 1995, sec. B, p. 8.

1. See Space Task Group, *The Post-Apollo Space Program: Directions for the Future* (Washington, D.C.: Executive Office of the President, 1969); National Commission on Space (Thomas O. Paine, chair), *Pioneering the Space Frontier* (New York: Bantam Books, 1986); Synthesis Group on America's Space Exploration Initiative (Thomas P. Stafford, chair), *America at the Threshold: Report of the Synthesis Group on America's Space Exploration Initiative* (Washington, D.C.: GPO, 1991).

2. Office of the White House Press Secretary (San Clemente, Calif.), The White House, Press Conference of Dr. James Fletcher and George M. Low, San Clemente Inn, California, 5 January 1972, NASA History Office, 8; Jules Crittenden, "NASA Chief Sets Sights on Manned Trip to Mars," *Boston Herald,* 11 July 1995; "Mission to Mars may be $300 billion cheaper than first estimates," *Aerospace Daily,* 9 March 1992, 389; Adam Rogers, "Son of the Shuttle," *Newsweek,* 1 July 1996, 69; Anne Eisele, "X-33 Competitors Wait Anxiously," *Space News,* 24 June 1996, 3; see also Carl Sagan, *Pale Blue Dot: A Vision of the Human Future in Space* (New York: Random House, 1994).

3. David B. Guralnik, ed., *Webster's New World Dictionary,* 2d College ed. (New York: Simon and Schuster, 1984).

4. "Now, my suspicion is that the universe is not only queerer than we suppose, but queerer than we *can* suppose." J. B. S. Haldane, *Possible Worlds and Other Papers* (New York: Harper & Brothers, 1927), 298.

5. John F. Kennedy, "Special Message to the Congress on Urgent National Needs," 25 May 1961, *Public Papers of the Presidents of the United States* (Washington, D.C.: GPO, 1962), 404; National Commission on Space, *Pioneering the Space Frontier,* opposite table of contents; see also Joseph J. Corn, *The Winged Gospel: America's Romance with Aviation, 1900–1950* (New York: Oxford University Press, 1983).

6. See also Frederick I. Ordway and Randy Liebermann, *Blueprint for Space: Science Fiction to Science Fact* (Washington, D.C.: Smithsonian Institution Press, 1992).

7. See W. Lance Bennett, *News: The Politics of Illusion,* 3d ed. (New York: Longman, 1996); James Fallows, *Breaking the News: How the Media Undermine American Democracy* (New York: Pantheon Books, 1996).

1. PRELUDE: THE EXPLORATION IDEAL

Epigraph: Michael Crichton, *Sphere* (New York: Alfred A. Knopf, 1987), 348.

1. NASA, "Minutes of Meeting of Research Steering Committee on Manned Space Flight," NASA Headquarters, Washington, D.C., 25–26 May 1959, 2.

2. NASA Office of Program Planning and Evaluation, "The Long Range Plan of the National Aeronautics and Space Administration," 16 December 1959, 28; NASA History Office, NASA Headquarters, Washington, D.C. (hereafter cited as NASA History Office).

3. Willy Ley, *Rockets: The Future of Travel Beyond the Stratosphere* (New York: Viking Press, 1944), chap. 1; Wernher von Braun and Frederick I. Ordway, *Space Travel: A History* (New York: Harper and Row, 1985), chap. 1; Johannes Kepler, *Somnium: The Dream* (Madison: University of Wisconsin Press, 1967).

4. Edward Everett Hale, *The Brick Moon* (1869; reprint, New York: Spiral Press, 1971); also see Frederick I. Ordway, "Dreams of Space Travel from Antiquity to Verne," in Ordway and Liebermann, *Blueprint for Space,* 35–48; and Ley, *Rockets,* chaps. 1 and 2.

5. Edgar Rice Burroughs, *A Princess of Mars* (Garden City, N.Y.: Doubleday, 1917).

6. William Shakespeare, *The Tempest* (New York: Oxford University Press, 1987); Nicholas Nayfack, *Forbidden Planet* (MGM, 1956).

7. Daniel Defoe, *Robinson Crusoe,* 1719 (New York: Knopf, 1992); C. S. Lewis, *The Lion, the Witch and the Wardrobe* (New York: Collier Books, 1950); Lewis Carroll, *Alice's Adventures in Wonderland and Through the Looking Glass* (New York: New American Library, 1960); L. Frank Baum, *The Wizard of Oz* (New York: Ballantine Books, 1979).

8. K. E. Tsiolkovskiy, "Exploration of the Universe with Reaction Machines" (1926) in *Collected Works of K. E. Tsiolkovskiy,* vol. 2, *Reactive Flying Machines,* ed. A. A. Blagonravov, NASA technical translation, NASA TT F-237 (Washington, D.C.: NASA, 1965), 212.

9. Jules Verne, *From the Earth to the Moon; All Around the Moon* (New York: Dover Publications, 1962); Ron Miller, "The Origin of the Rocket-Propelled Spaceship," *Quest* 4 (Winter 1995): 4–7.

10. "The Autobiography of K. E. Tsiolkovskii," in *Interplanetary Flight and Communication* vol. 3, no. 7, ed. N. A. Rynin (Jerusalem: Israel Program for Scientific Translations, 1971), 3; K. E. Tsiolkovskiy, "Investigation of Universal Space by Reactive Devices" (1926), in *Works on Rocket Technology,* ed. K. E. Tsiolkovskiy, NASA technical translation, TT F-243 (Washington, D.C.: NASA, 1965), 208–15; K. E. Tsiolkovskiy, *Beyond the Planet Earth* (novel), trans. Kenneth Syers (New York: Pergamon Press, 1960); see also Michael Stoiko, *Pioneers of Rocketry* (New York: Hawthorne Books, 1974).

11. Hermann Oberth, "From My Life," *Astronautics* (June 1959): 39.

12. Fritz Lang, *Frau im Mond,* 1929 (available through Foothill Video, Tujunga, California).

13. Willy Ley, *Rockets, Missiles, and Men in Space* (New York: Viking Press, 1968), 114.

14. Taken from Frank H. Winter, "Man, Rockets and Space Travel," *Mankind* 2 (December 1969): 23.

15. Esther C. Goddard and G. Edward Pendray, eds., *The Papers of Robert H. Goddard* (New York: McGraw-Hill, 1970), 1:7; see also von Braun and Ordway, *Space Travel,* 44; and R. H. Goddard to H. G. Wells, 20 April 1932, in J. D. Hunley, "Robert H. Goddard, Enigmatic Space Pioneer," unpublished paper, 1993, NASA History Office.

16. Quoted by Tom Crouch, "'To Fly to the Moon': Cosmic Voyaging in Fact and Fiction from Lucian to Sputnik," in *Science Fiction and Space Futures, Past and Present,* ed. Eugene Emme, AAS History series, vol. 5 (San Diego: American Astronautical Society, 1982), 8.

17. Goddard and Pendray, *Papers of Robert H. Goddard* 1:117, 419–20, 3:1612.

18. Goddard and Pendray, *Papers of Robert H. Goddard* 1:117, 121; Robert H. Goddard, *A Method of Reaching Extreme Altitudes,* Smithsonian Miscellaneous Collections, vol. 71, no. 2 (Washington, D.C.: Smithsonian Institution, 1919). See also John M. Logsdon, ed., *Exploring the Unknown: Selected Documents in the History of the U.S. Civil Space Program,* NASA SP-4218 (Washington, D.C.: NASA, 1995), 1:86–133.

19. "Topics of the Times," *New York Times,* 13 January 1920; Frank H. Winter, *Rockets into Space* (Cambridge: Harvard University Press, 1990), 19, 29.

20. Frank H. Winter, *Prelude to the Space Age: The Rocket Societies, 1924–1940* (Washington, D.C.: Smithsonian Institution Press, 1983).

21. *Bulletin,* American Interplanetary Society, June 1930, reproduced in Ordway and Liebermann, *Blueprint for Space,* 111.

22. G. Edward Pendray, "32 Years of ARS History," *Astronautics and Aerospace Engineering* 1 (February 1963): 124.

23. Willy Ley, "How It All Began," *Space World* 1 (June 1961): 48 and 50.

24. Woodford A. Helfin, "Who Said It First? 'Astronautics,'" *Aerospace Historian* 16 (Summer 1969): 44–47; see also Winter, *Prelude to the Space Age,* 25.

25. H. E. Ross, "The B.I.S. Space-ship," *Journal of the British Interplanetary Society* 5 (January 1939): 4–9; R. A. Smith, "The B.I.S. Coelostat," *Journal of the British Interplanetary Society* 5 (July 1939): 22–27; Arthur C. Clarke, "An Elementary Mathematical Approach to Astronautics," *Journal of the British Interplanetary Society* 5 (January 1939): 26–28; and other articles in the January 1939 and July 1939 issues of the journal. See also H. E. Ross, "The British Interplanetary Society's Astronautical Studies, 1937–39," in *First Steps Toward Space,* ed. Frederick C. Durant and George S. James, Smithsonian Annals of Flight no. 10 (Washington, D.C.: Smithsonian Institution Press, 1974), 209–16; Arthur C. Clarke, "We Can Rocket to the Moon—Now!" *Tales of Wonder* 7 (Summer 1939): 84–88.

26. Winter, *Prelude to the Space Age;* Frederick I. Ordway and Mitchell R. Sharpe, *The Rocket Team* (New York: Thomas Y. Crowell, 1979).

27. Ley, "How It All Began," 23–25, 48–52; *Current Biography 1953* (New York: H. W. Wilson, 1954), s.v. "Willy Ley," 356–59.

28. Walter Dornberger, quoted in Ordway and Sharpe, *Rocket Team,* 18.

29. Quoted in Daniel Lang, "A Reporter at Large: A Romantic Urge," *New Yorker* 27 (21 April 1951): 74.

30. Ordway and Sharpe, *Rocket Team,* 19.

31. Michael J. Neufeld, *The Rocket and the Reich: Peenemünde and the Coming of the Ballistic Missile Era* (New York: Free Press, 1995); Ordway and Sharpe, *Rocket Team,* chap. 1, 79 and 251–52.

32. Ernst Stuhlinger and Frederick I. Ordway, *Wernher von Braun: Crusader for Space* (Malabar, Fla.: Krieger Publishing, 1994), chap. 2.

33. Quoted in *Current Biography 1953,* s.v. "Willy Ley."

34. G. Edward Pendray, "The First Quarter Century of the American Rocket Society," *Jet Propulsion* 25 (November 1955): 587.

35. Pendray, "First Quarter Century of the American Rocket Society," 587.

36. The appendix to the novel, a technical treatise, was published in Germany and the United States. See Wernher von Braun, *The Mars Project* (Urbana: University of Illinois Press, 1953). Von Braun later published a fictional account of a trip to the Moon. See Wernher von Braun, *First Men to the Moon* (New York: Holt, Rinehart & Winston, 1958) and "First Men to the Moon," *Reader's Digest,* January 1961, 175–92.

37. See Steven Pyne, *The Ice* (Iowa City: University of Iowa Press, 1986); Henry M. Stanley, *How I Found Livingston* (Montreal: Dawson, 1872); David Mountfield, *A History of African Exploration* (New York: Hamlyn, 1976); Richard A. Van Orman, *The Explorers: Nineteenth Century Expeditions in Africa and the American West* (Albuquerque: University of New Mexico Press, 1984); Alfred Runte, *National Parks: The American Experience,* 2d ed. (Lincoln: University of Nebraska Press,

1987); George Kennan, "Announcement," National Geographic Society, Washington, D.C., October 1888; Sam Moskowitz, "The Growth of Science Fiction from 1900 to the Early 1950s," in Ordway and Liebermann, *Blueprint for Space,* 69–82.

38. Ley, *Rockets,* 3.

39. Philip E. Cleator, *Rockets Through Space: The Dawn of Interplanetary Travel* (New York: Simon and Schuster, 1936), 203.

40. Maurice K. Hanson, "The Payload on the Lunar Trip," *Journal of the British Interplanetary Society* 5 (January 1939): 16; "Report of the Technical Committee," *Journal of the British Interplanetary Society* 5 (July 1939): 19.

41. See, for example, George Pal, *Destination Moon* (Eagle Lion, 1950); and George Pal, *The Conquest of Space* (Paramount, 1955).

42. Daniel J. Boorstin, *The Discoverers* (New York: Random House, 1983), 278.

43. Lonora Foerstel and Angela Gilliam, *Confronting the Margaret Mead Legacy* (Philadelphia: Temple University Press, 1992); Joseph Conrad, *Heart of Darkness* (New York: Penguin Books, 1995).

2. MAKING SPACE FLIGHT SEEM REAL

Epigraph: "Man Will Conquer Space Soon," *Collier's,* 22 March 1952, 22–23.

1. George Gallup, ed., *The Gallup Poll: Public Opinion, 1935–1971* (New York: Random House, 1972), 2:875.

2. Walter A. McDougall, *. . . the Heavens and the Earth: A Political History of the Space Age* (New York: Basic Books, 1985), 12. McDougall expropriated this framework from sociologist Daniel Bell, "Technology, Nature, and Society," in *The Winding Passage,* by Bell (Cambridge: Abt Books, 1980).

3. Richard Adams Locke, *The Moon Hoax; or, A Discovery that the Moon Has a Vast Population of Human Beings* (Boston: Gregg Press, 1975); Roger Lancelyn Green, *Into Other Worlds: Space-Flight in Fiction, from Lucian to Lewis* (New York: Arno Press, 1975), chap. 7.

4. Edward Everett Hale, *The Brick Moon and Other Stories* (Freeport, N.Y.: Books for Libraries Press, 1970).

5. Rem Koolhaas, *Delirious New York* (New York: Oxford University Press, 1978).

6. Brian Horrigan, "Popular Culture and Visions of the Future in Space, 1901–2001," in *New Perspectives on Technology and American Culture,* ed. Bruce Sinclair (Philadelphia: American Philosophical Society, 1986), 52.

7. Marshall B. Tymn and Mike Ashley, *Science Fiction, Fantasy, and Weird Fiction Magazines* (Westport, Conn.: Greenwood Press, 1985).

8. Sam Moskowitz, *Science Fiction by Gaslight: A History and Anthology of Science Fiction in the Popular Magazines, 1891–1911* (Westport, Conn.: Hyperion Press, 1968); Paul A. Carter, *The Creation of Tomorrow: Fifty Years of Magazine Science Fiction* (New York: Columbia University Press, 1977).

9. Robert C. Dille, ed., *The Collected Works of Buck Rogers in the 25th Century* (New York: A&W Publishers, 1977).

10. John F. Kasson, *Amusing the Million: Coney Island at the Turn of the Century* (New York: Hill and Wang, 1978), 61; Frank H. Winter and Randy Liebermann, "A Trip to the Moon," *Air & Space* (November 1994): 62–76.

11. Gary Grossman, *Saturday Morning T.V.* (New York: Dell Publishing, 1981), 138.

12. E. E. "Doc" Smith, *The Skylark of Space* (New York: Pyramid Books, 1928).

13. The quote is taken from the "Writers Guide" to the "Star Trek" television series, 17 April 1967, lent by Gregory Jein to the National Air and Space Museum. The entire quote follows: "Let's go back to the days when some of us were working on the first television westerns. We did *not* recreate the Old West as it actually existed; instead we created a new Western form, actually a vast colorful backdrop *against which any kind of story could be told.*"

14. David Lasser, *The Conquest of Space* (New York: Penguin Press, 1931), 5.

15. *Bulletin,* the American Interplanetary Society, June 1930, reproduced in Ordway and Liebermann, *Blueprint for Space,* 111.

16. Jo Ranson, "Radio Dial-Log," *Brooklyn Daily Eagle*, 1 March 1932; Winter, *Prelude to the Space Age,* 76.

17. Cleator, *Rockets Through Space,* 163.

18. Willy Ley, *Rockets and Space Travel* (New York: Viking Press, 1947); Chesley Bonestell and Willy Ley, *The Conquest of Space* (New York: Viking Press, 1949).

19. Arthur C. Clarke, *The Exploration of Space* (New York: Harper and Brothers, 1951), 183–84; see also Neil McAleer, *Odyssey: The Authorised Biography of Arthur C. Clarke* (London: Victor Gollancz, 1992); and *Current Biography 1966,* s.v. "Arthur C. Clarke," 49–52.

20. Clarke, *Exploration of Space,* 61–62.

21. Ibid., 182.

22. Willy Ley, letter of invitation, First Annual Symposium on Space Travel, 1951, American Museum of Natural History, Hayden Planetarium Library, New York, N.Y. (hereafter cited as American Museum of Natural History).

23. Willy Ley, "Thirty Years of Space Travel Research," and American Museum of Natural History, "Space Travel Symposium Held at Hayden Planetarium," 12 October 1951, American Museum of Natural History.

24. Robert R. Coles, "The Conquest of Space," First Annual Symposium on Space Travel, 12 October 1951, American Museum of Natural History.

25. Robert R. Coles, "The Role of the Planetarium in Space Travel," Second Symposium on Space Travel, 13 October 1952, American Museum of Natural History.

26. Ordway and Sharpe, *Rocket Team.*

27. Wernher von Braun, "The Early Steps in the Realization of the Space Station," Second Symposium on Space Travel, 13 October 1952, American Museum of Natural History.

28. Joseph M. Chamberlain, "Introductory Remarks," Third Symposium on Space Travel, 4 May 1954, 3, American Museum of Natural History.

29. American Museum of Natural History, "Space-Travel Specialists Confer at Planetarium," 4 May 1954, American Museum of Natural History.

30. R. C. Truax, "A National Space Flight Program," Third Symposium on Space Travel, 4 May 1954, American Museum of Natural History.

31. Harry Hansen, ed., *The World Almanac and Book of Facts for 1952* (New York: New York World-Telegram and the Sun, 1952).

32. Fred L. Whipple, "Recollections of Pre-Sputnik Days," in Ordway and Liebermann, *Blueprint for Space,* 129.

33. "Man Will Conquer Space Soon," *Collier's,* 22 March 1952, cover and 22–23.

34. Wernher von Braun, "Man on the Moon: The Journey," *Collier's,* 18 October 1952, 51.

35. Wernher von Braun and Cornelius Ryan, "Baby Space Station," *Collier's,* 27 June 1953, 34.

36. Fred L. Whipple, "Is There Life on Mars?" *Collier's,* 30 April 1954, 21.

37. "Journey into Space," *Time,* 8 December 1952, 62.

38. "The Seer of Space: Lifetime of Rocket Work Gives Army's Von Braun Special Insight into the Future," *Life,* 18 November 1957, cover, 133–39.

39. David R. Smith, "They're Following Our Script: Walt Disney's Trip to Tomorrowland," *Future,* May 1978, 55.

40. Ward Kimball, *Man in Space* (Walt Disney, 1955).

41. Ibid.

42. Ward Kimball, *Man and the Moon* (Walt Disney, 1955). (Also titled *Tomorrow the Moon.*)

43. Ward Kimball, *Mars and Beyond* (Walt Disney, 1957).

44. Ryan A. Harmon, "Yesterday, Today, and Tomorrowland," *Disney News* 26 (Spring 1991): 18–23; Bruce Gordon and David Mumford, "Tomorrowland 1986: The Comet Returns," unpublished

manuscript, Walt Disney Archives, Burbank, California. Disney chose 1986 because it was the year that Halley's Comet would return.

45. "Blast Off!" *Vacationland* 5 (Spring 1961): 13, published by Disneyland, Inc., Anaheim History Room, Anaheim Public Library, Anaheim, California (hereafter cited as Anaheim History Room).

46. Kimball, *Tomorrow the Moon.*

47. "Space Travel: The Trip to the Moon Disneyland Style," *Disneyland News,* 10 March 1956, 10–11; "Feeling of Space Journey Accompanies Trip to Moon," *Disneyland News,* January 1957.

48. Smith, "They're Following Our Script," 56.

49. Harmon, "Yesterday, Today, and Tomorrowland," 19.

50. Walt Disney Productions, *Disneyland Diary: 1955–Today,* 1982; Jon C. A. DeKeles, "Disneyland Concordance," 1982; both in Anaheim History Room.

51. "Tomorrowland Built with Aid of Experts," *Disneyland News,* September 1955, 6.

52. David A. Hardy, *Visions of Space* (New York: Gallery Books, 1989).

53. James Nasmyth and James Carpenter, *The Moon* (London: J. Murray, 1874).

54. T. E. R. Phillips, *Splendour of the Heavens* (London: Hutchinson and Company, 1923).

55. *Astounding Science Fiction,* June 1938 and April 1939.

56. Arthur C. Clarke, "Space Future: Visions of Space," *Spaceflight* (May 1986): 201.

57. F. Barrows Colton, "News of the Universe," *National Geographic* 76 (July 1939): 1–32.

58. "Interview: Chesley Bonestell," *Space World* (December 1985): 9–12; also see Frederick C. Durant and Ron Miller, *Worlds Beyond: The Art of Chesley Bonestell* (Norfolk: Donning Company, 1983).

59. See Sagan, *Pale Blue Dot,* 115.

60. "Solar System," *Life,* 29 May 1944, 78–86; quote from 80.

61. "Trip to the Moon: Artist Paints Journey by Rocket," *Life,* 4 March 1946, 73–76.

62. Quoted from Durant and Miller, *World's Beyond,* 9. For one of the working drawings, see 7.

63. Willy Ley and Wernher von Braun, *The Exploration of Mars* (New York: Viking Press, 1956).

64. Cornelius Ryan, ed., *Across the Space Frontier* (New York: Viking Press, 1952); Cornelius Ryan, ed., *Conquest of the Moon* (New York: Viking Press, 1953).

65. Mike McIntyre, "Celestial Visions," *Air and Space* (August/September 1986): 86–92.

66. See Ron Miller, *Space Art* (New York: Starlog Magazine, 1978).

67. Clarke, *Exploration of Space;* R. A. Smith (text by Arthur C. Clarke), *The Exploration of the Moon* (London: Frederick Muller, 1954).

68. Jack Coggins and Fletcher Pratt, *Rockets, Jets, Guided Missiles and Space Ships* (New York: Random House, 1951); Jack Coggins and Fletcher Pratt, *By Space Ship to the Moon* (New York: Random House, 1952). For later artists, see Ben Bova, *Vision of the Future: The Art of Robert McCall* (New York: Harry N. Abrams, 1982); Robert McCall, *The Art of Robert McCall* (New York: Bantam Books, 1992); Ron Miller and William K. Hartmann, *The Grand Tour: A Traveler's Guide to the Solar System* (New York: Workman, 1981); and William K. Hartmann, Ron Miller, and Pamela Lee, *Out of the Cradle: Exploring the Frontiers Beyond Earth* (New York: Workman, 1984).

69. Gail Morgan Hickman, *The Films of George Pal* (New York: A. S. Barnes, 1977), 36–46; Robert A. Heinlein, *Rocket Ship Galileo* (New York: Charles Scribner's Sons, 1947).

70. Robert Heinlein, "Shooting Destination Moon," from Yoji Kondo, *Requiem: New Collected Works by Robert Heinlein* (New York: Tom Doherty Associates, 1992), 121.

71. Cobbett A. Steinberg, *Reel Facts* (New York: Vintage Books, 1978), 344–45.

72. David Wingrove, *The Science Fiction Film Source Book* (Harlow, U.K.: Longman, 1985), 303.

73. Pal, *Conquest of Space.*

74. Wingrove, *Science Fiction Film Source Book,* 304.

75. Gallup, *Gallup Poll: Public Opinion, 1935–1971* 2:875, 1306.

76. Ibid. 2:1521–22.

77. Robert A. Devine, *The Sputnik Challenge* (New York: Oxford University Press, 1993), 102–5.

78. Hugh L. Dryden, "Space Technology and the NACA," *Aeronautical Engineering Review* 17 (March 1958): 32–44. Dryden could not attend the institute's annual meeting and his statement was delivered by John Victory, executive secretary of the NACA. See also Michael H. Gorn, ed., *Hugh L. Dyden's Career in Aviation and Space,* monographs in aerospace history no. 5 (Washington, D.C.: NASA History Office, 1996).

79. National Advisory Committee for Aeronautics, "On the Subject of Space Flight," a resolution, 16 January 1958, NASA History Division, Nasa Headquarters, Washington, D.C.

80. Dryden, "Space Technology and the NACA," 33.

81. NASA, "Minutes of Meeting of Research Steering Committee on Manned Space Flight," 2.

82. Ibid., 9.

83. Ibid., 10.

84. NASA, "Minutes of Meeting of Research Steering Committee on Manned Space Flight," Ames Research Center, 25–26 June 1959, 6.

85. NASA Office of Program Planning and Evaluation, table 1.

86. Daniel Herman, Workshop on Automated Space Station, Washington, D.C., 18 March 1984, in U.S. Senate Committee on Appropriations, a subcommittee, *Department of Housing and Urban Development, and Certain Independent Agencies Appropriations for Fiscal Year 1985,* 98th Cong., 2d sess., 1984, 1266.

87. NASA, "NASA Long Range Planning Conference," concluding remarks by Dr. Paine, transcript of proceedings, Wallops Island, Virginia, 14 June 1970, 32; NASA History Office.

88. NASA, "Post-Apollo Space Program: Directions for the Future," summary of National Aeronautics and Space Administration's Report to the President's Space Task Group, September 1969; NASA, "America's Next Decades in Space," a report for the Space Task Group, September 1969; Space Task Group, *Post-Apollo Space Program;* all in NASA History Division.

89. National Commission on Space, *Pioneering the Space Frontier,* frontispiece.

90. Ibid., opposite Contents.

91. NASA, "NASA Establishes Office of Exploration," release 87–87, *NASA News,* 1 June 1987.

92. The White House, Office of the Press Secretary, "Fact Sheet: Presidential Directive on National Space Policy," 11 February 1988. See also Ronald Reagan, "1988 Legislative and Administrative Message: A Union of Individuals," *Weekly Compilation of Presidential Documents* 24 (25 January 1988): 119.

93. NASA, "Civil Space Exploration Initiative," undated, in Howard E. McCurdy, "The Decision to Send Humans Back to the Moon and On to Mars," Space Exploration Initiative History Project, March 1992, NASA History Division; also see Edward McNally interview, 7 August 1992, Anchorage, Alaska.

94. George Bush, "Remarks on the 20th Anniversary of the *Apollo 11* Moon Landing," 20 July 1989, *Public Papers of the Presidents of the United States* (Washington, D.C.: GPO, 1990), bk. 2, p. 991.

95. Ibid., 993.

3. THE COLD WAR

Epigraphs: Statement of Democratic Leader Lyndon B. Johnson to the Meeting of the Democratic Conference on 7 January 1958, statements of LBJ collection, box 23, Lyndon Baines Johnson Library, Austin, Texas; Clark R. Chapman and David Morrison, "Chicken Little Was Right," *Discover* (May 1991): 40–43.

1. The White House, *Introduction to Outer Space* (Washington, D.C.: GPO, 1958), 10.

2. Aerospace Industries Association of America, *Aerospace Facts and Figures, 1962* (Washington, D.C.: American Aviation Publications, 1962), 20. Funds available for missile development and production, fiscal years 1954 through 1961.

3. U.S. Bureau of the Budget, *The Budget of the United States Government,* fiscal year 1958 (Washington, D.C.: GPO, 1957), message of the president, sec. M, p. 45.

4. "Address to the Nation by President Eisenhower on the Cost of Government," 14 May 1957, reproduced in Robert L. Brannyan and Lawrence H. Larsen, *The Eisenhower Administration, 1953–1961: A Documentary History* (New York: Random House, 1971), 845.

5. "What Are We Waiting For?" *Collier's,* 22 March 1952, 23.

6. President's Science Advisory Committee, "Report of Ad Hoc Panel on Man-in-Space," December 1960, 8, NASA History Office. See also John Logsdon, *The Decision to Go to the Moon* (Cambridge, Mass.: MIT Press, 1970), 34–35.

7. "What Are We Waiting For?" 23.

8. Von Braun, "Early Steps in the Realization," 5.

9. Milton W. Rosen, *The Viking Rocket Story* (New York: Harper and Brothers, 1955).

10. Milton W. Rosen, "A Down-to-Earth View of Space Flight," Second Symposium on Space Travel, 13 October 1952, 1, American Museum of Natural History.

11. Ibid., 5.

12. William L. Laurence, "2 Rocket Experts Argue 'Moon' Plan," *New York Times,* 14 October 1952; Robert C. Boardman, "Space Rockets with Floating Base Predicted," *New York Herald Tribune,* 14 October 1952.

13. Laurence, "2 Rocket Experts Argue"; Boardman, "Space Rockets with Floating Base."

14. "Journey into Space," 62, 68.

15. Ibid., 70.

16. See "Chief Whip in Scientific Race," *Business Week,* 16 November 1957, 42–43; *Current Biography 1959,* s.v. "James R. Killian," 229–31.

17. White House, *Introduction to Outer Space.*

18. Ibid., 6.

19. Ibid., 10.

20. Ad Hoc Committee on Space (Jerome B. Wiesner, chair), "Report to the President Elect," 10 January 1961, 15–16, NASA History Office.

21. President's Science Advisory Committee, "The Next Decade in Space," February 1970, see 2–3, NASA History Office. See also NASA, *Outlook for Space: A Synopsis* (Washington, D.C.: NASA, 1976).

22. See Karlyn Keene and Everett Ladd, "Government as Villain," *Government Executive* (January 1988): 11–16.

23. Gallup, *Gallup Poll: Public Opinion, 1935–1971* 3:1720.

24. Media General/Associated Press Public Opinion Poll, June and July 1988, Media General Research, Richmond, Virginia; George Gallup, *The Gallup Poll: Public Opinion, 1989* (Wilmington, Del.: Scholarly Resources, 1990), 169–73; see also Herbert E. Krugman, "Public Attitudes Toward the Apollo Space Program, 1965–1975," *Journal of Communication* 27 (Autumn 1977): 87–93; Michael A. G. Michaud, "The New Demographics of Space," *Aviation Space* (Fall 1984): 46–47.

25. Dwight D. Eisenhower, "Farewell Radio and Television Address to the American People, January 17, 1961," in *Public Papers of the Presidents of the United States* (Washington, D.C.: GPO, 1961), 1038.

26. Stephen E. Ambrose, *Eisenhower,* vol. 2, *The President* (New York: Simon and Schuster, 1984), 257.

27. Dwight D. Eisenhower, "Why I Am a Republican," *Saturday Evening Post,* 11 April 1964, 19.

28. Technological Capabilities Panel (James Killian, chair), "Meeting the Threat of Surprise Attack," 14 February 1955, Jet Propulsion Laboratory, Pasadena, California.

29. R. Cargill Hall, "Origins of U.S. Space Policy: Eisenhower, Open Skies, and Freedom of Space," in *Exploring the Unknown: Selected Documents in the History of the U.S. Civil Space Program,* vol. 1, NASA SP-4407, ed. John M. Logsdon (Washington, D.C.: GPO, 1995). Also see Constance McLaughlin Green and Milton Lomask, *Vanguard: A History* (Washington, D.C.: Smithsonian Institution Press, 1971).

30. See Hall, "Origins of U.S. Space Policy," 1:213–29.

31. Hugh L. Dryden, memorandum for Dr. James R. Killian, 18 July 1958, NASA History Office; NASA, *Mercury Project Summary,* NASA SP-45 (Houston: Manned Spacecraft Center, 1963), 2; Lloyd S. Swenson, James M. Grimwood, and Charles C. Alexander, *This New Ocean: A History of Project Mercury,* NASA SP-4201 (Washington, D.C.: GPO, 1966), esp. 110–11.

32. Dwight D. Eisenhower, "Annual Budget Message to Congress," 16 January 1961, *Public Papers of the Presidents of the United States,* 972.

33. President's Science Advisory Committee, "Report of Ad Hoc Panel on Man-in-Space," 14 November 1960, NASA History Office, 5.

34. White House, *Introduction to Outer Space,* 6, 15.

35. Dwight D. Eisenhower, "Are We Headed in the Wrong Direction?" *Saturday Evening Post,* 11–18 August 1962, 24; see also Dwight R. Eisenhower, "Spending into Trouble," *Saturday Evening Post,* 18 May 1963, 19.

36. Eisenhower, "Farewell Radio and Television Address," 1038.

37. Eisenhower, "Why I Am a Republican," 19.

38. Robert L. Rosholt, *An Administrative History of NASA, 1958–1963,* NASA SP-4101 (Washington, D.C.: GPO, 1966), 79–81.

39. Ibid, 213–14.

40. Henry C. Dethloff, *Suddenly, Tomorrow Came: A History of the Johnson Space Center,* NASA SP-4307 (Houston: Johnson Space Center, 1993), chap. 3.

41. William J. Jorden, "Soviet Fires Earth Satellite into Space; It Is Circling the Globe at 18,000 M.P.H.; Sphere Traced in 4 Crossings over U.S.; 560 Miles High," *New York Times,* 5 October 1957; "Satellite Announcement Brings Mixed Reaction," *New York Times,* 5 October 1957; W. H. Lawrence, "Eisenhower Gets Missile Briefing," *New York Times,* 9 October 1957; see also "Satellites and Our Safety," *Newsweek,* 21 October 1957, 29–39; and E. Nelson Hayes, "Tracking Sputnik I," in *The Coming of the Space Age,* ed. Arthur C. Clarke (New York: Meredith Press, 1967), 11–12.

42. Harry Schwartz, "A Propaganda Triumph," *New York Times,* 6 October 1957, sec. A, p. 43. See also "Satellite: The World Takes a Second Look," *U.S. News & World Report,* 18 October 1957, 110.

43. "A Time of Danger," *Time,* 11 November 1957, 23.

44. Mr. Hagerty's News Conference, Saturday, 5 October 1957, Dwight D. Eisenhower Library, Abilene, Kansas; see also "Soviet Fires Earth Satellite into Space: Device is 8 Times Heavier Than One Planned by U.S.," *New York Times,* 5 October 1957, p. A1.

45. "Soviet Fires Earth Satellite into Space," A1; W. H. Lawrence, "Eisenhower Gets Missile Briefing," *New York Times,* 9 October 1957, p. A1.

46. Dwight D. Eisenhower, "The President's News Conference of October 9, 1957," *Public Papers of the Presidents of the United States* (Washington, D.C.: GPO, 1958), 73.

47. Dwight D. Eisenhower, "Radio and Television Address to the American People on Science in National Security," November 7, 1957, *Public Papers of the Presidents of the United States,* 794.

48. "Into Space: Man's Awesome Adventure," *Newsweek,* 14 October 1957, 38.

49. "Defense: The Race to Come," *Time,* 21 October 1957, 21. See also "Russia's Satellite, a Dazzling New Sight in the Heavens," *Life,* 21 October 1957, 19–35.

50. "Space Travel, Not in Five or Ten Years," *U.S. News & World Report,* 18 October 1957, 106; "Sputniks and Budgets," *New Republic,* 14 October 1957, 3.

51. Gallup, *Gallup Poll: Public Opinion, 1935–1971* 2:1467, 1522.

52. National Advisory Committee for Aeronautics to Dr. Killian's Office, 6 August 1958, NASA History Office.

53. Swenson, Grimwood, and Alexander, *This New Ocean,* 91–101; Dryden, "Space Technology and the NACA," 33.

54. U.S. Senate Armed Services Committee, Preparedness Investigating Subcommittee, *Inquiry into Satellite and Missile Programs,* 85th Cong., 1st and 2d sess., 1957 and 1958; Lyndon B. Johnson and Styles Bridges, Statement of the Senate Preparedness Subcommittee, 23 January 1958, from Bryce Harlow Papers, Dwight D. Eisenhower Library, Abilene, Kansas.

55. Robert A. Divine, *The Sputnik Challenge: Eisenhower's Response to the Soviet Satellite* (New York: Oxford University Press, 1993).

56. Gallup, *Gallup Poll: Public Opinion, 1935–1971* 3:1720.

57. George Reedy, memo to Lyndon B. Johnson, 17 October 1957, LBJ Library, Senate Papers, Box 421.

58. G. Edward Pendray, "Next Stop the Moon," *Collier's,* 7 September 1946, 12 (emphasis added).

59. Ibid., 77.

60. Robert S. Richardson, "Rocket Blitz from the Moon," *Collier's,* 23 October 1948, 24–25, 44–46.

61. Ibid., 25.

62. Caleb B. Laning and Robert A. Heinlein, "Flight into the Future," *Collier's,* 30 August 1947, 36.

63. Coggins and Pratt, *By Space Ship to the Moon,* 1–2.

64. Wernher von Braun, "Crossing the Last Frontier," *Collier's,* 22 March 1952, 74.

65. Ibid., 26.

66. "What Are We Waiting For?" 23. See also "Hello, Down There," *Collier's,* 16 September 1955, 82 and Ryan, *Across the Space Frontier,* xiii–xiv.

67. "What Are We Waiting For?" 23.

68. Pal, *Destination Moon.*

69. "Treaty on Principles Governing the Activities of States in the Exploration and Use of Outer Space, Including the Moon and Other Celestial Bodies," 10 October 1967, in U.S. Arms Control and Disarmament Agency, *Arms Control and Disarmament Agreements* (Washington, D.C.: Arms Control and Disarmament Agency, 1975), 46–55; Philip D. O'Neill, "The Development of International Law Governing the Military Use of Outer Space," in *National Interests and the Military Use of Space,* ed. William J. Durch (Cambridge, Mass.: Ballinger Publishing, 1984), 169–99.

70. William E. Burrows, *Deep Black: Space Espionage and National Security* (New York: Random House, 1987).

71. Paul B. Stares, *The Militarization of Space: U.S. Policy, 1945–1984* (Ithaca, N.Y.: Cornell University Press, 1985).

72. President's Science Advisory Committee, *Introduction to Outer Space,* 12.

73. Burrows, *Deep Black.*

74. "Journey into Space," 67, 73.

75. John 6. *The New American Bible* (New York: Benziger, 1970).

76. J. Gordon Melton, *The Encyclopedia of American Religions,* vol. 79 (Wilmington, N.C.: McGrath Publishing, 1978), 459–61; Charles H. Lippy and Peter W. Williams, eds., *Encyclopedia of the American Religious Experience,* vol. 2 (New York: Charles Scribner's Sons, 1988), 834–35.

77. Wm. H. Barton, "The End of the World," *Sky* 1 (October 1937): 3–14, 19; "The Talk of the Town," *New Yorker,* 9 July 1949, 11–13.

78. Max Wilhelm Meyer, *The End of the World* (Chicago: C. H. Kerr, 1905); Joseph McCabe, *The End of the World* (London: G. Routledge, 1920); George Gamow, *Biography of the Earth* (New York: Viking Press, 1941); Geoffrey Dennis, *The End of the World* (London: Eyre and Spottiswoode, 1930).

79. "Hayden Planetarium Shows Four Ways in Which the World May End," *Life,* 1 November 1937, 54–58; see also "Picture Show," *Coronet,* July 1947, 27–34; Barton, "End of the World," 19.

80. Lincoln Barnett, "The Earth is Born," *Life,* 8 December 1952, 87, 99–100.

81. Richard F. Dempewolff, "Five Roads to Doomsday," *Popular Mechanics,* February 1950, 83.

82. Albert Einstein, as told to Raymond Swing, "Atomic War or Peace," *Atlantic,* November 1947, 29.

83. Quoted from Kenneth Heuer, *The End of the World* (New York: Rinehart, 1953), 144–45.

84. *Life's Picture History of World War II* (New York: Time, 1950).

85. William L. Laurence, "How Hellish Is the H Bomb?" *Look,* 21 April 1953, 31–35.

86. John Lear, "Hiroshima, U.S.A.: Can Anything Be Done About It?" *Collier's,* 5 August 1950, 11–15.

87. Robert E. Sherwood, "The Third World War," *Collier's,* 27 October 1951, 29–31ff.

88. W. Warren Wager, *Terminal Visions: The Literature of Last Things* (Bloomington: Indiana University Press, 1982); Carl B. Yoke, ed., *Phoenix from the Ashes: The Literature of the Remade World* (New York: Greenwood Press, 1987); Eric S. Rabkin, Maertin H. Greenberg, and Joseph D. Olander, *The End of the World* (Carbondale: Southern Illinois University Press, 1983).

89. Ray Bradbury, *The Martian Chronicles* (New York: Bantam Books, 1950), 143–45.

90. Nevil Shute, *On the Beach* (New York: Ballantine, 1957); Stanley Kramer, *On the Beach* (United Artists, 1959); Cobbett S. Steinberg, *Film Facts* (New York: Facts on File, 1980).

91. David Weishart, *Them!* (Warner Brothers, 1954).

92. Gallup, *Gallup Poll: Public Opinion, 1935–1971,* 2:907, 1018, 1225, 1365, 1434–43, 1460, 1523; 3:1726.

93. Curtis Peebles, *Watch the Skies! A Chronicle of the Flying Saucer Myth* (Washington, D.C.: Smithsonian Institution Press, 1994).

94. Donald A. Keyhoe, "The Flying Saucers Are Real," *True,* January 1950, 11–13, 83–87; see also H. B. Darrach and Robert Ginna, "Have We Visitors from Space?" *Life,* 7 April 1952, 80–96.

95. Richard Kyle, *The Religious Fringe* (Downers Grove, Ill.: InterVarsity Press, 1993), 282–84; H. Taylor Buckner, "Flying Saucers Are for People," *Trans-action* 3 (May/June 1966): 10–13; Robert S. Ellwood, *Religious and Spiritual Groups in Modern America* (Englewood Cliffs, N.J.: Prentice-Hall, 1973), chap. 4; Peebles, *Watch the Skies!* chap. 7.

96. Julian Blaustein, *The Day the Earth Stood Still* (Twentieth Century–Fox, 1951).

97. Lawrence J. Tacker, *Flying Saucers and the U.S. Air Force* (New York: D. Van Nostrand, 1960).

98. David Michael Jacobs, *The UFO Controversy in America* (Bloomington: Indiana University Press, 1975).

99. Carl Sagan, "What's Really Going On?" *Parade,* 7 March 1993, 3–7; Joseph Klaits, *Servants of Satan: The Age of Witch Hunts* (Bloomington: Indiana University Press, 1985).

100. Edward U. Condon, *Scientific Study of Unidentified Flying Objects* (New York: E. P. Dutton, 1969), 514; "If You're Seeing Things in the Sky," *U.S. News and World Report,* 15 November 1957, 122–26.

101. Statement of Democratic Leader Lyndon B. Johnson, 3–4. See also Johnson, "The Vision of a Greater America," *General Electric Forum* (July–September 1962): 7–9.

102. Johnson, "Vision of a Greater America," 3.

103. Ibid.

104. "Man on the Moon: The Epic Journey of Apollo 11," CBS interview of Lyndon B. Johnson by Walter Cronkite, 21 July 1969.

105. W. Stuart Symington, Address, Veterans' Day, Jefferson City, Missouri, 11 November 1957, from Lee Saegesser, "High-Ground Advantage," NASA History Office.

106. John F. Kennedy, "If the Soviets Control Space," *Missiles and Rockets,* 10 October 1960, 12.

107. Peter J. Roman, *Eisenhower and the Missile Gap* (New York: Cornell University Press, 1995).

108. Homer A. Boushey, "Who Controls the Moon Control the Earth," excerpts from a speech delivered before the Aero Club of Washington, D.C., 28 January 1958, *U.S. News & World Report,* 7 February 1958, 54. Also see Boushey's testimony in Select Committee on Astronautics and Space Exploration, *Astronautics and Space Exploration,* 85th Cong., 2d sess., 1958, 521–26.

109. See the remarks of Thomas B. White, U.S. Air Force Chief of Staff, to the National Press Club of Washington, D.C., 29 November 1957; General James H. Doolittle, 1959; Eugene M. Zuckert, 22 April 1961; and General Thomas D. White, 30 June 1961; all in Saegesser, "High-Ground Advantage."

110. Paul Palmer, "Soviet Union vs. U.S.A.—What Are the Facts?" *Reader's Digest,* April 1958, 44.

111. President's Science Advisory Committee, *Introduction to Outer Space,* 12; Palmer, "Soviet Union vs. U.S.A.," 44.

112. T. R. B., "Washington Wire," *New Republic,* 25 November 1957, 2.

113. George R. Price, "Arguing the Case for Being Panicky," *Life,* 18 November 1957, 126. See also "Stepping Up the Pace," *Newsweek,* 21 October 1957, 30–34; "The Feat that Shook the Earth," *Life,* 21 October 1957, 19–35; and "World Will Be Ruled from Skies Above," *Life,* 17 May 1963, 4.

114. James R. Killian, *Sputnik, Scientists, and Eisenhower* (Cambridge, Mass.: MIT Press, 1977), 7.

115. Logsdon, *Decision to Go to the Moon,* 95–99, 126–29.

116. James R. Hansen, *Engineer in Charge: A History of the Langley Aeronautical Laboratory, 1917–1958,* NASA SP-4305 (Washington, D.C.: GPO, 1987), 376.

117. McCurdy interview with Max Faget, 9 November 1987, Houston.

118. Gallup, *Gallup Poll: Public Opinion, 1935–1971* 2:907, 1018, 1225, 1240–41, 1345, 1451, 1523, 1539; 3:1595, 1632, 1674, 1812, 1842, 1881, 1934, 1944, 1973; see also Vernon Van Dyke, *Pride and Power: The Rationale of the Space Program* (Urbana: University of Illinois Press, 1964).

119. See Leon Festinger, *When Prophecy Fails* (Minneapolis: University of Minneapolis Press, 1956).

120. Jeffrey Klein and Dan Stober, "The American Empire in Space," *San Jose Mercury News,* 2 August 1992; Jack Manno, *Arming the Heavens* (New York: Dodd, Mead, 1984); U.S. Space Command, Office of History, "The Role of Space Forces: Quotes from Desert Shield/Desert Storm," May 1993; and Thomas Karas, *The New High Ground* (New York: Simon and Schuster, 1983).

121. Chapman and Morrison, "Chicken Little Was Right," 40; see also David Morrison "Target Earth!" *Astronomy,* October 1995, 34–41; David Morrison, ed., "The Spaceguard Survey: Report of the NASA International Near-Earth-Object Detection Workshop," Jet Propulsion Laboratory, Pasadena, California, 11–12; and Duncan Steel, *Rogue Asteroids and Doomsday Comets* (New York: John Wiley and Sons, 1995).

122. George Pal, *When Worlds Collide* (Paramount, 1951).

123. Larry Niven and Jerry Pournelle, *Lucifer's Hammer* (Chicago: Playboy Press, 1977); Arthur C. Clarke, *The Hammer of God* (New York: Bantam Books, 1993).

124. Luis W. Alvarez, Walter Alvarez, Frank Asaro, and Helen V. Michel, "Extraterrestrial Cause for the Cretaceous-Tertiary Extinction," *Science* 208 (1980): 1095–108; John Noble Wilford, *The Riddle of the Dinosaur* (New York: Alfred A. Knopf, 1985), chaps. 14 and 15; R. Ganapathy, "Evidence for a Major Meteorite Impact on the Earth 34 Million Years Ago: Implication for Eocene Extinctions," *Science* 216 (May 1982): 885–86; see also Morrison, "Spaceguard Survey," 8–11.

125. John R. Spencer and Jacqueline Mitton, *The Great Comet Crash: The Impact of Comet Shoemaker-Levy 9 on Jupiter* (New York: Cambridge University Press, 1995).

126. NASA Solar System Exploration Division, "Report of the Near-Earth Objects Survey Working Group," NASA Office of Space Science, Washington, D.C., June 1995; NASA, "Near Earth Asteroid Rendezvous Press Kit," NASA Headquarters, Washington, D.C., February 1996, 5; Sagan, *Pale Blue Dot,* 327; see also "Cold War Weaponeers Gaze Skyward," *Astronomy,* October 1995, 41.

127. U.S. House Committee on Science, Space, and Technology, *Report to Accompany H.R. 5649: National Aeronautics and Space Administration Multiyear Authorization Act of 1990,* report 101–763, 101st Cong., 2d sess., 26 September 1990, 29–30; Morrison, "Spaceguard Survey," 49.

128. H.R. 4489, "An Act to Authorize Appropriations to the National Aeronautics and Space Administration," 103d Cong., 2d sess., sec. 225.

129. NASA, "Summary of the Findings of the Near-Earth Object (NEO) Interception Workshop," NASA Headquarters, Washington, D.C., n.d.

4. APOLLO: THE AURA OF COMPETENCE

Epigraph: Tom Horton, "On Environment: If America Could Send a Man to the Moon, Why Can't We . . . ?" *Baltimore Sun,* 22 July 1984.

1. Logsdon, *Decision to Go to the Moon.*

2. Total NASA appropriations increased from $524 million in fiscal year 1960 to $5.3 billion in fiscal year 1965. NASA, *Pocket Statistics,* 1993, sec. C, p. 16.

3. John F. Kennedy, "Special Message to Congress on Urgent National Needs," 25 May 1961, *Public Papers of the Presidents of the United States, John F. Kennedy,* 404.

4. Charles L. Schultze, memorandum for the President, 24 January 1966, 3, WHCF Ex FI 4, box 22, LBJ Library; Robert A. Divine, "Lyndon B. Johnson and the Politics of Space," in *The Johnson Years: Vietnam, the Environment, and Science,* ed. Robert A. Divine (Lawrence: University of Kansas Press, 1987), 238–39; see also Glen P. Wilson, "The Legislative Origins of NASA: The Role of Lyndon B. Johnson," *Prologue: Quarterly of the National Archives* 25 (Winter 1993): 363–73; Robert Dallek, "Johnson, Project Apollo, and the Politics of Space Program Planning," in *Spaceflight and the Myth of Presidential Leadership*, ed. Roger Launius and Howard E. McCurdy (Urbana: University of Illinois Press, 1997).

5. W. Henry Lambright, *Powering Apollo: James E. Webb of NASA* (Baltimore: Johns Hopkins University Press, 1995).

6. Paul. R. Abramson, *Political Attitudes in America* (San Francisco: W. H. Freeman, 1983), 12. See also Seymour Martin Lipset and William Schneider, *The Confidence Gap* (New York: Free Press, 1983).

7. NASA, *Pocket Statistics,* January 1988 (Washington, D.C.: NASA, 1988), sec. B, 32, 59–70.

8. See Richard P. Hallion, "The Development of American Launch Vehicles," in *Space Science Comes of Age: Perspectives in the History of the Space Sciences,* ed. Paul A. Hanle and Von del Chamberlain (Washington, D.C.: Smithsonian Institution Press, 1981), 115–34.

9. R. Cargill Hall, *Lunar Impact: A History of Project Ranger,* SP-4210 (Washington, D.C.: GPO, 1977); House Committee on Science and Astronautics, Subcommittee on NASA Oversight, *Investigation of Project Ranger,* 88th Cong., 2d sess., 1964.

10. Swenson, Grimwood, and Alexander, *This New Ocean,* 372–77.

11. Tom Wolfe, *The Right Stuff* (New York: Farrar, Straus, Giroux, 1979), 280–96. See also William J. Perkinson, "Grissom's Flight: Questions," *Baltimore Sun,* 22 July 1961.

12. "Month's Delay for Glenn Seen," *Washington Star,* 31 January 1962; Art Woodstone, "Television's $1,000,000 (When & If) Manshoot; Lotsa Prestige & Intrigue," *Variety,* 24 January 1962; Swenson, Grimwood, and Alexander, *This New Ocean,* 419–22.

13. Nate Haseltine, "Vanguard Fails, Burns in Test Firing; Hill Critics See Blow to U.S. Prestige," *Washington Post,* 7 December 1957; Joseph Alsop, "Making the Worst of It," *Washington Post,* 9 December 1957; Milton Bracker, "Vanguard Rocket Burns on Beach; Failure to Launch Satellite Assailed as Blow to U.S. Prestige," *New York Times,* 7 December 1957; "Sputternik," *New York Times,* 10 December 1957.

14. Ad Hoc Committee on Space, "Report to the President Elect," 10 January 1961, 16–17.

15. Swenson, Grimwood, and Alexander, *This New Ocean,* 272–79, 291–97, 326–28, 335–38.

16. James Barr, "Is Mercury Program Headed for Disaster?" *Missiles and Rockets,* 15 August 1960, 12.

17. Martin J. Collins and Sylvia K. Kraemer, *Space: Discovery and Exploration* (Washington, D.C.: Smithsonian Institution, 1993), 277.

18. See "Americans Jubilant over Shepard's Flight," *Washington Post,* 6 May 1961; Robert Conley, "Nation Exults over Space Feat," *New York Times,* 6 May 1961.

19. "The High Price of History," *Television,* April 1962, 65.

20. "8,000 Eyes in Orbit," *Life,* 2 March 1962, 2–3.

21. Joseph Arthur Angotti, "A Descriptive Analysis of NBC's Radio and Television Coverage of the First Manned Orbital Flight," Master's thesis, Indiana University, June 1965, 117–19; "Space: The New Ocean," *Time,* 2 March 1962, 14; Robert B. Voas, "John Glenn's Three Orbits in *Friendship 7,*" *National Geographic,* June 1962, 792–827.

22. See T. Keith Glennan, *The Birth of NASA: The Diary of T. Keith Glennan,* NASA SP-4105, ed. J. D. Hunley (Washington, D.C.: NASA, 1993), 20–21.

23. See James L. Kauffman, *Selling Outer Space: Kennedy, the Media, and Funding for Project Apollo, 1961–1963* (Tuscaloosa: University of Alabama Press, 1994).

24. Wolfe, *Right Stuff,* 24.

25. Ibid., 120.

26. "Space Voyagers Rarin' to Orbit," *Life,* 20 April 1959, 22.

27. Richard Slotkin, *Gunfighter Nation: The Myth of the Frontier in Twentieth-Century America* (New York: Atheneum, 1992).

28. Tom Wolfe, "The Last American Hero," in *The Kandy-Kolored Tangerine-Flake Streamline Baby,* by Wolfe (New York: Farrar, Straus, and Giroux, 1965).

29. Chuck Yeager and Leo Janos, *Yeager: An Autobiography* (New York: Bantam Books, 1985).

30. "The Adventures of Ozzie and Harriet," ABC, 3 October 1952 to 3 September 1966; "Father Knows Best," CBS, 3 October 1954 to 27 March 1955, 22 September 1958 to 17 September 1962, NBC 31 August 1955 to 17 September 1958, ABC, 30 September 1962 to 3 February 1967; "Leave It to Beaver," CBS, 11 October 1957 to 26 September 1958, ABC, 3 October 1958 to 12 September 1963; also see Steven Mintz and Susan Kellogg, *Domestic Revolutions: A Social History of American Family Life* (New York: Free Press, 1988), chap. 9.

31. John H. Glenn, "A New Era: May God Grant Us the Wisdom and Guidance to Use It Wisely," *Vital Speeches of the Day,* 15 March 1962, 324–26; Dora Jane Hamblin, "Applause, Tears and Laughter and the Emotions of a Long-ago Fourth of July," *Life,* 9 March 1962, 34.

32. Agreement among Malcolm Scott Carpenter et al. and C. Leo DeOrsey, 28 May 1959; agreement between Leo DeOrsey and Time Incorporated, 5 August 1959; both in NASA History Office.

33. *Life* magazine: 20 April, 14, 21 September, and 14 December 1959; 29 February, 21 March, 11 April, 9 May, 1 August, and 3 October 1960; 27 January, 3 March, 12, 19 May, 28 July, 4 August, and 8 December 1961; 2 February, 2, 9 March, 18 May, 1, 8 June, and 12, 26 October 1962; 24, 31 May, and 7 June 1963.

34. See Larry J. Sabato, *Feeding Frenzy: How Attack Journalism Has Transformed American Politics* (New York: Free Press, 1991).

35. Don A. Schanche to P. Michael Whye, 28 December 1976, NASA History Office.

36. See Robert Sherrod, "The Selling of the Astronauts," *Columbia Journalism Review* (May/June 1973): 16–25.

37. Dora Jane Hamblin to P. Michael Whye, 18 January 1977, NASA History Office.

38. "Backing up the Men, Brave Wives and Bright Children," *Life,* 20 April 1959, 24–25. See also "Seven Brave Women Behind the Astronauts," *Life,* 21 September 1959, 142–63.

39. Loudon S. Wainwright, "New Astronaut Team, Varied Men with One Goal, Poise for the Violent Journey," *Life,* 3 March 1961, 24–25.

40. "High Dreams for a Man and His Sons," *Life,* 8 June 1962, 38.

41. Jack Mann, "Rene Carpenter's Own Orbit," *Washington Post/Potomac,* 16 June 1974, pp. 8–19.

42. Hamblin to Whye, 18 January 1977.

43. See David Halberstam, *The Best and the Brightest* (New York: Random House, 1969), esp. 41.

44. U.S. Department of Commerce, *United States Science Exhibit: Seattle World's Fair,* final report (Washington, D.C.: GPO, 1963), 3.

45. Murray Morgan, *Century 21: The Story of the Seattle World's Fair, 1962* (Seattle: Acme Press, distributed by the University of Washington Press, 1963), 81–86.

46. U.S. Department of Commerce, *United States Science Exhibit: World's Fair in Seattle, 1962* (Seattle: Craftsman Press, 1962).

47. Ibid., 7.

48. Ibid.

49. NASA, "NASA Technical Exhibit: Space for the Benefit of Mankind," Century 21 Exposition, Seattle, Washington, 1962, from the private collection of Norman P. Bolotin, Redmond,

Washington; Washington State Department of Commerce and Economic Development, "Seattle World's Fair, 1962: Official Souvenir Program," 15, Seattle Center Foundation, Seattle, Washington.

50. Morgan, *Century 21,* 25.

51. "National Air and Space Museum—Historical Perspective," *Aerospace* 14 (June 1976): 2–3; "Freedom Seven Rests, at Last," *National Aeronautics,* April 1963, 2; Philip Hopkins, "The National Air Museum," *National Aeronautics,* June 1964, 4–5.

52. Flip Schulke, Debra Schulke, Penelope McPhee, and Raymond McPhee, *Your Future in Space: U.S. Space Camp Training Program* (New York: Crown Publishers, 1986); see also Edward O. Buckbee and Charles Walker, "Spaceflight and the Public Mind," in Ordway and Liebermann, *Blueprint for Space,* 189–97.

53. See, for example, "Space," *Time,* 2 March 1962, 11–18; Voas, "John Glenn's Three Orbits," 792–827; Jerry E. Bishop, "Gemini Team Maneuvers in Three Orbits of Earth While Probing for Knowledge to Aid U.S. Moon Flight," *Wall Street Journal,* 24 March 1965.

54. See Walter Sullivan, ed., *America's Race for the Moon* (New York: Random House, 1962); M. Scott Carpenter, L. Gordon Cooper Jr., John H. Glenn Jr., Virgil I. Grissom, Walter M. Schirra Jr., Alan B. Shepard Jr., and Donald K. Slayton, *We Seven* (New York: Simon and Schuster, 1962); Edgar M. Cortright, ed., *Apollo Expeditions to the Moon,* NASA SP-350 (Washington, D.C.: GPO, 1975).

55. See David Sanford, "Admen in Orbit," *New Republic,* 17 December 1966, 13–15.

56. Courtney G. Brooks, James M. Grimwood, and Loyd S. Swenson, *Chariots for Apollo: A History of Manned Lunar Spacecraft,* NASA SP-4205 (Washington, D.C.: NASA, 1979), 266.

57. See CBS Television, *February 14 15 16 17 18 19 20* (New York: Columbia Broadcasting System, n.d.)

58. "High Price of History," 65.

59. See, for example, "Networks in High Gear for G-T 6," *Broadcasting,* 25 October 1965, 78; Edwin Diamond, "Perfect Match: TV and Space," *Columbia Journalism Review* (Summer 1965): 18–20.

60. James Barron, "Jules Bergman," *New York Times,* 13 February 1987, sec. D, p. 20; Jules Bergman, "The Reluctant Astronaut, the Suborbital Water-Skier, Carpenter's Snafu," *TV Guide,* 2 March 1974, 8–13; "Bergman Gives Views on Science, Technology and Medicine," *Langley Researcher,* 20 May 1983, 4–5.

61. Diamond, "Perfect Match: TV and Space," 20.

62. Art Buchwald, "Countdown 1966," *Washington Post,* 26 August 1965.

63. See Blake Clark, "A Job for the Next Congress: Stop the Race to the Moon," *Reader's Digest,* January 1964, 75–79; Joe Alex Morris, "How Haste in Space Makes Waste," *Reader's Digest,* July 1964, 82–87; and Joe Alex Morris, "The Pork Barrel Goes into Orbit," *Reader's Digest,* August 1964, 87–92; or Amitai Etzioni, *The Moon-Doggle* (Garden City, N.Y.: Doubleday, 1964).

64. "The Cosmic Circus," *New York Herald Tribune,* 24 March 1965, p. 28.

65. "One of Our Finest Hours," *New York Times,* 21 February 1962, p. 44.

66. "Go!" *Washington Post,* 21 February 1962, sec. A, p. 24; "Jumping Gemini," *Washington Post,* 24 March 1965, sec. A, p. 20.

67. Brooks, Grimwood, and Swenson, *Chariots for Apollo,* chap. 3.

68. John F. Kennedy, "Address at Rice University in Houston on the Nation's Space Effort," 12 September 1962, *Public Papers of the Presidents of the United States* (Washington, D.C.: GPO, 1963), 669.

69. Kennedy, "Special Message to Congress on Urgent National Needs," 404.

70. See Lyndon B. Johnson, Memorandum for the President, "Evaluation of Space Program," 28 April 1961; and James E. Webb and Robert S. McNamara, "Recommendations for Our National Space Program: Changes, Policies, Goals," report to the Vice President, 8 May 1961; both in Logsdon, *Exploring the Unknown* 1:427, 443–44.

71. Logsdon, *Decision to Go to the Moon,* 118.

72. Arnold W. Frutkin Oral History, 4 April 1974, by Eugene M. Emme and Alex Roland, 28–29, and Arnold W. Frutkin Oral History, 30 July 1970, by John M. Logsdon, 17–18, both in NASA Historical Reference Collection, NASA Headquarters, Washington, D.C. See also Arnold W. Frutkin, *International Cooperation in Space* (Englewood Cliffs, N.J.: Prentice-Hall, 1965), chap. 3. There were numerous meetings between U.S. and Soviet representatives toward this end. See Arnold W. Frutkin, "Record of US-USSR Talks on Space Cooperation," 27 March–1 May 1962; Hugh L. Dryden, "Bilateral Meeting with Soviets on Outer Space," 30 May 1962, 1, 4, 6, 7 June 1962; James E. Webb to the President, 31 January 1964, all in NASA Historical Reference Collection, NASA History Office.

73. Dodd L. Harvey and Linda C. Ciccoritti, *U.S.-Soviet Cooperation in Space* (Miami: Center for Advanced International Studies, University of Miami, 1974), 78–79.

74. "Text of President Kennedy's Address on Peace Issues at U.N. General Assembly," *New York Times,* 21 September 1963, p. C6; see Yuri Karash, "The Price of Rivalry in Space," *Baltimore Sun,* 19 July 1994, p. 11A; McDougall, . . . *the Heavens and the Earth,* 394–96.

75. "Major Legislation—Appropriations," *Congressional Quarterly Almanac 1963* (Washington, D.C.: Congressional Quarterly, 1964), 170.

76. Charles L. Schultze, Memorandum for the President, 20 September 1966, with attached memorandum, subject: NASA's budget and Mr. Webb's letter, 1 September 1966, LBJ Library, WHCF Ex OS, box 2.

77. Schultze, Memorandum for the President, 24 January 1966; Charles L. Schultze, Memorandum for the President, 11 August 1967, LBJ Library, WHCF EX FI 4, box 30 (emphasis removed).

78. Lyndon B. Johnson to James Webb, 29 September 1967, LBJ Library, WHCF confidential file, box 43; James E. Webb, Memorandum for the President, 10 August 1967, WHCF EX FI 4, box 29.

79. Lyndon B. Johnson, "President's News Conference at the LBJ Ranch," 29 August 1965, *Public Papers of the Presidents of the United States* 2:945. See also Lyndon B. Johnson, "Remarks Following an Inspection of NASA's Michoud Assembly Facility Near New Orleans," 12 December 1967, *Public Papers of the Presidents of the United States* (Washington, D.C.: GPO, 1968), 2:1123.

80. Lyndon B. Johnson, "Remarks Upon Presenting the NASA Distinguished Service Medal to the Apollo 8 Astronauts," 9 January 1969, *Public Papers of the Presidents of the United States* (Washington, D.C.: GPO, 1970), 2:1247.

81. NASA, "Post-Apollo Space Program"; Space Task Group, *Post-Apollo Space Program,* 22; both in NASA History Office.

82. NASA, "Program Review: Manned Space Science and Advanced Manned Missions," Office of Programming, NASA Headquarters, 7 October 1965, 36; NASA, "Apollo Applications Program: Program Review Document," 15 November 1966, NASA History Office; Arnold S. Levine, *Managing NASA in the Apollo Era,* NASA SP-4102 (Washington, D.C.: GPO, 1982), 242–61. See also NASA, "Lunar Studies," 25 June 1964; and U.S. House Committee on Science and Astronautics, *Summary Report: Future Programs Task Group,* 89th Cong., 1st sess., 1965; all in NASA History Office.

83. Paul S. Boyer, *By the Bomb's Early Light* (New York: Pantheon, 1985).

84. See Norman Mailer, *The Armies of the Night* (New York: Signet, 1968), 135–51.

85. Rachel Carson, *Silent Spring* (Boston: Houghton Mifflin, 1962).

86. Paul Ehrlich, *The Population Bomb* (New York: Ballantine Books, 1968); see also Donella Meadows, Dennis L. Meadows, Jorgen Randers, and William W. Behrens III, *The Limits to Growth* (New Hyde Park, N.Y.: University Books, 1972); and Kirkpatrick Sale, *The Green Revolution: The American Environmental Movement, 1962–1992* (New York: Hill and Wang, 1993).

87. Joseph Heller, *Catch-22* (New York: Dell, 1961); Kurt Vonnegut, *Slaughterhouse Five* (New York: Delta Books, 1969).

88. Robert J. Samuelson, *The Good Life and Its Discontents* (New York: Times Books, 1995); Robert McNamara, *In Retrospect: The Tragedy and Lessons of Vietnam* (New York: Times Books, 1995).

89. John F. Kennedy, "Letter to the President of the Senate and to the Speaker of the House on Development of a Civil Supersonic Air Transport," 14 June 1963, *Public Papers of the Presidents of the United States,* 476.

90. See Mel Horwich, *Clipped Wings: The American SST Conflict* (Cambridge, Mass.: MIT Press, 1982).

91. Brooks, Grimwood, and Swenson, *Chariots for Apollo,* chap. 9.

92. Herbert E. Krugman, "Public Attitudes Toward the Apollo Space Program, 1965–1975," *Journal of Communication* 27 (Autumn 1977): 87–93; Gallup, *Gallup Poll: Public Opinion, 1935–1971* 3:1952.

93. Krugman, "Public Attitudes Toward the Apollo Space Program"; Bonestell and Ley, *Conquest of Space;* Brooks, Grimwood, and Swenson, *Chariots for Apollo,* 281.

94. "Columbuses of Space," *New York Times,* 22 December 1968; "Two Faces of Science," *New York Times,* 29 December 1968; Robert J. Donovan, "Moon Voyage Turns Men's Thoughts Inward," *Los Angeles Times,* 29 December 1968; "Apollo 8," *Philadelphia Sunday Bulletin,* 22 December 1968; "Footprints in the Dirty Sand," *Washington Post,* 28 December 1968.

95. "Apollo 8: 'Millennial' Event," *Los Angeles Times,* 29 December 1968.

96. Stanley Kubrick and Victor Lyndon, *Dr. Strangelove, or: How I Learned to Stop Worrying and Love the Bomb* (Columbia Pictures, 1963); Michael Gruskoff and Douglas Trumbull, *Silent Running* (Universal Pictures, 1971); Mort Abrahams, *Planet of the Apes* (APJAC/Twentieth Century–Fox, 1968); see also Dino DeLaurentis, *Barbarella* (Paramount Pictures, 1967) and Francis Ford Coppola and Lawrence Sturhahn, *THX 1138* (Warner Brothers, 1970); and Byron Kennedy, *Mad Max* (Orion Pictures, 1979).

97. Stanley Kubrick, *2001: A Space Odyssey* (MGM, 1968); Arthur C. Clarke, *2001: A Space Odyssey* (New York: New American Library, 1968).

98. Gene Roddenberry, "Star Trek," NBC, 8 September 1966 to 9 September 1969; Irwin Allen, "Lost in Space," CBS, 15 September 1965 to 11 September 1968; Stephen E. Whitfield and Gene Roddenberry, *The Making of Star Trek* (New York: Ballantine Books, 1968).

99. Michael Crichton, *The Andromeda Strain* (New York: Dell, 1969); Robert Wise, *The Andromeda Strain* (Universal Studios, 1971).

100. Gordon Carroll, David Giler, and Walter Hill, *Alien* (Brandywine-Shussell/Twentieth Century–Fox, 1979); see also Michael Deeley and Ridley Scott, *Blade Runner* (Warner Brothers, 1982).

101. Julia and Michael Phillips, *Close Encounters of the Third Kind* (Columbia Pictures, 1977); Steven Spielberg and Kathleen Kennedy, *E.T.: The Extraterrestrial* (Universal Studios, 1982); Lili Zanuck, *Cocoon* (TCF, 1985).

102. See Charles Murray and Catherine Bly Cox, *Apollo: The Race to the Moon* (New York: Simon and Schuster, 1989).

103. See Jim Lovell and Jeffrey Kluger, *Lost Moon: The Perilous Voyage of Apollo 13* (Boston: Houghton Mifflin, 1994).

104. Brian Grazer, *Apollo 13* (Universal, 1995); "Peerless Voyage: NASA's 'Finest Hour' Comes to the Silver Screen," *Spaceflight,* September 1995, 319–21. See also Norman Mailer, *A Fire on the Moon* (Boston: Little, Brown, 1969).

105. "The U.S. Space Program," Media General/Associated Press Public Opinion Poll.

106. See Howard E. McCurdy, *Inside NASA: High Technology and Organizational Change in the U.S. Space Program* (Baltimore: Johns Hopkins University Press, 1993), 146–55.

107. Michael A. G. Michaud, *Reaching for the High Frontier* (New York: Praeger, 1986), chap. 6; Michaud, "The New Demographics of Space," *Aviation Space,* Fall 1984, 46–47; Krugman, "Public Attitudes Toward the Apollo Space Program." See also Gallup, *Gallup Poll: Public Opinion, 1935–1971* 3:1952, 2184; Elizabeth H. Hastings and Philip K. Hastings, eds., *Index to International Public Opinion, 1979–1980* (Westport, Conn.: Greenwood Press, 1981), 73; Elizabeth H. Hastings and Philip K. Hastings, eds., *Index to International Public Opinion, 1982–1983* (Westport, Conn.: Greenwood Press, 1984), 183; "The Public's Agenda," *Time,* 30 March 1987, 37; "The U.S. Space

Program," Media General/Associated Press Public Opinion Poll 21, 22 June–2 July 1988; George Gallup, *The Gallup Poll: Public Opinion, 1988* (Wilmington, Del.: Scholarly Resources, 1989), 106–8.

108. Robert P. Mayo to Thomas O. Paine, 28 July 1969, NASA History Office; Space Task Group, *Post-Apollo Space Program;* Richard Witkin, "Agnew Proposes a Mars Landing," *New York Times,* 17 July 1969.

109. Caspar Weinberger, memo to the President, "Future of NASA," 12 August 1971, NASA History Office. See also Roger D. Launius, "NASA and the Decision to Build the Space Shuttle, 1969–72," *Historian* 57 (Autumn 1994): 17–34.

110. Weinberger, memo to the President, 12 August 1971; see also John Logsdon, "The Decision to Develop the Space Shuttle," *Space Policy* 2 (May 1986): 103–19; and Logsdon, "The Space Shuttle Decision: Technological and Political Choice," *Journal of Contemporary Business* 7, no. 3 (1978): 13–30.

111. Richard M. Nixon, "Space Shuttle Program," statement by the president announcing the decision to proceed with the development of the new Space Transportation System, *Weekly Compilation of Presidential Documents* 8, no. 2 (5 January 1972): 27–28; see also Nixon, "The Future of the United States Space Program," *Weekly Compilation of Presidential Documents* 6, no. 10 (7 March 1970): 328–31.

5. MYSTERIES OF LIFE

Epigraph: Gordon Freeman, *A Brief History of Time* (Paramount Home Video, 1993).

1. Joseph Campbell, *The Inner Reaches of Outer Space* (New York: Harper and Row, 1986); see also Joseph Campbell, *The Power of Myth* (New York: Doubleday, 1988).

2. Stephen Hawking, *A Brief History of Time* (New York: Bantam, 1988), 7–11, 174–75.

3. Pierre Simon Laplace, *The System of the World,* trans. J. Pond (London: R. Phillips, 1809); see also Agnes M. Clerke, *A Popular History of Astronomy During the Nineteenth Century* (London: Adam and Charles Black, 1902), chap. 9.

4. Isabel M. Lewis, "Life on Venus and Mars?" *Nature Magazine,* September 1934, 134.

5. Ley, *Rockets,* 34. See also A. Pannekoek, *A History of Astronomy* (New York: Barnes and Noble, 1969), chap. 35.

6. Kimball, *Man and the Moon.*

7. Clarke, *2001.*

8. Ley, *Rockets,* 37–39; Pannekoek, *History of Astronomy,* 381; Camille Flammarion, *Popular Astronomy* (London: Chatto and Windus, 1907), 370.

9. See John Noble Wilford, *Mars Beckons* (New York: Alfred A. Knopf, 1990), 22; Samuel Glasstone, *The Book of Mars,* NASA SP-179 (Washington, D.C.: GPO, 1968).

10. A. Lawrence Lowell, *Biography of Percival Lowell* (New York: Macmillan, 1935); see also William Graves Hoyt, *Lowell and Mars* (Tucson: University of Arizona Press, 1976).

11. Percival Lowell, *Mars* (New York: Houghton Mifflin, 1895); *Mars and Its Canals* (New York: Macmillan, 1907); and *Mars as the Abode of Life* (New York: Macmillan, 1908).

12. Lowell, *Mars,* 207–8.

13. Lowell, *Mars and Its Canals,* 377, 382.

14. Lowell, *Mars as the Abode of Life,* 215–14. See also Samuel Phelps Leland, *World Making: A Scientific Explanation of the Birth, Growth and Death of Worlds* (Chicago: Leland, 1906), 62–67.

15. Alfred Russel Wallace, *Is Mars Habitable?* (London: Macmillan, 1907). See also Wallace, *Man's Place in the Universe* (New York: McClure, Phillips, 1903) and William H. Pickering, "Recent Studies of the Martian and Lunar Canals," *Popular Astronomy,* February 1904.

16. H. G. Wells, *The Time Machine and War of the Worlds,* ed. Frank D. McConnell (New York: Oxford University Press, 1977), chap. 1, 125.

17. Frank D. McConnell, "Introduction," in Wells, *Time Machine and the War of the Worlds,* 3–10.

18. Wells, *Time Machine and War of the Worlds,* 124.

19. Lowell, *Mars as the Abode of Life,* 216.

20. See "Mars," in *The Encyclopedia of Science Fiction,* by John Clute and Peter Nicholls (New York: St. Martin's Press, 1993), 777–79.

21. Burroughs, *Princess of Mars.* The story first appeared in *All-Story* magazine in 1912 and was not released in book form until 1917. See also Richard A. Lupoff, *Barsoom: Edgar Rice Burroughs and the Martian Vision* (Baltimore: Mirage Press, 1976); Erling B. Holtsmark, *Edgar Rice Burroughs* (Boston: Twayne, 1986), chap. 2.

22. Hadley Cantril, *The Invasion from Mars: A Study of the Sociology of Panic* (Princeton, N.J.: Princeton University Press, 1940).

23. Kurt Neumann, *Rocketship X-M* (Lippert, 1950).

24. George Pal, *War of the Worlds* (Paramount, 1953).

25. See Arthur C. Clarke, *Childhood's End* (New York: Harcourt, Brace, and World, 1953).

26. Bradbury, *Martian Chronicles,* 32, 49–66, 102–10.

27. Kimball, *Mars and Beyond.*

28. Ray Bradbury, Arthur C. Clarke, B. Murray, Carl Sagan, and W. Sullivan, *Mars and the Mind of Man* (New York: Harper and Row, 1973), 22.

29. Boorstin, *Discoverers,* pt. 12.

30. T. H. White, ed., *The Bestiary: A Book of Beasts: Being a Translation from a Latin Bestiary of the Twelfth Century* (New York: Perigree, 1980), 24–25; Willy Ley, *Dawn of Zoology* (Englewood Cliffs, N.J.: Prentice-Hall, 1968), 97. For the origin of the goose story, see Robert Bartlett, *Gerald of Wales* (Oxford: Clarenndon Press, 1982), 136–37.

31. See Sully Zuckerman, *Great Zoos of the World: Their Origins and Significance* (Boulder, Colo.: Westview Press, 1980) and Marian Murray, *Circus! From Rome to Ringling* (Westport, Conn.: Greenwood Press, 1956).

32. White, *Bestiary,* 22–24.

33. John Ashton, *Curious Creatures in Zoology* (Detroit: Singing Tree Press, 1890), vi.

34. Josephine Waters Bennett, *The Rediscovery of Sir John Mandeville* (New York: Kraus Reprint, 1954).

35. Antonio Pigafetta, *Magellan's Voyage: A Narrative Account of the First Circumnavigation* (New Haven, Conn.: Yale University Press, 1969); Richard C. Temple, *The World Encompassed and Analogous Contemporary Documents Concerning Sir Francis Drake's Circumnavigation of the World* (New York: Cooper Square, 1969). No major narrative of the English discovery of North America exists. See James A. Williamson, *The Cabot Voyages and Bristol Discovery under Henry VII* (Cambridge: Cambridge University Press, 1962). See also Alexander de Humboldt and Aime Bonpland, *Personal Narrative of Travels to the Equinoctial Regions of the New Continent During the Years 1799–1804* (New York: AMS Press, 1966); John Bakeless, *The Eyes of Discovery: America as Seen by the First Explorers* (New York: Dover, 1950); Walter Raleigh, *The Discovery of the Large, Rich, and Beautiful Empire of Guiana* (Cleveland: World, 1966). Descriptions of races far exceed descriptions of beasts.

36. See Stephen E. Ambrose, *Meriwether Lewis, Thomas Jefferson and the Opening of the American West* (New York: Simon and Schuster, 1996).

37. Charles Darwin, *The Voyage of the Beagle* (New York: P. F. Collier and Son, 1909), chap. 17; Charles Darwin, *The Zoology of the Voyage of the Beagle* (New York: New York University Press, 1987).

38. Darwin, *Voyage of the Beagle,* chap. 10, 218.

39. See Steven J. Dick, *Plurality of Worlds: The Origins of the Extraterrestrial Life Debate from Democritus to Kant* (New York: Cambridge University Press, 1982) and Michael J. Crowe, *The Extraterrestrial Life Debate, 1750–1900* (New York: Cambridge University Press, 1986).

40. Aristotle, *On the Heavens,* trans. W. K. C. Guthrie (Cambridge: Harvard University Press, 1939), chap. 8; also see Friedrich Solmsen, *Aristotle's System of the Physical World* (Ithaca, N.Y.: Cornell University Press, 1960).

41. St. Thomas Aquinas, *Summa Theologica* (New York: Benziger Brothers, 1947), pt. 1, question 47, art. 3.

42. See "The Solar System: Exception or Rule?" in *We Are Not Alone: The Search for Intelligent Life on Other Worlds,* rev. ed., by Walter Sullivan (New York: McGraw-Hill, 1964), chap. 5; S. F. Dermott, *The Origin of the Solar System* (New York: John Wiley and Sons, 1978).

43. Immanuel Velikovsky, *Worlds in Collision* (Garden City, N.Y.: Doubleday, 1950); see also Immanuel Velikovsky, *Ages in Chaos* (Garden City, N.Y.: Doubleday, 1952) and *Earth in Upheaval* (Garden City, N.Y.: Doubleday, 1955). See also Ronald N. Bracewell, *The Galactic Club: Intelligent Life in Outer Space* (San Francisco: W. H. Freeman, 1974).

44. Genesis 1:27.

45. Bernard de Fontenelle, *Entretiens sur la Pluralité des Mondes* (1686; reprint, Berkeley and Los Angeles: University of California Press, 1990).

46. H. G. Wells, "The Things that Live on Mars," *Cosmopolitan* 44 (March 1908): 335–42.

47. Kenneth Heuer, *Men of Other Planets* (New York: Pellegrini and Cudahy, 1951), 150; see also Steven J. Dick, *The Biological Universe: The Twentieth Century Extraterrestrial Life Debate and the Limits of Science* (New York: Cambridge University Press, 1996). Gene Bylinsky, *Life in Darwin's Universe: Evolution and the Cosmos* (Garden City, N.Y.: Doubleday, 1981); David Milne, David Raup, John Billingham, Karl Niklaus, and Kevin Padian, eds., *The Evolution of Complex and Higher Organisms,* NASA SP-478 (Washington, D.C.: GPO, 1985); Willy Ley, "What Will 'Space People' Look Like," *This Week Magazine,* 10 November 1957; Alan D. Foster, *Alien Omnibus* (Warner Books, 1987).

48. I. S. Shklovskii and Carl Sagan, *Intelligent Life in the Universe* (San Francisco: Holden-Day, 1966), 329.

49. Bonnie Dalzell, "Exotic Bestiary for Vicarious Space Voyagers," *Smithsonian,* October 1974, 84–91.

50. See Wayne Douglas Barlowe and Ian Summers, *Barlowe's Guide to Extra-Terrestrials* (New York: Workman, 1979) and Issac Asimov, "Anatomy of a Man from Mars," *Esquire,* September 1965, 113–17, 200.

51. Howard Hawks, *The Thing* (RKO Radio Pictures, 1951).

52. Gordon Carroll, David Giller, and Walter Hill, *Alien* (Fox, 1979).

53. Gary Kurtz, *Star Wars* (Fox, 1977); also see Shane Johnson, *Star Wars Technical Journal* (New York: Ballantine Books, 1995).

54. *E.T.: The Extra-Terrestrial, Star Wars, Return of the Jedi,* and *The Empire Strikes Back.* Mark S. Hoffman, *The World Almanac and Book of Facts* (New York: Pharos Books, 1991), 308.

55. Henry Norris Russell, "Antropcentrism's Demise," *Scientific American* 169 (July 1943): 18–19.

56. H. Spencer Jones, *Life on Other Worlds* (New York: Macmillian, 1940), 244. See also "Life Beyond Earth?" *Time,* 7 October 1940, 62; and Bruce Bliven, "Is There Life on Other Planets," *Reader's Digest,* February 1955, 103–7.

57. See Crowe, *Extraterrestrial Life Debate,* chap. 9, 378–86.

58. Camille Flammarion, "Are the Planet's Inhabited?" *Harper's,* November 1904, 844.

59. Harlow Shapley, "Coming to Terms with the Cosmos," *Saturday Review,* 6 September 1958, 54. See also Harlow Shapley, *Of Stars and Men: The Human Response to an Expanding Universe* (Boston: Beacon Press, 1958).

60. See "Space Theology," *Time,* 19 September 1955, 81.

61. See Loren C. Eiseley, "Little Men and Flying Saucers," *Harper's,* March 1953, 86–91; Wyn Wachhorst, "Seeking the Center at the Edge: Perspectives on the Meaning of Man in Space," *Virginia*

Quarterly Review 69 (Winter 1993): 1–23; William Williams, *The Universe No Desert, the Earth No Monopoly* (Boston: James Munroe, 1855).

62. "Pictures Show Evidence of Life on Planet Mars," *Life,* 29 May 1944, 83; see also David Todd, "Professor Todd's Own Story of the Mars Expedition," *Cosmopolitan* 44 (March 1908): 343–51.

63. "Buck Rogers Baedeker," *Newsweek,* 28 February 1949, 48–49.

64. NASA, *Mariner-Mars 1964: Final Project Report,* NASA SP-139 (Washington, D.C.: GPO, 1967), 273–89; Glasstone, *Book of Mars.* Turn-of-the-century astronomers may have mapped these features. See William Sheehan, "Did Barnard & Mellish Really See Craters on Mars?" *Sky & Telescope,* July 1992, 23–25.

65. "An End to the Myths About Men on Mars," *U.S. New and World Report,* 9 August 1965, 4.

66. Lyndon B. Johnson, "Remarks Upon Viewing New Mariner 4 Pictures from Mars," 29 July 1965, *Public Papers of the Presidents of the United States,* 806.

67. NASA, *Mariner-Mars 1964: Final Project Report,* 318–22.

68. "Is There Life on Mars—or Earth?" *Time,* 7 January 1966, 44; Steven D. Kilston, Robert R. Drummond, and Carl Sagan, "A Search for Life on Earth at Kilometer Resolution," *Icarus* 5 (January 1966): 79–98.

69. Stuart Auerbach, "Mariner 7 Photographs Mysterious Mars Canals," *Washington Post,* 5 August 1969, p. 1.

70. Senate Aeronautical and Space Sciences Committee, *Future NASA Space Programs,* 91st Cong., 1st sess., 1969, 6–7, 68.

71. Stuart Auerbach, "Mars Pocked by Moon-Like Depressions," *Washington Post,* 1 August 1969, p. 1. See also "Surface of Mars Similar to Moon," *Baltimore Sun,* 1 August 1969, pp. 1, 4.

72. William K. Hartmann and Odell Raper, *The New Mars: The Discoveries of Mariner 9,* NASA SP-337 (Washington, D.C.: GPO, 1974).

73. Quoted from Henry S. F. Cooper, *The Search for Life on Mars* (New York: Holt, Rinehart and Winston, 1976), 78–79.

74. See Norman Horowitz, *To Utopia and Back* (New York: W. H. Freeman, 1986), chaps. 5 and 6; Wilford, *Mars Beckons.*

75. Kathy Sawyer, "NASA Prepares Craft for a Deep Encounter of the Martian Kind," *Washington Post,* 21 September 1992, sec. A, p. 3.

76. Horowitz, *To Utopia and Back,* xi.

77. See Carl Sagan, *Cosmos* (New York: Random House, 1980), 299–302; Carl Sagan and Frank Drake, "The Search for Extraterrestrial Intelligence," *Scientific American* 232 (May 1975): 80–89.

78. Robert Jastrow and Malcolm H. Thompson, *Astronomy: Fundamentals and Frontiers,* 4th ed. (New York: John Wiley and Sons, 1984), 71; Sagan, *Cosmos,* 299.

79. See "Life on a Billion Planets," *Time,* 3 March 1958, 42–42; "Anybody Out There?" *Time,* 23 November 1959, 84–84; "Advice from Space," *Time,* 29 December 1961, 26. Also see Frank D. Drake, *Intelligent Life in Space* (New York: Macmillan, 1962).

80. Isaac Asimov, *Foundation* (Garden City, N.Y.: Doubleday, 1951), *Foundation and Empire* (Garden City, N.Y.: Doubleday, 1952), and *Second Foundation* (Garden City, N.Y.: Doubleday, 1953).

81. Gene Roddenberry, "Star Trek," NBC, September 1966 to September 1969, seventy-nine episodes.

82. Asimov, *Foundation,* 3–6.

83. Quoted from Carl Sagan, *The Cosmic Connection* (New York: Doubleday, 1973), 25–26.

84. House Science and Technology Committee, *The Possibility of Intelligent Life Elsewhere in the Universe,* 94th Cong., 1st sess., 1975, 24–27.

85. Ibid., 32–53. See also Sagan and Drake, "Search for Extraterrestrial Intelligence," 80–89; and Frank Drake and Dava Sobel, *Is Anyone Out There? The Scientific Search for Extraterrestrial Intelligence* (New York: Delacorte Press, 1992), 182–84.

86. James C. Fletcher, "NASA and the 'Now' Syndrome," from an address to the National Academy of Engineering, Washington, D.C., November 1975, NASA brochure, 7; see also Fletcher, "Space: 30 Years into the Future," *Acta Astronautica* 19, no. 11 (1989): 855–57.

87. Joseph Smith, *The Holy Scriptures,* The Reorganized Church of Jesus Christ of Latter Day Saints (Independence, Mo.: Herald Publishing House, 1944), a revelation given to Joseph the Seer, para. 21; Roger D. Launius, "A Western Mormon in Washington, D.C.: James C. Fletcher, NASA, and the Final Frontier," *Pacific Historical Review* 64 (May 1995): 217–41.

88. See Philip Morrison, John Billingham, and John Wolfe, eds., *The Search for Extraterrestrial Intelligence (SETI),* NASA SP-419 (Washington, D.C.: GPO, 1977).

89. NASA, "SETI," National Aeronautics and Space Administration publication NP-114, June 1990.

90. Lance Frazer, "Small Change, High Gain," *Ad Astra,* September 1989, 19.

91. Rob Meckel, "Proxmire 'Fleeces' NASA over Communications," Proxmire biography file, NASA History Office. "Senate Rejects NASA Space Signal Plan," *Newport News Times-Herald,* 8 August 1978.

92. William Triplett, "SETI Takes the Hill," *Air and Space,* November 1992, 80–86; Lance Frazer, "Listening for Life," *Ad Astra,* September 1989, 16–22.

93. Dava Sobel, "Is Anybody Out There?" *Life,* September 1992, 14. See also Peter Bond, "Extra-Terrestrials Search Stepped Up," *Spaceflight,* January 1993, 6–7.

94. *Congressional Record,* 22 September 1993, S12151.

95. Sebastian von Hoerner, "Where Is Everybody?" in *The Quest for Extraterrestrial Life,* ed. Donald Goldsmith (Mill Valley, Calif.: University Science Books, 1980), 252; Michael H. Hart, "An Explanation of the Absence of Extraterrestrials on Earth," *Quarterly Journal of the Royal Astronomical Society* 16 (June 1975): 128–35; Eric M. Jones, "Colonization of the Galaxy," *Icarus* 28 (1976) 421–22; Freeman and Lampton, "Interstellar Archeology and the Prevalence of Intelligence," *Icarus* 25 (1975): 368–69; Sagan, *Cosmos,* chap. 12.

96. See Ralph Lapp, *Harper's,* March 1961, 63; "Advice from Space," *Time,* 29 December 1961, 26.

97. See Sagan, *Cosmos,* 301.

98. See Dole, *Habitable Planets for Man,* chap. 8; Neil F. Comins, *What If the Moon Didn't Exist?* (New York: HarperCollins, 1993).

99. Von Hoerner, "Where Is Everybody?" 251–53.

100. *Congressional Record,* 22 September 1993, S12152. See also Mariane K. Meuse, "Space Explodes! Alien Media Invades Earth," *Ad Astra,* January/February 1992, 42–46, 55.

101. See Nichio Kaku, *Hyperspace: A Scientific Odyssey Through Parallel Universes, Time Warps, and the Tenth Dimension* (New York: Oxford University Press, 1994), 196–201; George Smoot and Keay Davidson, *Wrinkles in Time* (New York: William Morrow, 1993), 83–86; NASA, *NASA Facts,* "COBE Observes Primeval Explosion," Goddard Space Flight Center, Greenbelt, Maryland, n.d.

102. Craig Covault, "Cosmic Background Explorer to Observe Big Bang Radiation," *Aviation Week & Space Technology,* 6 November 1989, 36–41.

103. Quoted from Paul Hoversten, "Relics of Universe's Birth Found," *USA Today,* 24 April 1992; and Thomas H. Maugh, "'Holy Grail' of the Cosmos," *Los Angeles Times,* 24 April 1992, Washington ed.

104. Malcolm W. Browne, "Despite New Data, Mysteries of Creation Persist," *New York Times,* 12 May 1992, sec. C, pp. 1, 10; NASA, "Cosmic Background Explorer Observes the Primeval Explosion," Goddard Space Flight Center, Greenbelt, Maryland, n.d.

105. Kathy Sawyer, "Big Bang 'Ripples' Have Universal Impact," *Washington Post,* 3 May 1992, sec. A, p. 1.

106. Hoversten, "Relics of Universe's Birth Found." Also see Billy Goodman, "Ancient Whisper," *Air & Space,* April/May 1992, 55–61.

107. Joseph Campbell, *The Power of Myth* (New York: Anchor Books, 1991).

108. Genesis 1:3–5.

109. Sawyer, "Big Bang 'Ripples' Have Universal Impact," sec. A, p. 20.

110. H. G. Wells, *The Time Machine* (New York: Berkley, 1963). George Pal released the motion picture version in 1960. Pal, *The Time Machine* (MGM, 1960).

111. See Paul J. Nahin, *Time Machines: Time Travel in Physics, Metaphysics, and Science Fiction* (New York: American Institute of Physics, 1993); Nicholls, *Science in Science Fiction,* chap. 5.

112. NASA, "Frontiers in Cosmology," Hubble Space Telescope Fact Sheet, Space Telescope Science Institute, Baltimore Maryland, n.d., NASA History Office.

113. Lockheed Missiles and Space Company, "Hubble: A Window into the Universe," 1986; see also NASA, "Hubble Space Telescope Media Reference Guide," published for NASA by Lockheed Missiles and Space Company, Sunnyvale, California, 1990; Joseph J. McRoberts, *Space Telescope,* NASA EP-166 (Washington, D.C.: GPO, n.d.); all in NASA History Office.

114. Steven Spielberg, Bob Gale, and Neil Canton, *Back to the Future* (Universal Studios, 1985).

115. Kaku, *Hyperspace,* x (emphasis removed); also see Kaku and Jennifer Trainer, *Beyond Einstein: The Cosmic Quest for the Theory of the Universe* (New York: Bantam Books, 1987); and Hawking, *Brief History of Time.*

116. Kip S. Thorne, *Black Holes and Time Warps* (New York: W. W. Norton, 1994); Barry Parker, *Cosmic Time Travel: A Scientific Odyssey* (New York: Plenum Press, 1991).

117. Michael S. Morris, Kip S. Thorne, and Ulvi Yurtsever, "Wormholes, Time Machines, and the Weak Energy Condition," *Physical Review Letters* 61 (26 September 1988): 1446–49; Michael S. Morris and Kip S. Thorne, "Wormholes in Spacetime and Their Use for Interstellar Travel," *American Journal of Physics* 56 (May 1988): 395–412; see also, Nahin, *Time Machines,* tech note 9.

118. Lewis Carroll, *Through the Looking Glass* (New York: Grosset and Dunlap, 1946); Lewis, *Lion, the Witch, and the Wardrobe.*

119. NASA, News Release 95-216, "Hubble Finds New Black Hole and Unexpected Mysteries," 4 December 1995.

120. Kaku, *Hyperspace,* chap. 10.

121. Patrick J. Kiger, "The New Galileo," *Baltimore Magazine,* February 1990, 107.

122. *Congressional Record,* 24 June 1993, H4057. See also *Congressional Record,* 23 June 1993, H3974-780, and 19 October 1993, H8114-24; and *Congressional Record,* 10 July 1991, S9430-43.

123. NASA News Release, "Jupiter's Europa Harbors Possible 'Warm Ice' or Liquid Water," release 96-164, NASA Headquarters, 13 August 1996.

124. See David C. Black, ed., *Project Orion: A Design Study of a System for Detecting Extrasolar Planets,* NASA SP-436 (Washington, D.C.: GPO, 1980); NASA, "TOPS: Toward Other Planetary Systems," Solar System Exploration Division, n.d.; D. DeFrees, *Exobiology in Earth Orbit,* NASA SP-500 (Washington, D.C.: GPO, 1989).

125. See Milne et al., *Evolution of Complex and Higher Organisms,* 153–54.

126. John Noble Wilford, "New Discoveries Turn Astronomers Toward Hunt for New Planets," *New York Times,* 23 January 1996; Wilford, "The Search for Solar Systems Accelerates Amid New Clues," *New York Times,* 21 April 1987; Malcolm Brown, "Clues Point to Young Planet Systems Nearby," *New York Times,* 12 June 1992; NASA, *NASA News,* release 92-226, "Hubble Discovers Protoplanetary Disks Around New Stars," 16 December 1992.

127. See Richard Corliss, "The Invasion Has Begun," *Time,* 8 July 1996, 58–64.

128. "Meteorite Find Incites Speculation on Mars Life," *Space News,* 5–11 August 1966, 2; David S. McKay, Everett K. Gibson Jr., Kathie L. Thomas-Keprta, Hojatollah Vali, Christopher S. Romanek, Simon J. Clemett, Xavier D. F. Chillier, Claude R. Maechling, and Richard N. Zare, "Search for Past Life on Mars: Possible Relic Biogenic Activity in Martian Meteorite ALH84001," *Science* 273 (16 August 1996): 924–30.

129. David Colton, "Discovery Would Equal Finding the New World," *USA Today,* 8 August 1996; NASA, *NASA News,* "Statement of Daniel S. Goldin, NASA Administrator," release 96-159, 6 August 1996; NASA, *NASA News,* "NASA Briefing Wednesday on Discovery of Possible Martian Life," note to editors N96-53, 6 August 1966; William J. Clinton, "NASA Discovery of Possible Life on Mars," *Weekly Compilation of Presidential Documents* (7 August 1996): 1417–18.

130. John Noble Wilford, "On Mars, Life's Getting Tougher (If Not Impossible)," *New York Times,* 22 December 1996.

6. THE EXTRATERRESTRIAL FRONTIER

Epigraph: Herbert F. Solow and Robert H. Justman, *Inside Star Trek* (New York: Pocket Books, 1996), 149.

1. See McCall, *Art of Robert McCall,* 28–29 for the mural; see also Stephen M. Fjellman, *Vinyl Leaves: Walt Disney World and America* (Boulder, Colo.: Westview Press, 1992), chap. 5; Richard R. Beard, *Walt Disney's Epcot* (New York: Harry N. Abrams, 1982), 136–63.

2. See Patricia Nelson Limerick, *The Legacy of Conquest: The Unbroken Past of the American West* (New York: W. W. Norton, 1987); Richard White, *"It's Your Misfortune and None of My Own": A New History of the American West* (Norman: University of Oklahoma Press, 1991); Donald Worster, *Rivers of Empire: Water, Aridity, and the Growth of the American West* (New York: Pantheon Books, 1985); and William Cronon, George Miles, and Jay Gitlin, eds., *Under the Open Sky: Rethinking America's Western Past* (New York: W. W. Norton, 1992).

3. See Slotkin, *Gunfighter Nation.*

4. "Day at Tranquility," *Washington Daily News,* 21 July 1969; "One Small Step—One Giant Leap," *Washington Post,* 21 July 1969.

5. Manfred van Ehrenfried, *Adventures on Santa Maria and Future Ships Sailing the Oceans of Space* (Glenside, Pa.: Custom Comic Services, 1991), 1.

6. National Commission on Space, *Pioneering the Space Frontier,* 8. See also Joseph F. Shea, "Manned Space Flight Program," address at the third national conference on the peaceful uses of space, Chicago, Illinois, 6 May 1963, NASA News release, NASA Headquarters, NASA History Office.

7. Buzz Aldrin, "The Mars Transit System," *Air & Space,* October/November 1990, 41, 42. See also Buzz Aldrin and John Barnes, *Encounter with Tiber* (New York: Warner Books, 1996).

8. "Excerpts of Remarks by Governor Ronald Reagan," America Legion State Convention, Sacramento, Calif., 26 June 1970, 4, Hoover Institution on War, Revolution, and Peace, Stanford, Calif.

9. Logsdon, *Decision to Go to the Moon,* 35; see also Ira C. Eaker, "Columbus and the Moon: Debates on Voyages Are Similar," *San Diego Union,* 22 September 1963; Daniel Goldin, "Celebrating the Spirit of Columbus," *National Forum* (Summer 1992): 8–9.

10. Howard E. McCurdy, *The Space Station Decision: Incremental Politics and Technological Choice* (Baltimore: Johns Hopkins University Press, 1990), 184; see also Johnson, "Remarks at Michoud Assembly Facility," 1697.

11. James M. Beggs, "Why the United States Needs a Space Station," remarks prepared for delivery at the Detroit Economic Club and Detroit Engineering Society, 23 June 1982, 2–3, NASA History Office; reprinted under the same title in *Vital Speeches,* 1 August 1982, 615–17.

12. James M. Beggs, "The Wilbur and Orville Wright Memorial Lecture," Royal Aeronautical Society, London, England, 13 December 1984, 2, NASA History Office; Michael Ryan, "Why They Come to the Ice," *Parade,* 11 July 1993, 5; see also James Beggs, "Space Tomorrow: The Antarctica Model," *IEEE Spectrum* 20 (September 1983): 89–90; and "A Terrestrial Testing Ground for Space Exploration" in NASA, *HQ Bulletin* (19 February 1991): 1, 3.

13. Ronald Reagan, "United States Space Policy," remarks on the completion of the fourth mission of the space shuttle *Columbia, Weekly Complication of Presidential Documents* 18 (4 July 1982): 870.

14. George Bush, "Remarks at the Texas A&I University Commencement Ceremony in Kingsville, Texas," *Weekly Compilation of Presidential Documents* 26, no. 20 (11 May 1990): 749.

15. NASA Grant Nsg-253-62; Bruce Mazlish, ed., *The Railroad and the Space Program: An Exploration in Historical Analogy* (Cambridge, Mass.: MIT Press, 1965).

16. Patricia Nelson Limerick, "Imagined Frontiers: Westward Expansion and the Future of the Space Program," in *Space Policy Alternatives,* ed. Radford Byerly (Boulder, Colo.: Westview Press, 1992), 249–61.

17. Frederick Jackson Turner, "The Significance of the Frontier in American History," from John M. Farager, *Rereading Frederick Jackson Turner* (New York: Henry Holt, 1994), 31–60.

18. Ibid., 32.

19. See George Rogers Taylor, ed., *The Turner Thesis: Concerning the Role of the Frontier in American History,* 3d ed. (Lexington: Mass.: D. C. Heath, 1972).

20. See for example Eric M. Jones, ed., "The Space Settlement Papers," in *Journal of the British Interplanetary Society* 39 (July 1986): 291–311.

21. Brad Darrach and Steve Petranek, "Mars: Our Next Home," *Life,* May 1991, 34.

22. "Go!" *Washington Post,* 21 February 1962, sec. A, p. 24. See also Donovan, "Moon Voyage Turns Men's Thoughts Inward"; "One Small Step—One Giant Leap," *Washington Post,* 21 July 1969; and "At Path, Not an End," *Washington Post,* 21 July 1969.

23. Sagan, *Pale Blue Dot,* xiv.

24. Ibid., xii; see also James A. Michener, "Space and the Human Quest," *National Forum* (Summer 1992): 3–5.

25. Sagan, *Pale Blue Dot,* 50. See Bryan Appleyard, *Understanding the Present: Science and the Soul of Modern Man* (London: Picador/Pan Books, 1992).

26. Sagan, *Pale Blue Dot,* 53.

27. James M. Beggs, "The Wilbur and Orville Wright Memorial Lecture," Royal Aeronautical Society, London, England, 13 December 1984, NASA History Office; Daniel Goldin, "Celebrating the Spirit of Columbus," *Phi Kappa Phi Journal* (Summer 1992): 8. See also Synthesis Group, *America at the Threshold,* iv.

28. Beggs, "Wilbur and Orville Wright Memorial Lecture," 2; also see Thomas O. Paine, "1969: A Space Odyssey," address at the American Institute of Aeronautics and Astronautics, Washington, D.C., 7 November 1968, 9–11; and Paine, "Thomas A. Edison Memorial Lecture," address at the Naval Research Laboratory, Washington, D.C., 11 March 1969, 11–12; all in NASA History Office.

29. Kennedy, "Address at Rice University," 12 September 1962, 373.

30. Sagan, *Pale Blue Dot,* 371, 382; Fletcher, "NASA and the 'Now' Syndrome," 3; Goddard and Pendray, *Papers of Robert H. Goddard* 3:1612; Richard Rhodes, "God Pity a One-Dream Man," *American Heritage,* June/July 1980, 32.

31. Walter J. Hickel, "In Space: One World United," in *Lunar Bases and Space Activities of the 21st Century,* ed. W. W. Mendell (Houston: Lunar and Planetary Institute, 1985), 15, 17; Beggs, "Wilbur and Orville Wright Memorial Lecture," 2.

32. Paine, "Thomas A. Edison Memorial Lecture," 11.

33. Ibid.

34. Ibid., 10.

35. Paul D. Lowman, "Lunar Bases and Post-Apollo Lunar Exploration: An Annotated Bibliography of Federally-Funded American Studies, 1960–82," Geophysics Branch, Goddard Space Flight Center, October 1984; NASA, "NASA Initiates Lunar Base Study Program," news release no. 63-91, 6 May 1963; both in NASA History Office.

36. Boeing Company, "Initial Concept of Lunar Exploration Systems for Apollo (LESA)," D2-1000057, NASW 792, Boeing Company Aero-Space Division; see also House Subcommittee on Manned Space Flight, *1965 NASA Authorization,* 88th Cong., 2d sess., 1964, pt. 2, 587–626; and House Committee on Science and Astronautics, *Future National Space Objectives,* 89th Cong., 2d sess., 1966; NASA, "Apollo Applications Program," 3–4.

37. Lockheed, "Study of Mission Modes and System Analysis for Lunar Exploration (MIMOSA), Final Report," Lockheed Missiles and Space Company, NAS 8-20262, Sunnyvale, California; see also NASA, Future Programs Task Group, "Summary Report," NASA Headquarters, January 1965, 47–48, NASA History Office.

38. Clarke, *2001,* 61–65.

39. Paul D. Lowman, "Lunar Bases: A Post-Apollo Evaluation," in Mendell, *Lunar Bases and Space Activities,* 38, 42–43; F. Nozette, C. L. Lichtenberg, P. Spudis, R. Bonner, W. Ort, E. Malaret, M. Robinson, E. M. Shoemaker, "The Clementine Bistatic Radar Experiment," *Science* 274 (29 November 1996): 1495–98.

40. Gerard K. O'Neill, "The Colonization of Space," *Physics Today* 27 (September 1974): 32–40.

41. Gerard K. O'Neill, *The High Frontier: Human Colonies in Space* (New York: William Morrow, 1976), app. 1.

42. Michaud, *Reaching for the High Frontier,* 65.

43. O'Neill, "Colonization of Space," 36; Gerard K. O'Neill, "A Lagrangian community?" *Nature* 250 (23 August 1974): 636.

44. Ibid., 37, 39; O'Neill, *High Frontier,* 64.

45. O'Neill, "Colonization of Space," 37.

46. Ibid., 37.

47. The Saturn 5 launch vehicle cost a total of $255 million in hardware and operational costs to place a lunar module weighing fifteen hundred kilograms (thirty-three hundred pounds) on the Moon. Howard E. McCurdy, "The Cost of Space Flight," *Space Policy* 10 (November 1994): 277–89.

48. O'Neill, *High Frontier,* 138–41.

49. Stewart Brand, *Space Colonies* (New York: Penguin Books, 1977), 15; see also O'Neill, "Colonization of Space," 37–38. See also Jerry Grey, ed., *Space Manufacturing Facilities (Space Colonies)* (New York: American Institute of Aeronautics and Astronautics, 1977).

50. Tsiolkovskiy, *Beyond the Planet Earth,* chaps. 12 and 13; J. D. Bernal, *The World, the Flesh, and the Devil* (London: Methuen, 1929; Bloomington: University of Indiana Press, 1969); J. N. Leonard, *Flight into Space* (New York: Signet, 1954), chap. 22; Dandridge M. Cole and Donald W. Cox, *Islands in Space: The Challenge of the Planetoids* (New York: Chilton, 1964); Dandridge M. Cole, "Extraterrestrial Colonies," *Navigation* 7 (Summer–Autumn 1960): 83–98; Arthur C. Clarke, *Islands in the Sky* (New York: Holt, Rinehart, and Winston, 1954); Krafft A. Ehricke, "Extraterrestrial Imperative," *Bulletin of the Atomic Scientists* (November 1971): 18–26; see also Michaud, *Reaching for the High Frontier,* chap. 4.

51. See Brand, *Space Colonies;* Michaud, *Reaching for the High Frontier,* chap. 5; and O'Neill, *High Frontier,* 253–61. Richard D. Johnson, *Space Settlements: A Design Study,* NASA SP-413 (Washington, D.C.: GPO, 1977).

52. Quoted from Michaud, *Reaching for the High Frontier,* 86.

53. Buckbee and Walker, "Spaceflight and the Public Mind," 193–96; Michaud, *Reaching for the High Frontier,* chaps. 3 and 5.

54. Michaud, *Reaching for the High Frontier,* chap. 7.

55. "The Mars Declaration," in "The Way to Mars," (Pasadena, Calif.: Planetary Society).

56. Trudy E. Bell, "Space Activists on Rise," *Insight* (National Space Institute) (August/September 1980).

57. Michaud, *Reaching for the High Frontier,* chaps. 8 and 9.

58. National Commission on Space, *Pioneering the Space Frontier,* 3, 140–42, 189–90.

59. Ibid., 2, 69, 86–87, 121, 135.

60. Thomas O. Paine, "Mars Colonization," *Phi Kappa Phi Journal* (Summer 1992): 26. See also National Commission on Space, *Pioneering the Space Frontier,* 89; Benton Clark, "Chemistry of the Martian Surface: Resources for the Manned Exploration of Mars," in *The Case for Mars,* vol. 57, science and technology series, ed. Penelope U. Boston (San Diego: American Astronautical Society, 1984), 197–208; and Christopher P. McKay, *The Case for Mars II,* vol. 62, science and technology series, ed. Christopher P. McKay (San Diego: American Astronautical Society, 1985), chap. 7 ("Utilizing Martian Resources").

61. National Commission on Space, *Pioneering the Space Frontier,* 72. See also *Planetary Explorer,* a publication of the General Dynamics Space Systems Division, San Diego, California, 1988.

62. James Fletcher, "Excerpts from Remarks Prepared for Delivery, National Space Symposium," 14 April 1988.

63. Sally K. Ride, *Leadership and America's Future in Space,* Report to the Administrator (Washington, D.C.: NASA, 1987), 29–31.
64. Leonard David, "Moon Fever," *Space World,* August 1988, 6–8.
65. Mendell, *Lunar Bases;* Craig Covault, "Manned U.S. Lunar Station Wins Support," *Aviation Week & Space Technology,* 19 November 1984, 73–86; Darren L. Burnham, "Back to the Moon with Robots?" *Spaceflight* 35 (February 1993): 54–57; David, "Moon Fever," 7.
66. Boston, *Case for Mars,* x.
67. McKay, *Case for Mars II,* ix–x; Carol R. Stoker, *The Case for Mars III: Strategies for Exploration—General Interest and Overview,* vol. 74, science and technology series (San Diego: American Astronautical Society, 1989).
68. Duke B. Reiber, *The NASA Mars Conference,* vol. 71, science and technology series (San Diego: American Astronautical Society, 1988); NASA, "NASA Establishes Office of Exploration."
69. Ride, *Leadership and America's Future in Space,* 32–35; "Journey to Mars," transparency in NASA, Office of Exploration, "Exploration Initiative: A Long-Range, Continuing Commitment," January 1990; NASA, "Report of the 90-Day Study on Human Exploration of the Moon and Mars," NASA Headquarters, November 1989; Synthesis Group, *America at the Threshold,* 21.
70. Alcestis and James Oberg, "The Future of Mars," in Boston, *Case for Mars,* 311–13.
71. Ross Rocklynne, "Water for Mars," *Astounding Stories,* April 1937, 10–46;
72. Will Stewart, "Collision Orbit," *Astounding Science Fiction,* July 1942.
73. Robert A. Heinlein, *Farmer in the Sky* (New York: Ballantine, 1950).
74. Carl Sagan, "The Planet Venus," *Science* 133 (24 March 1961): 857–58.
75. M. M. Averner and R. D. MacElroy, eds., *On the Habitability of Mars: An Approach to Planetary Ecosynthesis,* NASA SP-414 (Springfield, Va.: National Technical Information Service, 1976); also see Christopher P. McKay, "Living and Working on Mars," in Reiber, *NASA Mars Conference,* 522; "Terraforming: Making an Earth of Mars," *Planetary Report* 7 (November/December 1987).
76. Alcestis R. Oberg, "The Grass Roots of the Mars Conference," in Boston, *Case for Mars,* ix–xii.
77. See James Edward Oberg, *New Earths: Transforming Other Planets for Humanity* (Harrisburg, Pa.: Stackpole Books, 1981).
78. Brad Darrach and Steve Petranek, "Mars: Our Next Home," *Life,* May 1991, 32–35.
79. Sagan, *Pale Blue Dot,* 339; see also James B. Pollack and Carl Sagan, "Planetary Engineering," in *Near-Earth Resources,* ed. J. Lewis and M. Matthews (Tucson: University of Arizona Press, 1992).
80. White House, "Remarks by the President," 20 July 1989; George Bush, "Remarks at the Texas A&I University Commencement Ceremony in Kingsville, Texas," 750.
81. Sagan, *Pale Blue Dot,* 264.
82. Andrew Lawler, "Newsmaker Forum: Barbara Mikulski," *Space News,* 13 November 1989; George Bush, "Remarks to Employees of the George C. Marshall Space Flight Center in Huntsville, Alabama," *Weekly Compilation of Presidential Documents* 26, no. 25 (20 June 1990): 981–83; see also "Bush Goes on Counterattack Against Mars Mission Critics," *Congressional Quarterly Weekly Report,* 23 June 1990, 1958; see also Gregg Easterbrook, "The Case Against Mars," in Stoker, *Case for Mars III,* 49–54.
83. See Patricia Nelson Limerick, *The Legacy of Conquest: The Unbroken Past of the American West* (New York: W. W. Norton, 1987), 323–24; Richard White, *"It's Your Misfortune and None of My Own,"* chap. 21; James R. Grossman, ed., *The Frontier in American Culture* (Berkeley and Los Angeles: University of California Press, 1994).
84. Limerick, "Imagined Frontiers," 251.
85. Quoted from Brian W. Dippie, "The Winning of the West Reconsidered," *Wilson Quarterly* 14 (Summer 1990): 73.
86. Thomas O. Paine, "Head of NASA Has New Vision of 1984," *New York Times,* 17 July 1969. Konstantin Tsiolkovskiy, quoted by Nicholas Daniloff, *The Kremlin and the Cosmos* (New York: Knopf, 1972), 20; O'Neill, *High Frontier,* 219.

87. Arthur E. Bestor, *Backwoods Utopias,* 2d ed. (Philadelphia: University of Pennsylvania Press, 1970).

88. See Limerick, *Legacy of Conquest,* chap. 3.

7. STATIONS IN SPACE

Epigraph: House Science and Technology Committee, Subcommittee on Space Science and Applications, *NASA's Space Station Activities,* 98th Cong., 1st. sess, 1983, 4.

1. Ronald Reagan, "The State of the Union: Address Delivered Before a Joint Session of the Congress," *Weekly Compilation of Presidential Documents* 20 (25 January 1984): 90.

2. Ley, *Rockets,* 223; Fritz Sykora, "Guido von Pirquet: Austrian Pioneer of Astronautics" (paper presented at the Fourth History Symposium of the International Academy of Astronautics, Constance, German Federal Republic, October 1970), NASA History Office; Willy Ley, *Rockets, Missiles, and Space Travel* (New York: Viking Press, 1951), 317; Willy Ley, *Rockets and Space Travel* (New York: Viking Press, 1948), 284. See also Cornelius Ryan, ed., *Man on the Moon* (London: Sidgwick and Jackson, 1953), 24.

3. Willard B. Robinson, *American Forts: Architectural Form and Function* (Urbana: University of Illinois Press, 1977); Robert B. Roberts, *Encyclopedia of Historic Forts: The Military, Pioneer, and Trading Posts of the United States* (New York: Macmillan, 1988).

4. Roland Huntford, *The Last Place on Earth* (New York: Atheneum, 1983).

5. John Hunt, *The Ascent of Everest* (Seattle: Mountaineers, 1993); Micheline Morin, *Everest: From the First Attempt to the Final Victory* (New York: John Day, 1955); see also Galen Rowell, *In the Throne Room of the Mountain Gods* (San Francisco: Sierra Club Books, 1977).

6. See Adam Gruen, "The Port Unknown: A History of the Space Station Freedom Program," unpublished manuscript, NASA History Office, 1–2; also Gruen, "The Port Unknown," a dissertation submitted in partial fulfillment for the requirements for the degree of Doctor of Philosophy in the Department of History in the Graduate School of Duke University, 1989.

7. Hermann Oberth, *Man into Space: New Projects for Rocket and Space Travel,* translated from the German by G. P. H. De Freville (London: Weidenfeld and Nicolson, 1957), 73; Hermann Oberth, *Ways to Spaceflight,* NASA technical translation TT F-622 (Washington, D.C.: NASA, 1972), chap. 20; Ley, *Rockets,* 220; von Braun, "Crossing the Last Frontier," 28–29; von Braun, "Man on the Moon," 52.

8. See also Robert Gilruth, "Manned Space Stations," *Spaceflight* (August 1969): 258; S. Fred Singer, *Manned Laboratories in Space* (New York: Springer-Verlag, 1969).

9. See Oscar Schachter, "Who Owns the Universe?" *Collier's,* 22 March 1952, 36, 70–71; Senate Aeronautical and Space Sciences Committee, *Legal Problems of Space Exploration,* 87th Cong., 1st sess., 1961; Myres S. McDougal, Harold D. Lasswell, and Ivan A. Vlasic, *Law and Public Order in Space* (New Haven, Conn.: Yale University Press, 1963); and Clive Cussler, *Cyclops* (New York: Simon and Schuster, 1986).

10. "What Are We Waiting For?" 23.

11. Oberth, *Man into Space,* 61, 107–8.

12. Ley, *Rockets,* 227.

13. Von Braun, "Crossing the Last Frontier."

14. See Curtis Peebles, "The Manned Orbiting Laboratory," a three-part article in *Spaceflight* 22 (April 1980): 155–60, 22 (June 1980): 248–53, and 24 (June 1982): 274–77.

15. See Stares, *Militarization of Space;* and Curtis Peebles, *Battle for Space* (New York: Beaufort Books, 1983).

16. Hans Mark, *Space Station: A Personal Journey* (Durham, N.C.: Duke University Press, 1987), 50; see also Brooks, Grimwood, and Swenson, *Chariots for Apollo,* chap. 3; and Wernher von Braun, "Concluding Remarks by Dr. Wernher von Braun about Mode Selection for the Lunar Landing

Program," given to Dr. Joseph F. Shea, Deputy Director (Systems), Office of Manned Space Flight, 7 June 1962, NASA History Office.

17. Reinhold Messner, *Antarctica: Both Heaven and Earth* (Seattle: Mountaineers, 1991).

18. Testimony of George Mueller, Senate Aeronautical and Space Sciences Committee, *NASA Authorization for Fiscal Year 1970,* 91st cong., 1st sess., 1969; "What Are We Waiting For?" 23.

19. See Hale, *Brick Moon;* Kimball, *Man and the Moon;* Anatoly Andanov and Gennady Maximov, "Space Stations of the Future—A Soviet View," *Spaceflight* (August 1969): 264–65; Clarke, *Islands in the Sky.*

20. Von Braun, "Crossing the Last Frontier," 29, 72.

21. Willy Ley, "A Station in Space," *Collier's,* 22 March 1952, 30–31.

22. The USS *Kitty Hawk,* for example, is 1,065 feet long and holds a crew of fifty-three hundred when the air wing is on board; Clarke, *2001.*

23. Gene Roddenberry, *Star Trek: The Motion Picture* (Paramount, 1979); see also Gene Roddenberry, *Star Trek: The Motion Picture, A Novel* (New York: Simon and Schuster, 1979).

24. NASA, *Space Station: Key to the Future* (Washington, D.C.: GPO, n.d.); William Nromyle, "NASA Aims at 100-Man Station," *Aviation Week & Space Technology,* 24 February 1969, 16–17.

25. David Baker, "Space Station Situation Report—1: The North American Rockwell Proposal," *Spaceflight* 13 (September 1971): 318–34; Baker, "Space Station Situation Report—2: The McDonnell Douglas Proposal," *Spaceflight* 13 (September 1971): 344–51; McDonnell Douglas Astronautics Company, *Space Station* MSFC-DRL-160, Executive Summary, Contract NAS8-25140, August 1970; Nieson S. Himmel, "Advanced Space Station Concepts," *Aviation Week & Space Technology,* 22 September 1969, 100–113; Irving Stone, "NASA Launches Space Station Task," *Air Force/Space Digest* (July 1969): 79–82; John Logsdon, "Space Stations: A Policy History," prepared for the Johnson Space Center, NASA contract NAS9-16461, George Washington University, Washington, D.C., n.d.; Ray Hook, "Historical Review," *Journal of Engineering for Industry* 4 (November 1984): 276–86.

26. NASA, "NASA Announces Baseline Configuration for Space Station," *NASA News,* release 86–61, 14 May 1986.

27. NASA, *The Space Station: A Description of the Configuration Established at the Systems Requirements Review (SRR)* (Washington, D.C.: NASA Office of Space Station, 1986), 30.

28. Linda N. Ezell, *NASA Historical Data Book: Programs and Projects, 1969–1978,* SP-4012 (Washington, D.C.: GPO, 1988), 94.

29. James M. Beggs, NASA Administrator, to Craig L. Fuller, Assistant to the President for Cabinet Affairs, 12 April 1984, NASA History Office; Reagan, "State of the Union," 25 January 1984, 90.

30. Von Braun, "Crossing the Last Frontier," 22 March 1952, 26.

31. McCurdy, *Space Station Decision,* 149–50.

32. Robert W. Smith, *The Space Telescope: A Study of NASA, Science, Technology, and Politics* (New York: Cambridge University Press, 1989).

33. See Oberth, *Man into Space,* 67–71; NASA Space Station Task Force, *Program Description Document: Mission Description Document* (Washington, D.C.: NASA, 1984), bk. 2, sec. 3, 5–12.

34. Von Braun, "Crossing the Last Frontier," 72.

35. House Science and Technology Committee, Space Science and Applications Subcommittee, *NASA's Space Station Activities,* 98th Cong., 1st sess, 1983, 6; see also Senate Commerce, Science, and Transportation Committee, Science, Technology, and Space Subcommittee, *Civil Space Station,* 98th Cong., 1st sess., 1983, 30; and McCurdy, *Space Station Decision,* 146–49.

36. "Ivory Tower in Space," *Nature* 307 (5 January 1984): 2.

37. Oberth, *Man into Space,* 63; Hermann Oberth, *Rockets in Planetary Space* (1923), NASA TT F-9227 (Washington, D.C.: NASA, 1964), 93–94; see also Space Science Board, Committee on Space Biology and Medicine, *A Strategy for Space Biology and Medical Science for the 1980s and 1990s* (Washington, D.C.: National Academy Press, 1987).

38. See John McLucas, *Space Commerce* (Cambridge: Harvard University Press, 1991).

39. See McCurdy, *Space Station Decision,* 179–80; Craig Covault, "Reagan Briefed on Space Station," *Aviation Week & Space Technology,* 8 August 1983, 16–18; James R. Asker, "No Windfalls Yet, But Space Commerce Advances," *Aviation Week & Space Technology,* 19 April 1993, 26–27; and Camille M. Jernigan and Elizabeth Penetecost, *Space Industrialization Opportunities* (Park Ridge, N.J.: Noyes, 1985).

40. NASA, *Space Station,* 8–9.

41. McLucas, *Space Commerce,* 191.

42. Henry S. F. Cooper, "Annals of Space," *New Yorker,* 2 September 1991, 44–51.

43. Eliot Marshall, "Space Stations in Lobbyland," *Air & Space,* December 1988/January 1989, 54–61; Gruen, "Port Unknown," 284–86; Kathy Sawyer, "Commercial Space Laboratory Not Needed, Expert Panel Says," *Washington Post,* 12 April 1989; T. A. Heppenheimer, "Son of Space Station," *Discover,* July 1988, 64–66.

44. NASA, "The Post-Apollo Space Program: Directions for the Future," summary of NASA's report to the President's Space Task Group, September 1969, 2, 9; see also Thomas Paine, Memorandum to the President, 26 February 1969, in Logsdon, *Exploring the Unknown,* 517–18.

45. Clarke Covington and Robert O. Piland, "Space Operations Center: Next Goal for Manned Space Flight?" *Astronautics & Aeronautics* 18 (September 1980): 30–37; NASA, "Space Operations Center: A Concept Analysis," vol. 1, Summary, Johnson Space Center, 29 November 1979.

46. Covington and Piland, "Space Operations Center," 33.

47. See McCurdy, *Space Station Decision,* 77–90.

48. Von Braun, "Crossing the Last Frontier," 25.

49. Senate Commerce, Science, and Transportation Committee, *Civil Space Station,* 27.

50. NASA, *Space Station,* 34; NASA, "NASA Announces Baseline Configuration for Space Station," *NASA News,* release 86-61, 14 May 1986.

51. NASA, *The Space Station,* 2, 34.

52. Ibid., 3, 36; McCurdy, "Cost of Space Flight," 277–89.

53. NASA, "NASA Announces Baseline Configuration for Space Station," 2; NASA, *Space Station.*

54. House Committee on Science and Technology, *NASA's Space Station Activities,* 4; Mitchell Waldrop, "Space City: 2001 It's Not," *Science 83* (October 1983); see also John Noble Wilford, "When Man Has Stations in Space," *New York Times,* 19 October 1969.

55. NASA, "NASA Proceeding Toward Space Station Development," *NASA News,* release 87–50, 3 April 1987; NASA, "NASA Issues Requests for Proposals for Space Station Development," *NASA News,* release 87–65, 24 April 1987; see also Andrew J. Stofan, "Space Station: A Step into the Future," NASA Headquarters, n.d.; Stofan, "Preparing for the Future," *Aerospace America* 25 (September 1987): 16–22.

56. Ride, *Leadership and America's Future in Space,* 43.

57. Advisory Committee on the Future of the U.S. Space Program (Norman Augustine, chair), *Report of the Advisory Committee* (Washington, D.C.: GPO, 1990), 29; NASA, "Report to Congress on the Restructured Space Station," 20 March 1991; Technical and Administrative Services Corporation, "Space Station Freedom Media Handbook," May 1992, 23.

58. NASA Space Station Redesign Team, "Final Report to the Advisory Committee on the Redesign of the Space Station," June 1993, 31; see also Advisory Committee on the Redesign of the Space Station (Charles M. Vest, chair), *Final Report to the President* (Washington, D.C.: GPO, 1993); Warren E. Leary, "Clinton Plans to Ask Congress to Approve Smaller, Cheaper Space Station," *New York Times,* 18 June 1993; Daniel S. Goldin to John H. Gibbons, Cost Report for Space Station Alpha, 20 September 1993.

59. NASA, "Alpha Station: Addendum to Program Implementation Plan," 1 November 1993; White House, Office of the Vice President, "Joint Statements on Space Cooperation, Aeronautics and Earth Observation," 2 September 1993; White House, Office of the Vice President, "United

States–Russian Joint Commission on Energy and Space, Joint Statement on Cooperation in Space," 2 September 1993; Steven A. Holmes, "U.S. and Russians Join in New Plan for Space Station," *New York Times,* 3 September 1993; NASA, "Space Station Transition Status Report #1," 26 July 1993; NASA, "International Space Station: Creating a World-Class Orbiting Laboratory," January 1995; NASA, "International Space Station Fact Book," 1 June 1995.

60. Advisory Committee on the Redesign of the Space Station, *Final Report to the President,* 39–43.

61. "What Are We Waiting For?" 23.

62. James M. Beggs to James A. Baker, 24 August 1983; Peggy Finarelli to OMB/Bart Borrasca, 8 September 1983; both in NASA History Office; John Hodge interview, 10 July 1985; NASA Office of Comptroller, "Aerospace Price Deflator," NASA Headquarters, 1994.

63. James C. Miller, memorandum for the President, 10 February 1987, NASA History Office; see also U.S. House Committee on Government Operations, Subcommittee on Government Activities and Transportation, *Cost, Justification, and Benefits of NASA's Space Station,* 102d Cong., 1st sess., 1991; and McCurdy, "Cost of Space Flight," 277–89.

64. U.S. Senate Commerce, Science, and Transportation Committee, Subcommittee on Science, Technology, and Space, *NASA Authorization for Fiscal Year 1984,* 98th Cong., 1st sess., 1983, 51.

65. Gruen, "The Port Unknown: A History," 131–33; Philip E. Culbertson to Neil B. Hutchinson, "Space Station Program Cost Estimates," 14 August 1985, NASA History Office.

66. Von Braun, "Crossing the Last Frontier," 24–25; Clarke, *2001,* 41–48.

67. Presidential Commission on the Space Shuttle Challenger Accident (William P. Rogers, chair), *Report to the President* (Washington, D.C.: GPO, 1986), 164.

68. NASA, "Report to Congress on the Restructured Space Station," 20 March 1991, 5.

69. NASA, *Space Station,* 26–27; Richard DeMeis, "Fleeing Freedom," *Aerospace America* 27 (May 1989): 38–41; Andrew Lawler, "NASA: No Permanent Station Crew Without Escape Vehicle," *Space News,* 12 February 1990.

70. "What Are We Waiting For?" 23.

71. U.S. Senate Committee on Appropriations, *Department of Housing and Urban Development, and Certain Independent Agencies,* Appropriations for Fiscal Year 1985, 98th Cong., 2d sess., 1984, esp. 1266.

72. Kathy Sawyer, "Astronauts Express Fears over Space Station," *Washington Post,* 19 July 1986, Craig Covault, "Launch Capability, EVA Concerns Force Space Station Redesign," *Aviation Week & Space Technology,* 21 July 1986, 18–20.

73. NASA Office of Space Station, "Proceedings of the Space Station Evolution Workshop, Williamsburg, Virginia, 10–13 September 1985," NASA History Office.

74. Synthesis Group, *America at the Threshold,* 6 and 83; Advisory Committee on the Future of the U.S. Space Program, *Report of the Advisory Committee*, 29.

8. SPACECRAFT

Epigraph: Boyce Rensberger, "The Prophet in His Orbit," *Washington Post,* 7 November 1985, sec. C., p. 6.

1. Corn, *Winged Gospel.*

2. Stephen Pendo, *Aviation in the Cinema* (Metuchen, N.J.: Scarecrow Press, 1985).

3. Office of the White House Press Secretary, Press Conference of Dr. James Fletcher and George M. Low; NASA, "Space Shuttle," 1972; NASA, "Fact Sheet: The Economics of the Space Shuttle," July 1972, all in NASA History Office; also see House Committee on Science and Technology, Subcommittee on Space Science and Applications, *Operational Cost Estimates: Space Shuttle,* 94th Cong., 2d sess., 1976.

4. President's Science Advisory Committee, Space Science and Technology Panel, "The Next Decade in Space," NASA History Office, March 1970.

5. See Wolfe, *Right Stuff;* and Joseph D. Atkinson and Jay M. Shafritz, *The Real Stuff* (New York: Praeger, 1985).

6. Sylvia D. Fries, *NASA Engineers and the Age of Apollo* SP 4104 (Washington, D.C.: GPO, 1992); McCurdy, *Inside NASA,* 78–89.

7. Mark Sullivan, *Our Times: The United States 1900–1925* (New York: Charles Scribner's Sons, 1927), 2:556, 599. See also Roger E. Bilstein, *Flight in America: From the Wrights to the Astronauts,* rev. ed. (Baltimore: Johns Hopkins University Press, 1994) and Bilstein, "The Airplane, the Wrights, and the American Public," in *The Wright Brothers: Heirs of Prometheus,* ed. Richard P. Hallion (Washington, D.C.: Smithsonian Institution Press, 1978), 39–51.

8. Hallion, *Wright Brothers,* 75–87; Tom D. Crouch, *The Bishop's Boys: A Life of Wilbur and Orville Wright* (New York: W. W. Norton, 1986); Arthur G. Renstrom, *Wilbur & Orville Wright: A Chronology* (Washington, D.C.: Library of Congress, 1975).

9. Robert Scharff and Walter S. Taylor, *Over Land and Sea: A Biography of Glenn Hammond Curtiss* (New York: David McKay, 1968).

10. Corn, *Winged Gospel,* 12–13; see also Don Dwiggins, *The Barnstormers: Flying Daredevils of the Roaring Twenties* (New York: Grosset and Dunlap, 1968).

11. Orville Wright, "Future of the Aeroplane," *Country Life,* January 1909; Sullivan, *Our Times* 2:558.

12. See Walter H. G. Armytage, *A Social History of Engineering* (Cambridge, Mass.: MIT Press, 1961), 268–70.

13. Bilstein, *Flight in America.*

14. Robert J. Serling, *Wrights to Wide-Bodies: The First Seventy-five Years* (Washington, D.C.: Air Transport Association, 1978).

15. "An Airplane in Every Garage?" *Scribner's,* September 1935, 179–82; see also Corn, *Winged Gospel,* chap. 5.

16. Douglas J. Ingells, *Tin Goose: The Fabulous Ford Trimotor* (Fallbook, Calif.:, Aero Publishers, 1968); Fred E. Weick, "Development of the Ercoupe, an Airplane for Simplified Private Flying," *SAE Journal* 44 (December 1941): 520–31; Max Karant, "The Unbelievable Truth About Hammond's Experimental Plane," *Popular Aviation* 19 (October 1936): 56; Corn, *Winged Gospel,* chap. 5.

17. See Henry S. F. Cooper, "Annals of Space," *New Yorker,* 2 September 1991, 41–69; Stephan Wilkinson, "The Legacy of the Lifting Body," *Air & Space,* April/May 1991, 50–62.

18. Barton C. Hacker and James M. Grimwood, *On the Shoulders of Titans: A History of Project Gemini,* SP-4203 (Washington, D.C.: GPO, 1977), 139.

19. National Academy of Sciences, "Review of Project Mercury," *IG Bulletin* 80 (February 1964): 5.

20. Tom Wolfe, "Columbia's Landing Closes a Circle," *National Geographic,* October 1981, 475; Tom Wolfe, "Everyman vs. Astropower," *Newsweek,* 10 February 1986, 41.

21. Hacker and Grimwood, *On the Shoulders of Titans,* 19–20, 123–25, 144–48, 170–73.

22. Bilstein, *Flight in America,* 24; Hallion, *Wright Brothers,* 80; Crouch, *Bishop's Boys,* 375–76, 434–35; Scharff and Taylor, *Over Land and Sea,* 215.

23. Bilstein, *Flight in America,* 57–58, 101.

24. Ibid.

25. H. G. Wells, *A Critical Edition of "The War of the Worlds,"* with introduction and notes by David Y. Hughes and Harry M. Geduld (Bloomington: Indiana University Press, 1993), 53, 200.

26. Verne, *From the Earth to the Moon;* Verne, *Round the Moon;* see also Ron Miller, "The Spaceship as Icon: Designs from Verne to the Early 1950s," in Ordway and Liebermann, *Blueprint for Space,* 51; Ron Miller, *The Dream Machines: An Illustrated History of the Spaceship in Art, Science, and Literature* (Malabar, Fla.: Keieger, 1993), 47–54.

27. Lang, *Frau im Mond,* 1929.

28. Clarke, *2001,* 46.

29. Ibid., 56–57.

30. Ibid., chap. 7; House Science and Astronautics Committee, Subcommittee on Manned Space Flight, *1974 NASA Authorization,* 93d Cong., 1st sess., 1973, 1274.

31. Alex Raymond, *Flash Gordon* (New York: Nostalgia Press, 1974); see Miller, "Spaceship as Icon," 65.

32. Ley, *Rockets, Missiles, and Men in Space,* 506; see also Peter G. Cooksley, *Flying Bomb: The Story of Hitler's V-Weapons in World War II* (New York: Charles Scribner's Sons, 1979), 165.

33. U.S. Atomic Energy Commission, *Nuclear Propulsion for Space* (Oak Ridge, Tenn.: U.S. Atomic Energy Commission, 1967).

34. Synthesis Group, *America at the Threshold,* 66–68.

35. Kurtz, *Star Wars;* see also George Lucas, *Star Wars: A New Hope* (previously titled *Star Wars: The Adventures of Luke Skywalker*) (New York: Ballantine Books, 1976), 101, 110–11.

36. Lucas, *Star Wars,* 115.

37. See Peter Nicholls, *The Science in Science Fiction* (New York: Alfred A. Knopf, 1983).

38. Smith, *Skylark of Space.*

39. George Mueller, "Space: The Future of Mankind," *Spaceflight* 27 (March 1985): 105; Mueller, "Antimatter & Distant Space Flight," *Spaceflight* 25 (May 1983): 207.

40. John H. Mauldin, *Prospects for Interstellar Travel,* vol. 80 Science and Technology Series (San Diego: American Astronautical Society, 1992); Brice N. Cassenti, "A Comparison of Interstellar Propulsion Methods, *Journal of the British Interplanetary Society* 35 (March 1982): 116–24; Alan Bond and Anthony Martin, "Project Daedalus—The Final Report on the BIS Starship Study," *Journal of the British Interplanetary Society,* supplement (1978): 5–8, 37–42; and L. D. Jaffe, C. Ivie, J. C. Lewis, R. Lipes, H. N. Norton, J. W. Stearns, L. D. Stimpson, and P. Weissman, "An Interstellar Precursor Mission," *Journal of the British Interplanetary Society* 33 (January 1980): 3–26.

41. Mueller, "Antimatter & Distant Space Flight"; Mauldin, *Prospects for Interstellar Travel;* Nicholls, *Science in Science Fiction,* 78–79.

42. Robert L. Forward, "Antimatter Propulsion," *Journal of the British Interplanetary Society* 35 (September 1982): 391–95.

43. Rick Sternbach and Michael Okuda, *Star Trek: The Next Generation Technical Manual* (New York: Pocket Books, 1991), 60–61, 67; E. F. Mallove, R. L. Forward, Z. Paprotny, and J. Lehmann, "Interstellar Travel and Communication: A Bibliography," *Journal of the British Interplanetary Society* 33 (June 1980): 201–48.

44. Sternbach and Okuda, *Star Trek,* 55.

45. Disneyworld, "Journey to Mars," opened 1975.

46. Mauldin, *Prospects for Interstellar Travel,* ix.

47. Space Task Group, *Post-Apollo Space Program,* 15.

48. Wernher von Braun, "The Spaceplane That Can Put YOU in Orbit," *Popular Science,* July 1970, 37; Rick Gore, "When the Space Shuttle Finally Flies," *National Geographic,* March 1981, 317.

49. Michael Collins, "Orbiter Is First Spacecraft Designed for Shuttle Runs," *Smithsonian,* May 1977, 38.

50. Cooper, "Annals of Space," 64.

51. Quoted from Wolfe, "Everyman vs. Astropower," 41; see also James A. Michener, "Manifest Destiny," *Omni,* April 1981, 48–50, 102–4.

52. Quoted from McCurdy, *Inside NASA,* 87.

53. Thomas O. Paine, "Head of NASA Has a New Vision of 1984," *New York Times,* 17 July 1969.

54. President's Science Advisory Committee, "The Next Decade in Space," Executive Office of the President, Office of Science and Technology, March 1970, 38.

55. House Committee on Science and Technology, Subcommittee on Space Science and Applications, *Operational Cost Estimates: Space Shuttle,* 94th Cong., 2d sess., 1976, 11; NASA, "Space Shuttle Economics Simplified," 26 January 1972, 3, NASA History Office; J. S. Butz, "The Coming Age of the Economy Flight into Space," *Air Force/Space Digest,* December 1969, 42.

56. Center for Aerospace Education Development, "Space Shuttle: A Space Transportation System Activities Book," Civil Air Patrol, U.S. Air Force, n.d.; James J. Haggerty, "Space Shuttle: Next Giant Step for Mankind," *Aerospace* 14 (December 1976): 3, 4. See also Florence S. Steinberg, *Aboard the*

Space Shuttle (Washington, D.C.: NASA Division of Public Affairs, 1980); and Wernher von Braun, "The Reusable Space Transport," *American Scientist* 60 (November–December 1972): 730–38.

57. T. O. Paine to the President, 26 March 1970, NASA History Office; James C. Fletcher and William P. Clements, "NASA/DOD Memorandum of Understanding on Management and Operation of Space Transportation System," 14 January 1977, 8, NASA History Office; see also Howard E. McCurdy, "The Costing Models of the Early 1970s and the Launching of the Space Shuttle Program," in *L'Ambition Technologique: Naissance d'Ariane,* ed. Emmanuel Chadeau (Paris: Institut d'Histoire de l'Industrie, 1995).

58. NASA, "Space Shuttle Economics" from "Space Shuttle: Appendix to Space Shuttle Fact Sheet," February 1972, NASA History Office; Klaus P. Heiss and Oskar Morgenstern, "Economic Analysis of the Space Shuttle System: Executive Summary," study prepared for NASA under contract NASW-2081, 31 January 1972.

59. See Brian O'Leary, "The Space Shuttle: NASA's White Elephant in the Sky," *Bulletin of the Atomic Scientists* (February 1983): 36–43; Logsdon, "Decision to Develop the Space Shuttle," 103–19; and Testimony of Ralph Lapp, Senate Aeronautical and Space Sciences Committee, *NASA Authorization for Fiscal Year 1973,* 92nd Cong., 2d sess., 1972, 1069–86.

60. See Alex Roland, "The Shuttle: Triumph or Turkey?" *Discover,* November 1985, 14–24; John M. Logsdon, "The Space Shuttle Program: A Policy Failure?" *Science* 232 (30 May 1986): 1099–1105; Roger A. Pielke and Radford Byerly, "The Space Shuttle Program: Performance versus Promise," in Byerly, *Space Policy Alternatives,* 223–45.

61. NASA, Office of Space Flight, "Shuttle Launch Cost," March 2, 1983; Ed Campion to Lee Saegesser, 21 September 1990; both in NASA History Office; NASA, Office of Space Flight, "Budget Control Package in Support of FY 93 Budget to Congress: Shuttle Average Cost Per Flight," 1992; U.S. General Accounting Office, "Space Transportation: The Content and Uses of Shuttle Cost Estimates," GAO/NSIAD-93-115, January 1993.

62. See Logsdon, "Space Shuttle Decision," 13–30; Logsdon, "Decision to Develop the Space Shuttle."

63. W. R. Lucas, Program Development memorandum to Dr. Rees, 16 June 1970; NASA, "Space Shuttle," February 1972; both in NASA History Office.

64. NASA News, "Space Shuttle Decisions," release no. 72-61, 15 March 1972, NASA History Office.

65. See Cooper, "Annals of Space."

66. G. Harry Stine, "The Sky is Going to Fall," *Analog Science Fiction/Science Fact,* August 1983, 75. See also Ben Bova, "The Shuttle, Yes," *New York Times,* 4 January 1982.

67. Rudy Abramson, "NASA to Study 2,000 Safety-Critical Parts," *Los Angeles Times,* 18 March 1986; Space Transportation System, "Return to Flight Status: Critical Item Waiver Status," Lyndon B. Johnson Space Center, August 1988; L Systems, Inc., "Risk Analysis of Space Transportation During the Space Station Era," prepared under contract NAS8-38076 for George C. Marshall Space Flight Center, 15 December 1989; Marcia Dunn, "Space Safety," Associated Press, 27 February 1995; NASA History Office; see also Kevin McKean, "They Fly in the Face of Danger," *Discover,* April 1986, 48–58.

68. Michael Collins, "Riding the Beast," *Washington Post,* 30 January 1986, sec. A, p. 25.

69. R. P. Feynman, "Personal Observations on Reliability of Shuttle," appendix F, in Presidential Commission on the Space Shuttle Challenger Accident, *Report of the Presidential Commission* (Washington, D.C.: GPO, 1986). See also *Report of the Presidential Commission,* chap. 8; Feynman, "An Outsider's View of the Challenger Inquiry," *Physics Today* 41 (February 1988): 26–37; Feynman, *What Do You Care What Other People Think?* as told to Ralph Leighton (New York: W. W. Norton, 1988); and Diane Vaughan, *The Challenger Launch Decision: Risky Technology, Culture, and Deviance at NASA* (Chicago: University of Chicago Press, 1996).

9. LIFE ON EARTH

Epigraph: NASA, *Why Man Explores,* symposium held at Bechman auditorium, California Institute of Technology, Pasadena, Calif., 2 July 1976 (Washington, D.C.: GPO, 1977), 11.

1. George Orwell, *1984: A Novel* (New York: Harcourt, Brace, 1949); Aldous Huxley, *Brave New World* (New York: Harper and Brothers, 1946); Frederick A. Hayek, *The Road to Serfdom* (Chicago: University of Chicago Press, 1944).

2. Gail S. Davidson, "Packaging the New: Design and the American Consumer, 1925–1975," Cooper-Hewitt Museum of Design, New York, 8 February–14 August 1994.

3. David Gelernter, *1939: The Lost World of the Fair* (New York: Free Press, 1995).

4. Donald J. Bush, *The Streamlined Decade* (New York: George Braziller, 1975), 3.

5. Daniel Goldin, "Celebrating the Spirit of Columbus," in "American at 500: Pioneering the Space Frontier," *National Forum* (Summer 1992): 8–9.

6. Gordon and Mumford, "Tomorrowland 1986," 2. See also "Tomorrowland: Show World of Future," *Disneyland News,* July 1955; and "Disneyland's New Tomorrowland," *Vacationland,* Summer 1967.

7. Willy Ley, "Inside the Moon Ship," *Collier's,* 18 October 1952, 56. See also Willy Ley, "Station in Space," *Collier's,* 22 March 1952, 30–31; Willy Ley, "Inside the Lunar Base," *Collier's,* 25 October 1952, 46–47.

8. Daniel S. Goldin, "Space Station: Built It for America," *Washington Post,* 28 July 1992, sec. A, p. 19. See also NASA, *Spinoff* (Washington, D.C.: GPO, 1990, also 1994); and Tom Alexander, "The Unexpected Payoff of Project Apollo," *Fortune,* July 1969, 114–15.

9. Evert Clark, "Satellite Spying Cited by Johnson," *New York Times,* 17 March 1967; quoted from Chalmers Roberts, *The Nuclear Years* (New York: Praeger, 1970), 87.

10. See Michael L. Smith, "Selling the Moon: The U.S. Manned Space Program and the Triumph of Commodity Scientism," in *The Culture of Consumption: Critical Essays in American History, 1880–1980,* ed. Richard W. Fox and T. J. Jackson Lears (New York: Pantheon Books, 1983).

11. See McCurdy, *Inside NASA,* 73.

12. David Riesman, *The Lonely Crowd* (New Haven, Conn.: Yale University Press, 1950).

13. Rockefeller Brothers Fund, *Prospect for America: The Rockefeller Panel Reports* (Garden City, N.Y.: Doubleday, 1961), xix.

14. Smith, "Selling the Moon," 195.

15. William H. Whyte, *The Organization Man* (New York: Simon and Schuster, 1956); Vance Packard, *The Waste Makers* (New York: D. McKay, 1960); William J. Lederer, *A Nation of Sheep* (New York: Norton, 1961).

16. Wolfe, "Columbia's Landing Closes a Circle," 475.

17. U.S. Congress, Select Committee on Astronautics and Space Exploration, *Astronautics and Space Exploration,* 85th Cong., 1st sess., 1958, 117, 951; Select Committee on Astronautics and Space Exploration, *The National Space Program,* report of the select committee (Washington, D.C.: GPO, 1958), 4.

18. "Space: The New Ocean," *Time,* 2 March 1962, 11; also see Robert Holz, "Man in Space," *Aviation Week,* 5 March 1962, 13; "Cooperation in Space," *New Republic,* 5 March 1962, 3.

19. Alex Roland, "NASA's Manned-Space Nonsense," *New York Times,* 4 October 1987, sec. 4, p. 23; James A. Van Allen, "Space Science, Space Technology and the Space Station," *Scientific American* 254 (January 1986): 37.

20. Van Allen, "Space Science, Space Technology and the Space Station," 37; William D. McCann, "Mars Viewed as Last Stop for Manned Space Flights," *Plain Dealer,* 27 December 1969; Bill Green, "Earth to NASA," *New York Times,* 27 August 1989.

21. Boyd, "In Space: Instruments or Man?" 70; also see "Space: The New Ocean," *Time,* 2 March 1962, 12; Swenson, Grimwood, and Alexander, *This New Ocean,* 428.

22. James Van Allen, "Space Station and Manned Flights Raise NASA Program Balance Issues," *Aviation Week & Space Technology,* 25 January 1988, 153.

23. Daniel S. Greenberg, "Robots in Space Are Less Costly," *Philadelphia Inquirer,* 18 September 1987; Van Allen, "Space Science, Space Technology and the Space Station."

24. R. L. F. Boyd, "In Space: Instruments or Man?" *International Science and Technology,* May 1965, 75.

25. Homer E. Newell, *The Mission of Man in Space* (Washington, D.C.: GPO, 1963), 5; James C. Fletcher to Edward M. Kennedy, 24 September 1971, NASA History Office; Robert Jastrow, "Man in Space or Chip in Space?" *New York Times Magazine,* 31 January 1971, 63.

26. President's Science Advisory Committee, "The Next Decade in Space," February 1970, 36.

27. Robert Glatzer, *The New Advertising* (New York: Citadel Press, 1970); Vance Packard, *The Hidden Persuaders* (New York: D. McKay, 1957).

28. Smith, "Selling the Moon"; Jane and Michael Stern, *Auto Ads* (New York: Random House, 1979).

29. Voas, "John Glenn's Three Orbits"; Linda Neuman Ezell, *NASA Historical Data Book: Programs and Projects 1958–1968,* NASA SP-4012 (Washington, D.C.: GPO, 1988), 2:173; Grimwood, *This New Ocean,* 314, 575.

30. Advisory Committee on the Future of the U.S. Space Program, *Report of the Advisory Committee,* 6.

31. Jessica Mathews, "Romance of the Manned Space Missions," *Washington Post,* 23 December 1990; James W. McCulla, "Why We Need Humans in Outer Space," *Washington Post,* 4 January 1991; Dean Rusk, "Just Say No to Mars," *Washington Post,* 22 September 1989; Paul Recer, "Scientists vs. Engineers: They Fued over Moon Flights," 12 October 1969.

32. Apollo 11 Crew Pre-Mission Press Conference, 5 July 1969, 2:00 P.M., NASA History Office. See also John Noble Wilford, "Humans and Machines in Space: A Vision of Our Space Future," *Space Times,* March–April 1991, 15.

33. Van Allen, "Space Science, Space Technology and the Space Station," 32.

34. Corn, *Winged Gospel,* 88.

35. Quoted from W. E. Debnam, "Women's Place in Aviation as Seen by Endurance Fliers," *Southern Aviation* 4 (December 1932): 11. See also Claudia M. Oakes, *United States Women in Aviation, 1930–1939* (Washington, D.C.: Smithsonian Institution Press, 1985).

36. Mary S. Lovell, *Straight on Toward Morning: The Biography of Beryl Markham* (New York: St. Martin's Press, 1987); Beryl Markham, *West with the Night* (San Francisco: North Point Press, 1983).

37. Anne Morrow Lindbergh, *North to the Orient* (New York: Harcourt, Brace, 1935); Deborah G. Douglas, *United States Women in Aviation, 1940–1985,* Smithsonian Studies in Air and Space no. 7 (Washington, D.C.: Smithsonian Institution Press, 1990).

38. Claudia M. Oakes, *United States Women in Aviation, 1930–1939,* Smithsonian Studies in Air and Space, no. 6 (Washington, D.C.: Smithsonian Institution Press, 1985); Doris L. Rich, *Amelia Earhart: A Biography* (Washington, D.C.: Smithsonian Institution, 1989).

39. Pamela Sargent, ed., *Women of Wonder: The Classic Years* (San Diego: Harcourt Brace and Company, 1995).

40. Solow and Justman, *Inside Star Trek.*

41. Pal, *Conquest of Space;* Nayfack, *Forbidden Planet.*

42. Speech by Wernher von Braun given at Mississippi State College, 19 November 1962, 6, NASA History Office; U.S. House Committee on Science and Astronautics, Special Subcommittee on the Selection of Astronauts, *Qualifications for Astronauts,* 87th Cong., 2d sess., 1962, 5, 58.

43. See Hugh L. Dryden to Jacqueline Cochran, 18 June 1962, NASA History Office.

44. U.S. House Committee on Science and Astronautics, *Qualifications for Astronauts,* 81; Jerrie Cobb with Jane Rieker, *Woman into Space: The Jerrie Cobb Story* (Englewood Cliffs, N.J.: Prentice-Hall, 1963), 149.

45. See paper presented by Jerrie Cobb, Aviation/Space Writers Association, twenty-third annual meeting and news conference, 1 May 1961, NASA History Office; "A Lady Proves She's Fit for Space Flight," *Life,* 29 August 1960, 72–76.

46. Lyndon B. Johnson to Miss Cobb, 23 April 1962; Jerrie Cobb to James E. Webb, 7 August 1962; Jerrie Cobb to the President, 10 February 1964; NASA History Office.

47. House Science and Astronautics Committee, *Qualifications for Astronauts,* 5.

48. Jacqueline Cochran to Jerrie Cobb, 23 March 1962; see also Cochran to James E. Webb, 14 June 1962; memo to James Webb and Hugh Dryden, 1 August 1962; and Joseph D. Atkinson and Jay M. Shafritz, *The Real Stuff: A History of NASA's Astronaut Recruitment Program* (New York: Praeger, 1985), chap. 5; letters and memos in NASA History Office.

49. "13 Women Triumphing Vicariously," *New York Times,* 5 February 1995, "Collins Fulfills Dreams for Mercury 13 Women," *NASA Headquarter Bulletin,* 21 February 1995.

50. Elizabeth S. Bell, *Sisters of the Wind: Voices of Early Women Aviators* (Pasadena, Calif.: Trilogy Books, 1994); Margery Brown, "Flying is Changing Women," *Pictorial Review,* June 1930.

51. Media General/Associate Press Public Opinion Poll, "The U.S. Space Program," poll no. 21, 22 June–2 July 1988; see also Gallup, *Gallup Poll: Public Opinion, 1989,* 171; George Gallup, *The Gallup Report,* March 1986, 11; Elizabeth H. Hastings and Philip K. Hastings, *Index to International Public Opinion, 1985–1986* (Westport, Conn.: Greenwood Press, 1987), 469–70.

52. Atkinson and Shafritz, *Real Stuff.*

53. Sagan, *Pale Blue Dot,* 6.

54. Archibald MacLeish, "A Reflection: Riders on Earth Together, Brothers in Eternal Cold," *New York Times,* 25 December 1968; Jimmy Carter, "Remarks at the Congressional Space Medal of Honor Awards Ceremony," Kennedy Space Center, *Weekly Compilation of Presidential Documents* 14 (1 October 1978): 1685.

55. Alvin Toffler, *The Third Wave* (New York: Bantam Books, 1980), 408; see also Toffler, "The Space Program's Impact on Society," in Paula Korn, ed., *Humans and Machines in Space: The Payoff* (San Diego: American Astronautical Society, 1992), 87.

56. Meadows et al., *Limits to Growth,* 24.

57. Ibid., 187, see also chap. 5; H. S. D. Cole, *Models of Doom: A Critique of the Limits to Growth* (New York: Universe Books, 1973).

58. See Lynn Margulis and Dorion Sagan, *Microcosmos: Four Billion Years of Microbial Evolution* (New York: Summit Books, 1986).

59. Lawrence E. Joseph, *Gaia: The Growth of an Idea* (New York: St. Martin's Press, 1990), 1–2. See also James Lovelock, *Ages of Gaia: A Biography of Our Living Earth* (New York: W. W. Norton, 1988).

60. Ride, *Leadership and America's Future in Space,* 23; Burton I. Edelson, "Mission to Planet Earth," *Science* 227 (25 January 1985): 367; Craig Covault, "Major Space Effort Mobilized to Blunt Environmental Threat," *Aviation Week & Space Technology,* 13 March 1989, 36–44; James R. Asker, "Earth Mission Faces Growing Pains," *Aviation Week & Space Technology,* 21 February 1994, 36.

61. Alvin Toffler, *Future Shock* (New York: Random House, 1970), chap. 4.

62. Sagan, *Pale Blue Dot.*

63. Theresa M. Foley, "NASA Prepares for Protests over Nuclear System Launch on Shuttle in October," *Aviation Week & Space Technology,* 26 June 1989; "Court Rejects Activists' Bid to Halt Galileo/Shuttle Launch," *Aviation Week & Space Technology,* 16 October 1989, 21; Charles Perrow, "The Habit of Courting Disaster," *Nation,* 11 October 1986; and Perrow, *Normal Accidents: Living with High-Risk Technologies* (New York: Basic, 1984).

CONCLUSION: IMAGINATION AND THE POLICY AGENDA

Epigraph: Sign in a store displaying Indian artifacts, author unknown, the Clotheshorse Trading Company, Seattle, Washington.

1. See Corn, *Winged Gospel.*

2. Robert Wohl, *A Passion for Wings: Aviation and the Western Imagination 1908–1918* (New Haven, Conn.: Yale University Press, 1994), 1.

3. Charles Murray, *Losing Ground: American Social Policy, 1950–1980* (New York: Basic Books, 1984).

4. James E. Anderson, *Public Policymaking* (Boston: Houghton Mifflin, 1994).

5. See, for example, Roderick Nash, *Wilderness and the American Mind* (New Haven, Conn.: Yale University Press, 1967); Jackson Lears, *Fables of Abundance: A Cultural History of Advertising in America* (New York: Basic Books, 1994); Kristin Ross, *Fast Cars, Clean Bodies* (Cambridge, Mass.: MIT Press, 1995); Nicholas B. Dirks, Geoff Eley, and Sherry B. Ortner, *Culture/Power/History* (Princeton, N.J.: Princeton University Press, 1994).

6. See Kenneth Clark, *Civilisation: A Personal View* (New York: Harper and Row, 1969), 322–23.

7. Charles Dickens, *Oliver Twist* (1838; reprint, New York: Bantam Books, 1982).

8. See Roderick Nash, *Wilderness and the American Mind,* 3d ed. (New Haven, Conn.: Yale University Press, 1982).

9. See Runte, *National Parks.*

10. Anne R. Morand, Joni L. Kinsey, and Mary Panzer, *Splendors of the American West: Thomas Moran's Art of the Grand Canyon and Yellowstone* (Birmingham, Ala.: Birmingham Museum of Art, 1990). See also Barbara Novack, *Nature and Culture: American Landscape, 1825–1865* (New York: Oxford University Press, 1980).

11. See Paul Brooks, *Speaking for Nature: How Literary Naturalists from Henry Thoreau to Rachel Carson Have Shaped America* (Boston: Houghton Mifflin, 1980).

12. See G. Edward White, *The Eastern Establishment and the Western Experience* (New Haven, Conn.: Yale University Press, 1989).

13. Frank R. Baumgartner and Bryan D. Jones, *Agendas and Instability in American Politics* (Chicago: University of Chicago Press, 1993).

14. Harriet Beecher Stowe, *Uncle Tom's Cabin,* ed. with an introduction by Philip van Doren Stern (New York: Paul S. Eriksson, 1851).

15. Moira Davidson Reynolds, *Uncle Tom's Cabin and Mid-Nineteenth Century United States* (Jefferson, N.C.: McFarland and Company, 1985), 7.

16. Stowe, *Uncle Tom's Cabin,* 560.

17. Reynolds, *Uncle Tom's Cabin and Mid-Nineteenth Century United States,* 12.

18. Also see Thomas F. Gossett, *Uncle Tom's Cabin and American Culture* (Dallas: Southern Methodist University Press, 1985), 314.

19. See John Tytell, *Ezra Pound* (New York: Anchor Press, 1987); William Perlberg, *Miracle on 34th Street* (Fox, 1947); and J. D. Salinger, *Franny and Zooey* (Boston: Little, Brown, 1961).

20. Ken Kesey, *One Flew Over the Cuckoo's Nest* (New York: Viking Press, 1962).

21. Paul McHugh, "Psychiatric Misadventures," *American Scholar* 61 (Autumn 1992): 498. Also see Erving Goffman, *Asylums: Essays on the Social Situation of Mental Patients and Other Inmates* (Garden City, N.Y.: Doubleday Anchor, 1961) and Thomas S. Szasz, *The Myth of Mental Illness* (New York: Harper and Row, 1961).

22. See Myron Magnet, *The Dream and the Nightmare: The Sixties Legacy to the Underclass* (New York: William Morrow, 1993).

23. See Charles Perrow, *Complex Organizations* (Glenview, Ill.: Scott, Foresman, 1972); Charles T. Goodsell, *The Case for Bureaucracy,* 3d ed. (Chatham, N.J.: Chatham House, 1994).

24. George Orwell, *1984: A Novel* (New York: Harcourt, Brace, 1949); Heller, *Catch-22;* Ivan Reitman, *Ghostbusters* (Columbia Pictures, 1984); also see Joe Queenan, "Evil Empire on the Potomac," *Washington Post,* 13 November 1994, sec. C, p. 5.

25. Alexis de Tocqueville, *Democracy in America* (New York: Schocken Books, 1835).

26. Mark Twain, *Huckleberry Finn* (New York: Harcourt, Brace & World, 1961).

27. Slotkin, *Gunfighter Nation.*

28. Gareth Morgan, *Imaginization: The Art of Creative Management* (Newbury Park, Calif.: Sage Publications, 1993).

29. See C. Northcote Parkinson, *Parkinson's Law and Other Studies in Administration* (Boston: Houghton Mifflin, 1957) and Anthony Downs, *Inside Bureaucracy* (Boston: Little, Brown, 1967).

30. Upton Sinclair, *The Jungle* (New York: New American Library, 1905).

31. Ibid., 349.

32. E. E. Schattschneider, *The Semi-Sovereign People* (New York: Holt, Rinehart and Winston, 1960), 3.

33. W. Lance Bennett and David L. Paletz, *Taken by Storm: The Media, Public Opinion, and U.S. Foreign Policy in the Gulf War* (Chicago: University of Chicago Press, 1994).

34. William Sims Bainbridge, *The Spaceflight Revolution: A Sociological Study* (New York: John Wiley and Sons, 1976), 13.

35. John Calvin Batchelor, in *What is the Value of Space Exploration? A Symposium,* sponsored by the Mission from Planet Earth Study Office, Office of Space Science, NASA Headquarters, and the University of Maryland at College Park, 18–19 July 1994, 31.

36. Paul Theroux, *Sailing Through China* (Boston: Houghton Mifflin, 1984), 23; see also Dwayne A. Day, "Paradigm Lost," *Space Policy* 11 (August 1995): 153–59.

37. See Samuelson, *The Good Life and Its Discontents.*

38. Ryan A. Harmon, "Predicting the Future," *Disney News,* Fall 1991, 35.

39. Lewis L. Strauss, "Remarks Prepared by Lewis L. Strauss, Chairman, United States Atomic Energy Commission, for Delivery at the Founders' Day Dinner, National Association of Science Writers," 16 September 1954, New York, Department of Energy History Division; also see Brian Balogh, *Chain Reaction: Expert Debate and Public Participation in American Commercial Nuclear Power, 1945–1975* (New York: Cambridge University Press, 1991), 113.

40. See Corn, *Winged Gospel,* chap. 5.

41. Richard M. Nixon, "Space Shuttle Program," *Weekly Compilation of Presidential Documents* (5 January 1972): 27; Office of the White House Press Secretary, Press Conference of Dr. James Fletcher and George M. Low, 8.

42. See Festinger, *When Prophecy Fails;* Leon Festinger, *A Theory of Cognitive Dissonance* (Stanford, Calif.: Stanford University Press, 1957).

43. See Jack Lemming, "The Future of Space Activities," *Spaceflight,* April 1994, 110–11; Dethloff, *Suddenly, Tomorrow Came,* chap. 16.

44. Corn, *Winged Gospel.*

INDEX

Page numbers in bold indicate illustrations